Systems Analysis Techniques

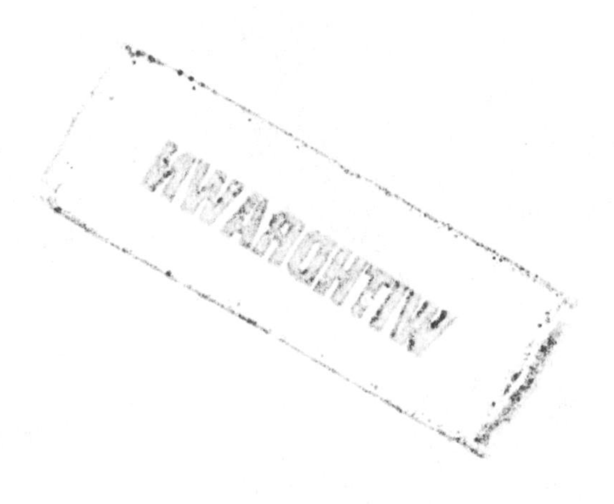

Systems Analysis Techniques

Barbara Robinson
Mary Prior

INTERNATIONAL THOMSON COMPUTER PRESS
I(T)P An International Thomson Publishing Company

London • Bonn • Boston • Johannesburg • Madrid • Melbourne • Mexico City • New York • Paris • Singapore
Tokyo • Toronto • Albany, NY • Belmont, CA • Cincinnati, OH • Detroit, MI

Systems Analysis Techniques

Copyright © 1995 Barbara Robinson and Mary Prior

I(T)P A division of International Thomson Publishing Inc.
The ITP logo is a trademark under licence.

For more information, contact:

International Thomson Computer Press
Berkshire House
168-173 High Holborn
London WC1V 7AA
UK

International Thomson Computer Press
20 Park Plaza
Suite 1001
Boston, MA 02116
USA

Imprints of International Thomson Publishing

International Thomson Publishing GmbH
Königswinterer Straße 418
53227 Bonn
Germany

Thomas Nelson Australia
102 Dodds Street
South Melbourne, 3205
Victoria
Australia

Nelson Canada
1120 Birchmount Road
Scarborough, Ontario
Canada M1K 5G4

International Thomson Publishing South Africa
PO Box 2459
Halfway House
1685 South Africa

International Thomson Publishing Asia
60 Albert Street #15-01
Albert Complex
Singapore 189969

International Thomson Publishing Japan
Hirakawacho Kyowa Building, 3F
2-2-1 Hirakawacho
Chiyoda-ku, 102 Tokyo
Japan

International Thomson Editores
Seneca, 53
Colonia Polanco
11560 Mexico D. F. Mexico

International Thomson Publishing France
Tours Maine-Montparnasse
33 avenue du Maine
75755 Paris Cedex 15
France

All rights reserved. No part of this work which is copyright may be reproduced or used in any form or by any means – graphic, electronic, or mechanical, including photocopying, recording, taping or information storage and retrieval systems – without the written permission of the Publisher, except in accordance with the provisions of the Copyright Designs and Patents Act 1988.

Products and services that are referred to in this book may be either trademarks and/or registered trademarks of their respective owners. The Publisher/s and Author/s make no claim to these trademarks.

Whilst the Publisher has taken all reasonable care in the preparation of this book the Publisher makes no representation, express or implied, with regard to the accuracy of the information contained in this book and cannot accept any legal responsibility or liability for any errors or omissions from the book or the consequences thereof.

British Library Cataloguing-in-Publication Data
A catalogue record for this book is available from the British Library

Library of Congress Cataloging-in-Publication Data
A catalog record for this book is available from the Library of Congress

First Printed 1995
Reprinted 1996

ISBN 1-85032-183-3

Typeset in the UK by Columns Design & Production Services Ltd, Reading
Printed in the UK by The Alden Press, Oxford

http://www.itcpmedia.com

Contents

Preface

This book is about the techniques used in structured systems analysis. It presents models used to analyse an information system from the three perspectives of process, data and time. The emphasis is on the analysis of business information systems. Reference is made to major structured methods such as SSADM, however, the approach is general and is not tied to a single methodology; we demonstrate the main techniques that are used in most structured systems development methods.

The book has a practical emphasis and is intended to aid students and others who are learning the practice of structured systems analysis. Descriptions of techniques and the illustration of a completed model offer little help in how to arrive at the finished product. Hence we provide guidelines for the development of each model, worked examples to illustrate the application of the guidelines and exercises with solutions to provide the reader with an opportunity to practise the techniques.

One major case scenario has been used for the worked examples throughout the text, and another for the exercises, supplemented by smaller case scenarios where necessary. The intention is to demonstrate how an information system may be viewed from three perspectives, and how the models of those three views interrelate. Using a small system to illustrate the application of the techniques allows us to demonstrate all of the aspects of the analysis process within a reasonable amount of time and space. Given the limits of space within a book of this size and the limits of time in the vast majority of teaching contexts, the use of a constrained scenario allows the student to grasp the basics of the techniques being taught without becoming too distracted by the complexities of a large-scale project.

We would like to thank Professor David Howe of De Montfort University for his helpful advice and comments, and for his support and encouragement; also, our colleagues Simon Bennett for reading and commenting on the book and Terry Sherwood for his help in the preparation of a number of the diagrams. Any errors in the text are, of course, our own.

Chapter 1

Introduction

OBJECTIVES

This chapter:

- explains the scope of this book;
- presents the case scenarios that will be used in the subsequent chapters.

Structured systems analysis

Systems analysis involves the investigation of a system to gain a thorough understanding of how it works, what are its present problems and what are its requirements. The output of systems analysis is a detailed specification of the requirements for a new system.

This book is about the techniques used in structured systems analysis to analyse an information system. Structured techniques for systems analysis and design have been in existence since the 1970s, following upon the heels of structured programming techniques.

See Gane and Sarson (1979) and DeMarco (1979); see also the discussion in Chapter 3 ('Other approaches to process modelling').

The structured approach to systems analysis and design is characterized by the use of top-down decomposition and modelling techniques within discrete phases defined by the systems development lifecycle. The typical life cycle is illustrated in Figure 1.1.

In concentrating on the modelling techniques used during the systems analysis phase of the lifecycle, this book does not cover those activities that must precede the production of models, such as fact finding, nor those that are complementary to it such as the production of a feasibility report or a requirements specification, nor issues such as project management. Techniques used in other phases such as systems design are also excluded. These topics are well covered in other texts.

See Blethyn and Parker (1990), Senn (1989), Skidmore (1994), Skidmore and Wroe (1990), Daniels and Yeates (1988), Kendall and Kendall (1992) and Howe (1989).

Systems development methodologies

In recent years, a number of systems development methodologies have evolved. There is no standard definition of a methodology, and there is wide

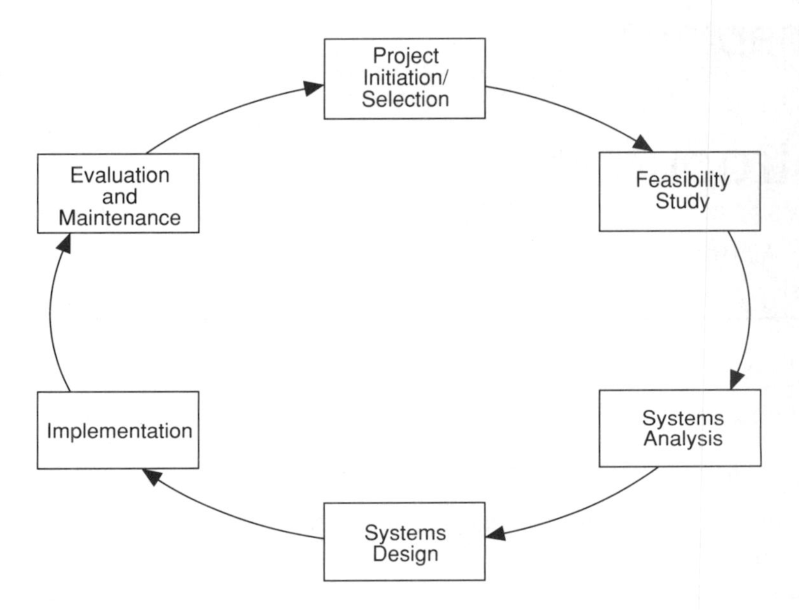

Figure 1.1 The systems development lifecycle.

For further discussion of methodologies see Avison and Fitzgerald (1988) and Flynn (1992).

See below for a discussion of CASE tools.

See Skidmore (1994).

See Ward and Mellor (1985) and Goldsmith (1993) for the modelling of real-time systems.

variation between what is offered by proprietary products claiming to be methodologies. Typically, we would expect a systems development methodology to provide more than just a set of modelling techniques. It should define the stages into which a systems development project should be broken down; specify the tasks to be carried out and the outputs to be produced at each stage; provide guidelines for the management and control of the project, and be underpinned by a philosophy regarding the approach it advocates to systems development. The potential user of the methodology should be able to acquire adequate manuals, training and the support of a CASE tool.

This book does not attempt to teach one systems development methodology; we have taken the 'toolkit' approach advocated by Skidmore and made reference to specific methodologies where appropriate. The techniques covered, or variants of them, are used in most structured systems methodologies. They are applicable to the analysis of business information systems. Some structured methodologies incorporate enhanced notation and additional techniques for modelling aspects of real-time systems; these have not been included in this book.

Prototyping

The linear progression implied by the systems development lifecycle is not reflected in practice when a certain amount of overlap and iteration between the phases is unavoidable, even desirable. In particular, recent emphasis on maximizing user involvement to ensure that user requirements are met, together with the availability of software tools to speed up application development, have led to the introduction of prototyping. Prototyping is seen by some as a systems development methodology in its own right; however, we take the view that it can be incorporated into a project that uses a structured approach, and reference is made in Chapter 8 as to how this may be achieved.

See Maude and Willis (1991) and Vonk (1990).

Hard and soft approaches

Approaches to systems development can be characterized as being 'hard' or 'soft' (Flynn 1992). Soft approaches emphasize the human and social aspects of information systems within their environment, and may be directed towards problem solving without any reference to computerization. An example is the Soft Systems Methodology developed by Checkland.

See Checkland (1981) and Checkland and Scholes (1990).

In contrast to this, hard approaches emphasize the application of techniques in a systematic manner to produce a new (usually computerized) system; most structured development approaches, and this book, fall into this category. There have been attempts to combine these two approaches, for example Multiview uses ideas from Soft Systems Methodology, ETHICS and structured methods to advocate a contingency approach that advises the application of the most appropriate techniques, soft or hard, depending upon the circumstances. Like the authors of Multiview, we regard soft approaches as complementary to, rather than a replacement for, the hard approach of this book.

See Avison and Wood-Harper (1990) and Mumford (1983).

Object oriented approaches

There has recently been a rapid increase in the development of object oriented technologies, including a proliferation of object oriented analysis and design methods. We have not attempted to include any object oriented analysis techniques in this book, partly due to constraints of space, but more importantly, because an object oriented approach demands a complete shift in perception, which requires a book to itself to present. The majority of object oriented analysis and design methods are in their infancy, with only limited experience of their application; it is our view that, despite the importance of these developments, there will remain a place for structured systems analysis techniques for some time to come.

See Graham (1994), Sully (1993) and Patel, Hayes and Sun (1994) for discussion of object oriented analysis and design.

CASE tools

The availability of Computer-Aided Software Engineering (CASE) tools has increased dramatically over the past decade. These tools provide software

See Chapter 4 ('Use of a CASE tool') and Chapter 8 ('Use of CASE tools and prototyping').

support for the systems development process. As with methodologies, there is wide variation between the facilities offered by different CASE tools. They may support one or more phases of the lifecycle and be tailored to support a particular methodology. A CASE tool to support systems analysis should provide the facilities to create the various models required, to check for errors of syntax and consistency both within a single model and between models, and to create and maintain a data dictionary.

The use of a CASE tool will save the systems analyst much tedious work in checking models, aid the production of documentation and improve communication both with the user and between colleagues working on the same project. The use of structured systems analysis techniques involves the construction of several models, each of which must be internally consistent and all of which must be consistent with each other. The use of structured techniques, particularly in large systems development projects, is impractical without the support of a CASE tool. The overall result of the use of a CASE tool should be an improvement in the quality of the delivered information system.

The implementation of a CASE tool within an organization is not a trivial matter, and will involve consideration of the methodology in use (if any) and its support by the proposed CASE tool, changes in work practice that may accompany the introduction of a new tool and/or methodology, the training that development staff will need and the selection of staff and projects for piloting the use of the tool. Management support is essential as there are considerable resource implications not only for the purchase of the CASE tool and for staff training, but also because the productivity of development teams may suffer at first; the gains will be medium to long term rather than instant.

Outline of contents

The remainder of this book presents models that may be used to analyse an information system from the three perspectives of process, data and time. In practice, the type of information system under investigation may suggest that more emphasis needs to be placed upon one view and less on another. For example, an information system may involve complex data structures and require relatively straightforward processing, in which case most effort will need to be expended in preparing the data model. In another case, the data structures may be uncomplicated but there may be a complex combination of events and processing required, in which case more effort will need to be spent on the process and time models. Nevertheless, it is essential for all three views to be thoroughly understood and mastered before such decisions can sensibly be made.

Chapters 2 and 3 discuss how to model the process view of an information system using data flow diagrams; Chapters 5 and 6 deal with the data view, modelled using an entity-relationship model; Chapter 7 presents the time view, modelled using entity life histories. Chapter 4 introduces the data dictionary, which is an important component underpinning all of the other models. Chapter 8 discusses issues involved in cross-checking the three views of the information system. Chapter 9 provides some suggestions on how to approach a systems analysis assignment or examination.

Case scenarios

The following case scenarios, Somerleyton Animal Park and Albany Hotel, will be used for the worked examples and exercises, respectively, supplemented by smaller scenarios which are introduced in the text as they are required.

Somerleyton Animal Park

Introduction

Somerleyton Animal Park was established by animal-lover Chris Lyon in 1980. He sees its primary aim as being to contribute to the conservation of endangered species. To this end, the park has an active breeding programme, collaborating with other animal parks and zoos to breed animals and reintroduce them to the wild. Somerleyton depends for its financial survival upon the many visitors that it now welcomes each year. Chris intends their visit to be an educational experience as well as a recreational one, and this is especially true of parties of school children. Thus the second important aim of the park is to extend the awareness of its visitors and the public at large to the threats facing the world's wildlife today.

Since its foundation, the park has had a number of breeding successes and has expanded considerably. The staffing establishment currently includes 30 specialists to care for the animals, three full time clerical assistants to help with the office work, and a number of other staff to run the entrance and visitor facilities such as the cafeteria and shop.

Most of the animal park's records are maintained manually; for example, a file for each animal is kept in a filing cabinet, and the ordering system is also organized on a manual basis. However, prompted by the enthusiasm of Jo Wilson, the headkeeper responsible for the large cats, Chris bought a microcomputer about three years ago. Jo set about putting some of the large cat species information, currently kept in a card file, onto a database package, and started to produce word-processed notes for distribution to parties of school children. Thinking this to be a good idea, a year later Tim Price, the headkeeper of the reptile house, followed suit and started to set up a database of information about the animals under his charge. One or two other keepers have also been expressing a wish to start files on their animals.

However, a recent incident has alerted Chris to the problems inherent in a number of his staff setting up independent systems. A party of children was due to visit the animal park, and Chris asked Jo to prepare one of her information sheets about big cats and also, since Tim was on holiday, one about reptiles. However, Jo was unable to do the latter, since Tim had gone his own way and set up (so he claimed) a superior system, and Jo couldn't figure out how to use it.

The lack of a computerized database of animals is beginning to put the park at a disadvantage when it comes to the exchange of animals for breeding purposes. Contacting other parks individually to find a suitable mate for an animal, or to offer a possible mate, is a time-consuming and therefore expensive activity. Most other parks have computerized systems and/or contribute to a national database of information, thus speeding up the process of finding suitable potential mates for their animals. Chris would like to increase the efficiency of this aspect of his work.

The expansion of the animal park has been accompanied by an increase in the work of the office staff. There is now a permanent backlog of paperwork. The writing out of orders for supplies for the animals takes time, and often delivery notes arrive before the corresponding order has been filed, making checking a time-consuming task. The office staff have been requesting more clerical help to get rid of the backlog.

Chris thought it was about time to review the use of information technology in the office. Consequently, he has called in a firm of consultants to advise him.

Current organization of animal records

Disliking the small cages provided by traditional zoos, Chris has allocated as much space as possible for each of his specimens. The animal park is divided into a number of areas (for example, an area for big cats and another for reptiles). Each area contains a number of enclosures, and each animal is confined to one enclosure. However, an enclosure could contain more than one animal, for example a mother and her offspring. Each area is under the care of a headkeeper, though more than one keeper helps to look after the larger areas. A keeper only works on one area at any one time.

Species information is held centrally on a card file (see Figure 1.2 for a sample card). These are filed alphabetically under the species name, and they indicate how many animals of that type are represented in the animal park.

Each animal has a folder devoted to it, which is kept in the filing cabinet under the animal's name (for example, Fred the gorilla). This includes data such as the animal's date of birth and its breeding history. The animal park employs its own full-time veterinary surgeon, and each time an animal is treated an entry is made in the animal's folder. If the animal is lent to another animal park, a note of the date sent and of the destination, together with any comments, is made in the animal's folder. A temporary folder is

Species Card

Species Name: ______________________________

Specimens Held:

Animal Name	Place and Date of Birth	Parents (if known)
____________	____________	____________
____________	____________	____________
____________	____________	____________

Figure 1.2 Species card.

made for animals that are on loan to Somerleyton from elsewhere. See Figure 1.3 for a sample animal record.

When Jo and Tim set up their computerized files, they added brief notes about each species in addition to the bare facts included on the original card file. Chris would like to see the card file replaced by a system which contains information about all types of animal, so that educational notes could be produced for children and, indeed, booklets could be produced for sale to all visitors.

Ordering supplies

Each headkeeper maintains a record of the sorts of food that each animal in his or her area should be fed, and in what quantities. This information is written on sheets of paper attached to a clipboard which is kept within the appropriate area, so that if the headkeeper is away, the animals will be correctly fed. Each animal may be given more than one type of food, and each type of food may be fed to a number of animals.

The headkeepers are responsible for ensuring that there are adequate supplies of food in stock. They inform the office staff each week how much of each type of food needs to be ordered for their areas. The office staff then co-ordinate requests for the same food type. A number of suppliers are used, and their names, addresses and telephone numbers are kept in a looseleaf file. Because of the large quantities required and the difficulty of

Animal Record

Animal Name: ____________________

Species: ____________________

Birth Date/Place: ____________________

Parents: ____________________

____

Breeding History:

Date	Place	Number and Names of Offspring

Veterinary Record:

Date	Description

Figure 1.3 Animal record.

obtaining some foods at certain times of the year, there is more than one possible supplier for each type of food. Most of the ordering is done via the telephone. One of the clerical assistants phones a supplier to see if they can provide, for example, 20 kilos of bananas the following week. If the first supplier cannot provide the whole order, the assistant will try others until the complete quantity is ordered. A standard order form is then written out

Animal Record (reverse)

Exchange History:

Date	Loaned to/from	Comments

Reintroduction to Wild:

Date	Location	Comments

Figure 1.3 Continued.

(see Figure 1.4), one for each supplier, and filed by date in an 'orders placed' file. Other necessary supplies such as straw are ordered in the same way.

When deliveries are received, the keepers check the delivery note against the goods received, amend it if necessary and pass it on to the office, where it is checked against the orders placed file. If they tally, the order form and

Order Form

Order Number: 27585

Date: 6/7/94

Supplier: Waltons

Item	Quantity	Price
Bananas	50 Kilos	£25
Apples	40 Kilos	£24
Oranges	35 Kilos	£21

Figure 1.4 Somerleyton Animal Park order form.

the delivery note are stapled together and placed in an 'awaiting payment' file alphabetically by the supplier's name. Any discrepancies are taken up with the supplier, and the supplier's response is noted on the order form before the two documents are filed. Most suppliers send an invoice each month. When an invoice is received it is matched up with the documents in the awaiting payment file, the payment is made, the invoice is stapled to the other documents and all three placed in a back orders file.

Staff in the shop and the cafeteria order their own goods themselves, using separate procedures.

Requirements

1. To record data on animals, including their breeding history and treatment details to facilitate the exchange of animals for breeding purposes. Details of animals on loan from elsewhere should also be recorded.
2. Facility to extract information on individual animals or species by request.
3. To provide educational notes on all species for school party visits.
4. To maintain the data needed to produce booklets for all visits. No decision has been made yet regarding the printing of the booklets.
5. To maintain a booking system for school party visits.
6. To provide the office staff with a weekly list of food requirements and possible suppliers. The selection of the actual supplier to be used is to remain as a manual task. For the time being, it is envisaged that the ordering will still be done by telephone. The determination of food requirements per animal is to remain the responsibility of the headkeepers and as a manual task. The ordering of food for the shop and cafeterias is beyond the scope of this system.
7. To progress orders through placement, delivery of food supplies and payment.

Albany Hotel

The Albany is a 90 room hotel in Leicester. In addition to the attractions of the city centre, it is also within reach of Rutland Water, which provides excellent facilities for sailing, fishing and bird watching.

The hotel offers both single and double rooms, most with bathrooms but some with washbasins only.

Bookings may be made either by telephone or by the use of a booking form (see Figure 1.5).

The hotel receptionists respond to any telephone enquiries, providing information to potential guests who may or may not already have made bookings. They have a large wall chart which shows room availability, a leaflet showing the room rate charges (which depend upon the room size and facilities), and a file containing a card for each guest or potential guest, together with any correspondence. A receptionist is able to respond to a telephone booking request immediately by checking the room chart for availability. The potential guest can be informed immediately if the accommodation requested is not available, otherwise the accommodation charge is then calculated and the potential guest given a provisional booking over the

Albany Hotel, Shire Street, Leicester, Tel. 0116 299 8888

Booking Form

Name: J. Smith

Address: 500 High Street
Anytown
Middlesex AE1 6PY

Tel. no: 0800 998899

Room(s) required: 1 double(s) with bathroom/~~washbasin~~* £40 per night
1 single(s) with bathroom/~~washbasin~~* £22 per night

Dates required: From 22-4-95 to 24-4-95 inclusive.

I wish to pay by cheque/~~credit card~~*.

My credit card number is: ________ Expiry date: ________

I agree to abide by the terms and conditions of the Albany Hotel (set out overleaf)

J. Smith (please sign).

* Please delete as appropriate.

Hotel use only

Total cost: £________

Deposit required: £________

Figure 1.5 Albany Hotel booking form.

telephone. The receptionist will follow this up by completing a booking form which is sent to the potential guest together with a request for a 20% deposit or a credit card number. A file is opened for the potential guest at that time.

On return of the signed booking form and either the deposit or credit card number, the receptionist will check the form and, if necessary, request further information. If the deposit is enclosed, the booking is confirmed both on the hotel records and by letter to the potential guest. If the guest intends to pay by credit card, the number is checked with the credit card company. Again, if the guest's credit is good, the booking is confirmed but if not, the guest is requested to offer an alternative method of payment. If the potential guest uses a booking form then he/she should include his/her deposit or credit card number with other details. Again, the receptionist will check the booking form, make a provisional booking and confirm it only when the payment method has been checked out.

For unexpected arrivals, the usual information is taken from the guest together with the credit card number, if applicable, and entered onto a booking form. The credit card number will then be checked and the guest(s) informed if an alternative method of payment is required.

It is usual for guests to pay on departure. Any additional charges, such as meals, drinks and minibar sales, will be added to their accommodation charge by the receptionist. Copies of signed dockets are sent to reception from the restaurant and coffee shop after each meal. The room service staff make a daily count of stock in the minibar kept in each guest room, calculate the minibar sales for each room and inform the receptionist.

The receptionist processes the settlement of the account and issues a receipt. All payments received are passed to the cashier, who then posts the amounts to the sales ledger.

Requirements

1. To maintain accurate and efficient guest booking, including changes to bookings and cancellations.
2. To maintain guest accounts.
3. To archive and keep guest details for 18 months (cancellations may be ignored).
4. To send Christmas cards to guests who stayed in the hotel for a week or longer in the current calendar year.
5. To post guest details to the sales ledger.
6. To respond quickly to potential guest/guest enquiries about availability, prices and current bookings.

Chapter 2

Process Modelling 1

OBJECTIVES

In this chapter you will learn:

- □ how a data flow diagram (DFD) may be used to model the process view of a system;
- □ the components of a DFD;
- □ using guidelines, how to develop:
  - a current physical top level DFD
  - a context diagram
  - a current logical top level DFD.

Introduction

Most business-oriented systems suggested for development are based on a system already in operation. There may be known changes, additional requirements and the consideration of anticipated future needs, but the fundamentals of the existing system remain an important consideration. A business user, when questioned in detail about his/her operation, will usually focus on the processes of that operation. A process may be defined as an action or series of actions which produce a change or development. The process view of a system may be modelled by a **data flow diagram** (DFD).

Data Flow Diagrams – basic concepts

A DFD is a diagrammatic representation of the passage of data through a system, and depicts any changes made to that data and what data is stored. Figure 2.1 is an example of a DFD.

A DFD comprises the following components or elements:

- process;
- data flow;
- data store; and
- terminator.

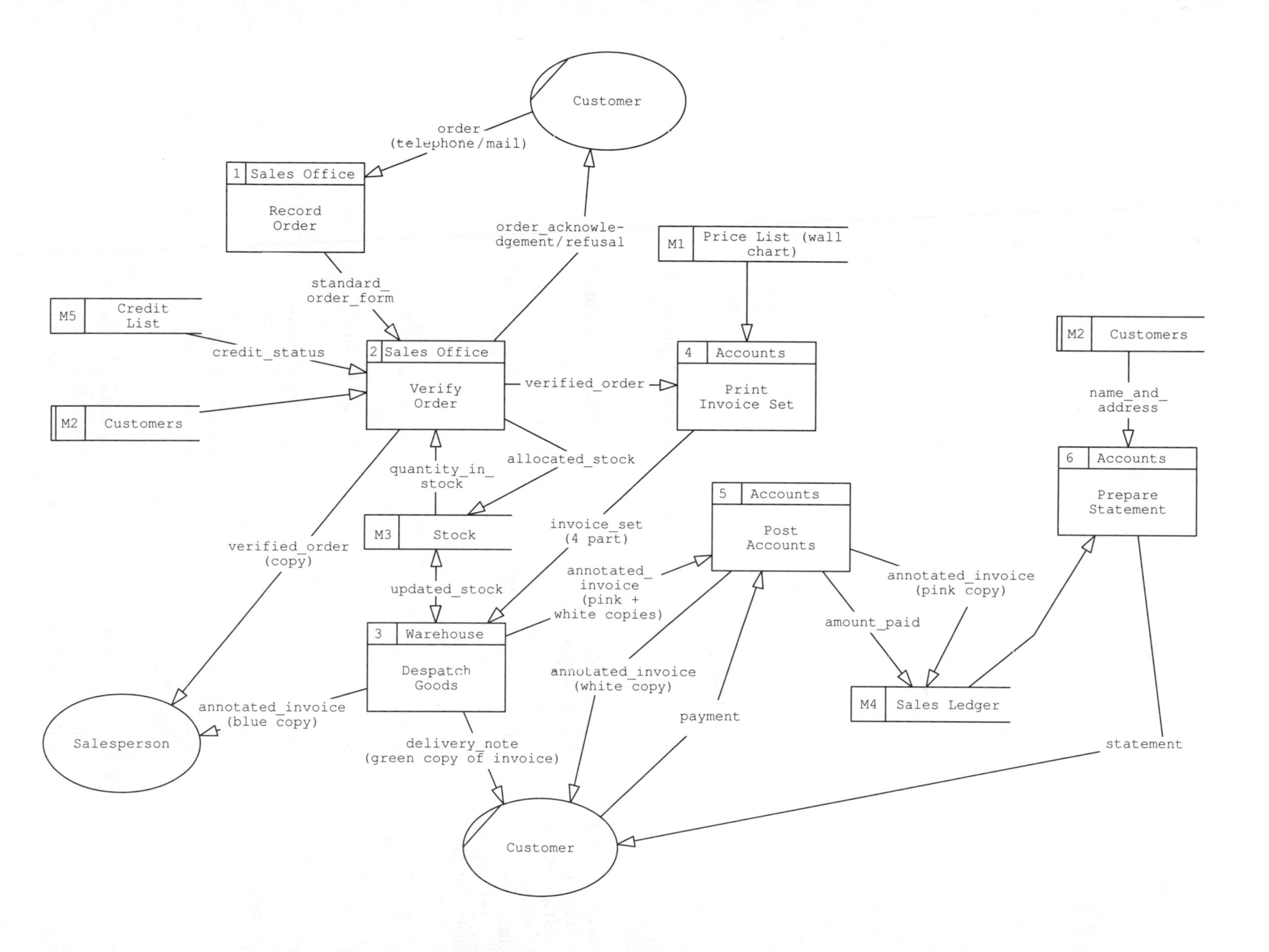

Figure 2.1 Example of a physical DFD – order processing system.

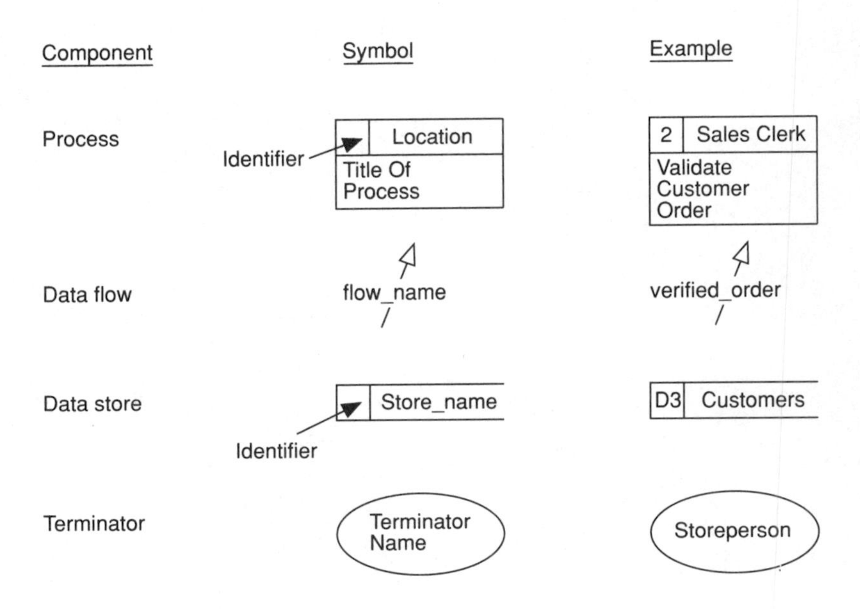

Figure 2.2 DFD symbols.

The symbols to be used for each component are shown in Figure 2.2.

The numbering sequence does *not* indicate the order in which the processes will be activated. Avoid using the verb 'process' where possible. The shape of the symbol indicates that it is a process, so it is possible to give more information about what the process actually does. For example, 'Verify Order' is more explicit than 'Process Order'.

Process

A **process** is used to transform data from one form or state to another. Each process is identified by a number, starting at 1.

The **location** shows who performs the process or where it is performed. The process title should be meaningful, and is written as an active verb followed by one or more objects.

Data Flow

Data passing to and from processes is indicated by a continuous line, with an arrowhead showing the direction of the flow and with a label for identification. This is known as a **data flow,** and defined as 'data in motion'. A data flow comprises one or more pieces or elements of data. The data that is moved is sometimes referred to as a 'packet of data'. The packet of data and its contents are usually defined in a data dictionary (discussed in Chapter 4).

A packet of data that enters and leaves a process may on first examination seem to be the same, and therefore be identified by the same name. However, a process must produce a change or development; for example, in Figure 2.1, the data flows `standard_order_form` and `verified_order` have the same data elements but, for the data flow leaving the process the status of the data has changed – it is now verified. Therefore, the data flows need different names. Similarly, data flow names should be different if the form of their data elements changes on entry to and exit from a process.

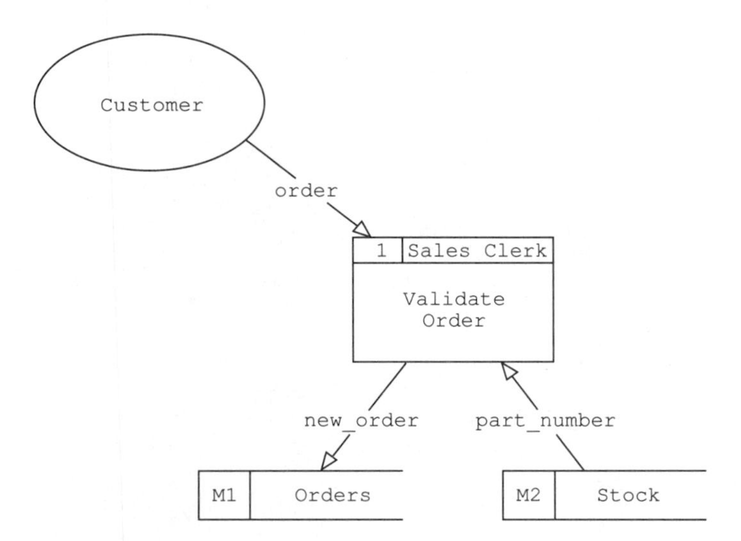

Figure 2.3 Example of the use of a data store.

Using the same approach as in Figure 2.4, the data flow new_order shown in Figure 2.3 may also be omitted. In this text, data flows to and from data stores have been omitted where it is felt that their meaning is obvious, in order to make the diagrams less 'cluttered'. However, if it is felt that clarification is helpful they are included.

Data Store

Data held in a file, a pile of documents of one type, a wall chart, filing cabinet or database is shown within a **data store.**

Sometimes referred to simply as a **store**.

This is known as 'data at rest'. A data store usually comprises a collection of packets of data each of which may consist of a number of individual data elements; for example, the data store Stock may contain for each different stocked item a packet of data with data elements part_number, description, quantity_in_stock and allocated_stock. The definition of this data is held in the data dictionary (discussed in Chapter 4).

Typically, the name chosen to identify the data store is written in the plural form.

Figure 2.3 shows the data element part_number moving between the data store and the process. The data element is made available to the process, but remains in the data store and is unchanged. Data moving from a data store to a process is often referred to as data being read or accessed.

Figure 2.3 shows the packet of data new_order being written to the data store Orders. Data moving from a process to a data store is often described as a write, or may be an update or a delete.

Remember that it is the process that specifies the action to be taken. The process and data flow names give an indication of the action being taken and the data being moved, but the data dictionary (discussed in Chapter 4) should be examined for precise definitions.

Figure 2.4 shows the data element quantity_in_stock moving between the process Issue Part and the data store Stock. A double-headed arrow (◁—▷) may be used when data is both examined and updated. Data must not be passed from data store to data store without passing through a process.

Figure 2.4 shows the packet of data deleted_part moving from the process Delete Part to the data store Stock. The DFD indicates that the packet of data in the data store for the specified stocked item is deleted. This action may be confirmed by the examination of the data dictionary (discussed in Chapter 4).

If an entire instance of a packet of data is transferred to or from a data store the data flow label may be left blank; for example in Figure 2.4 the unlabelled data flow between the process Record New Part and the data store

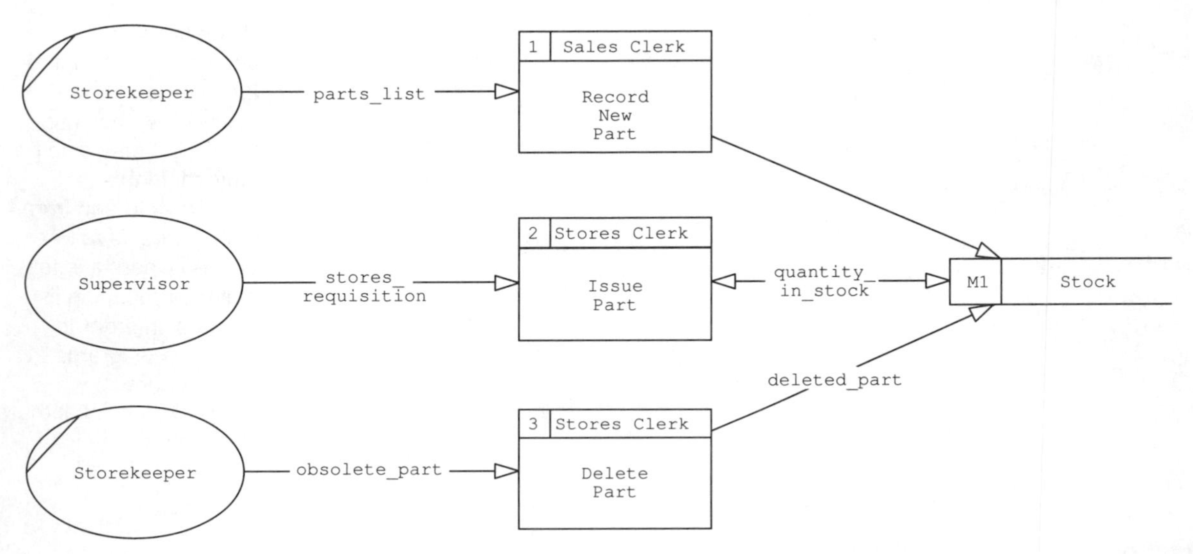

Figure 2.4 Further example of the use of a data store.

See Figure 2.1, 3.7 and 3.8. M = Manual Data Store (physical DFDs only) D = Main Data store (logical DFDs only). The numbers are unique. On lower level diagrams, the identifier indicates the process within whose boundary it exists; for example, Figure 3.7 data store `Prices` D3/1 belonging to process 3. Other names for terminator are external entity, source (originator) or sink (recipient). Typically, a terminator is defined in the singular form.

`Stock` indicates that the entire packet of data for an item of stock is transferred (or written) to the data store.

The way that data is passed or held may be shown on the diagram typically in brackets after the data flow or data store to which it relates; for example, in Figure 2.1 `order (telephone/mail)`, `annotated_invoice (pink copy)` and `Price List (wall chart)`.

Each data store is uniquely identified by a letter followed by a number.

Terminator

Data originating from or finishing outside the boundary of the defined system is shown by a **terminator**. This may be a person or department, another system or another organization.

Approach to drawing a current physical DFD

Useful systems analysis texts for fact finding include Blethyn and Parker (1990), Senn (1989) and Skidmore (1994).

As stated previously, most business-oriented systems suggested for development are based on a system already in operation. Therefore, in the majority of cases it is impractical to ignore totally the existing system. For the systems analyst, examining the existing system offers an opportunity to become familiar with current practice, and also to gain the user's confidence. The development of a DFD will assist with this understanding. It is anticipated that the systems analyst will be working from agreed terms of reference so that the scope of the project has been accepted. This means that the initial boundary of the system may readily be identified. The systems analyst will obtain information about the existing system using well-understood fact finding techniques.

A current physical DFD shows what happens to data within the existing (current) system, how it is processed, where, sometimes when and by whom. The following guidelines will be used to develop a current physical DFD for the worked example. The resulting diagram is known as a top level DFD.

Guidelines for drawing a current physical top level DFD

1. Identify the inputs and outputs of the system – for a clerical system, these are likely to be documents which will become data flows.
2. Identify the sources and recipients of the inputs and outputs – these will become terminators.
3. Draw the terminators and associated data flows identified in (1) and (2) round the outside of the page.
4. For each data flow on the diagram, identify and draw a process within the system which will receive/generate the data.
5. For each of the processes identified in (4), draw any associated data stores with their data flows.
6. Add any further processes that transform data produced totally within the system.
7. Add any further data flows and data stores required by those processes.
8. Add any internal data flows required between processes.
9. Check the diagram for consistency and completeness. For example, check for processes that have an input and no output; check for processes that have an output and no input; check that the process identifiers and their titles are both unique; check that unlabelled data flows do represent an entire instance of a packet of data being transferred to or from a data store; examine any duplicate data flows to ensure that the form and state of the data are the same; and check that the capability exists to create and delete entire instances of packets of data to and from a data store. These issues are addressed in Worked example 2.1 guideline step (9).

You might find it easier to work on (1) and (2) at the same time. For a large system, you might need to return and repeat these and subsequent steps before all the inputs and outputs are identified.
You might find it easier to work on (4) and (5) at the same time.

Process locations need not be unique.

Some of this checking may be undertaken by the use of a data dictionary (discussed in Chapter 4) and/or a CASE Tool (discussed in Chapter 8).

During the preparation of a current physical DFD, the systems analyst often realizes that the information obtained is incomplete or inconsistent. Further fact finding should then take place with the user. Modelling the current system allows the systems analyst to ensure that, as far as possible, the facts are complete and consistent. This approach is also possible for you when working in a role-play situation. When you are working from a scenario where the information supplied is incomplete, assumptions need to be made. It is useful to prepare an 'assumptions sheet' in parallel with the development of the DFD. However, sometimes it is difficult to provide a suitable assumption. Then, an appropriate query would be stated and no changes made to the DFD. This makes the 'assumptions sheet' a combination of queries and assumptions – perhaps a 'quassump sheet'!

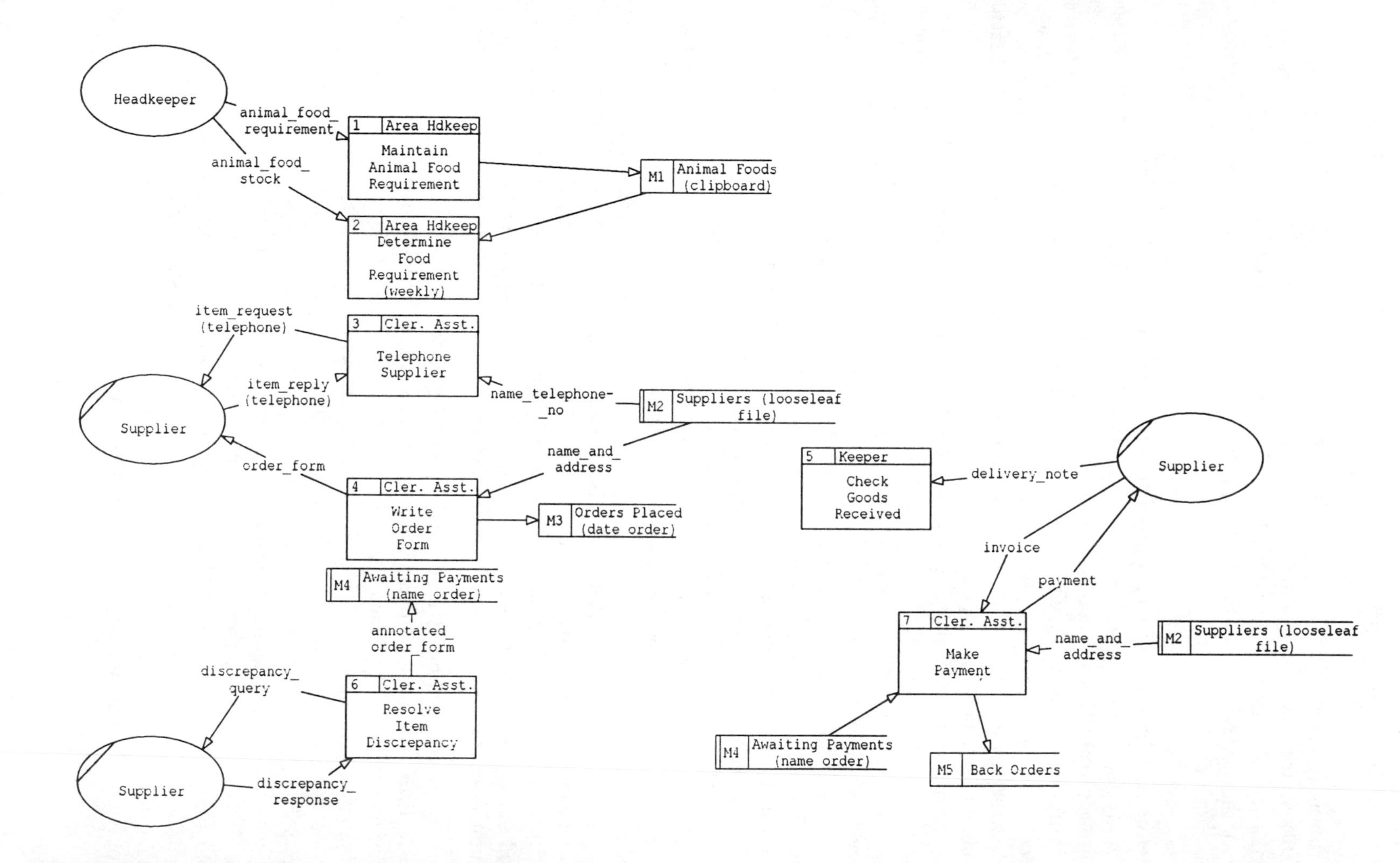

Figure 2.5 Worked example 2.1 DFD after guideline step 5.

Worked example 2.1

Using the specified guidelines, develop a current physical top level data flow diagram for the ordering supplies section (subsystem) of the Somerleyton Animal Park case scenario.

The end result is shown in Figure 2.6. We now show how this DFD is built up step by step.

Solution: For the purpose of this and subsequent worked examples, the user will not be available for consultation, so a 'queries and assumptions sheet' will probably need to be prepared.

Guideline steps 1 and 2:

Inputs and outputs	*Sources*	*Recipients*
animal_food_requirement	Headkeeper	
animal_food_stock	Headkeeper	
item_request		Supplier
item_reply	Supplier	
order_form		Supplier
delivery_note	Supplier	
discrepancy_query		Supplier
discrepancy_reply	Supplier	
invoice	Supplier	
payment		Supplier

The identification of the terminator Supplier and associated inputs and outputs is relatively straightforward. However, it is probably expected that the headkeepers are in fact part of the system. It is possible that a person might perform some activities within the system and some beyond its scope. In this case, the headkeeper has been identified as a terminator who will provide the system with the animal food requirements and the levels of animal food in stock.

These animal food requirements will then be recorded within the system, as shown in Figure 2.5, by process 1 storing the information in the Animal Foods (clipboard) data store.

Guideline step 3: Terminators are usually placed around the outside of the DFD. This is difficult for the terminator Supplier as it has many associated data flows. A duplicate terminator may be drawn with a diagonal line indicator marked on each occurrence of that terminator. After this guideline step, the DFD consists of those parts of Figure 2.5 comprising just the terminators (Headkeeper, and the duplicate Supplier terminators) and their associated data flows (animal_food_requirement, animal_food_stock and the eight data flows for Supplier).

Guideline steps 4 and 5: See Figure 2.5.

Queries and assumptions:

1. The checks made of the delivery note against the goods received may be conducted by the keeper or the headkeeper, depending upon availability.

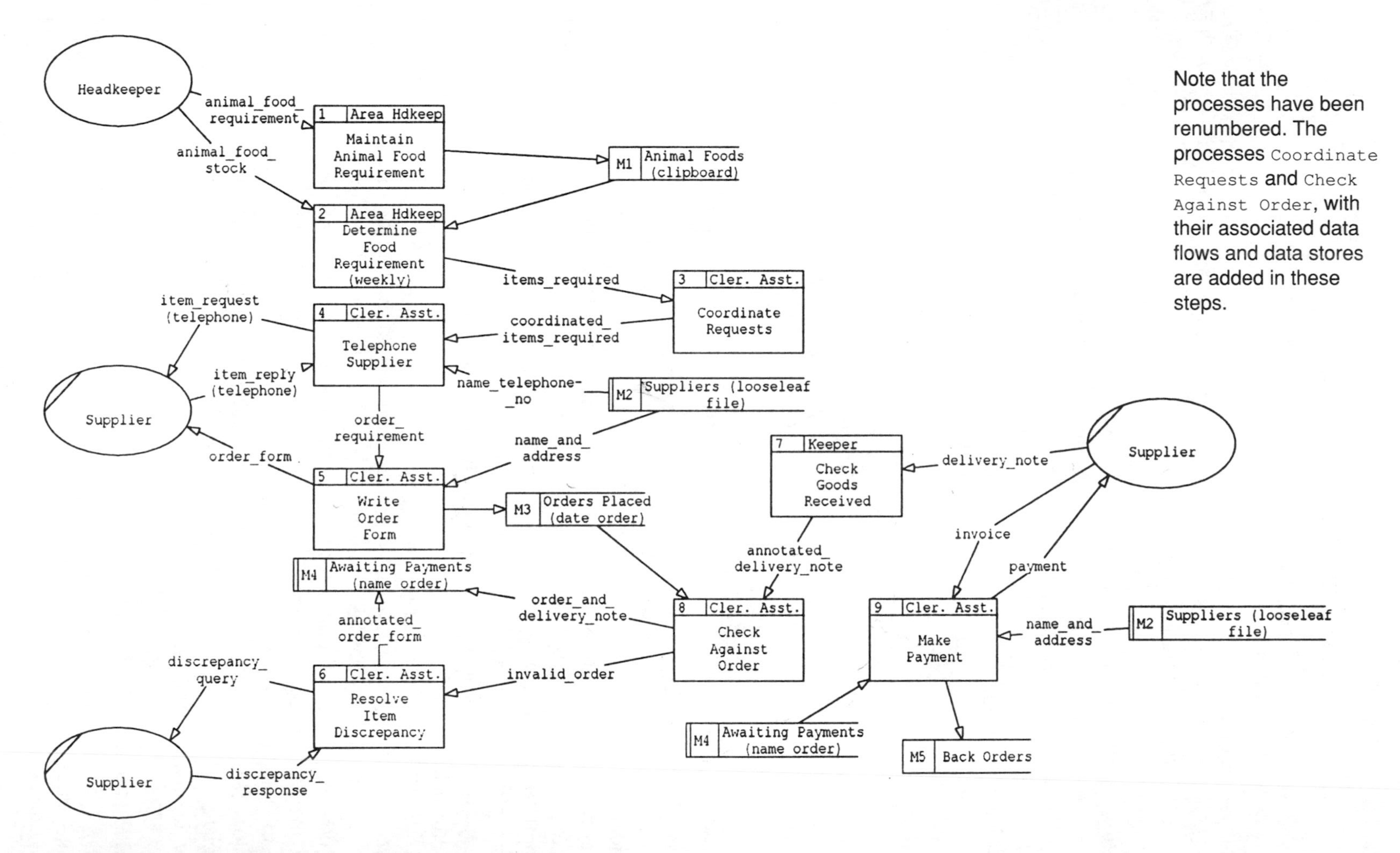

Note that the processes have been renumbered. The processes `Coordinate Requests` and `Check Against Order`, with their associated data flows and data stores are added in these steps.

Figure 2.6 Worked example 2.1 DFD after guideline step 8.

Keeper has been entered in the process location to cover both eventualities.

2. The scenario states that the discrepancies are taken up with the supplier. It is assumed that this task is performed by a clerical assistant.
3. The scenario states that after discrepancies have been checked the 'two documents are filed'. It is assumed that they are filed in the `Awaiting Payments` data store.
4. The scenario states that invoices are received and checked, and payments made, but not by whom! It is assumed that the payments are made by a clerical assistant.

Guideline steps 6, 7 and 8: See Figure 2.6.
Queries and assumptions continued:

5. The scenario states that the 'office staff' coordinate requests for item ordering. It is assumed that this is performed by a clerical assistant.
6. It is assumed that 'items' and 'other necessary' `supplies` are ordered in the same way, so no extension to the DFD is necessary to deal with the latter category.

Guideline step 9, Checks:

- There are no processes that have an input and no output.
- There are no processes that have an output and no input.
- All process identifiers are unique.
- All process titles are unique.
- There are no unlabelled data flows between processes.
- Unlabelled data flows do represent an entire instance of a packet of data being transferred to or from a data store.
- Duplicate data flows have been checked and do contain data in the same form and state.

Queries and assumptions continued:

7. `Suppliers` is a read-only data store – that is, there is no capability to add details for a new supplier, or to change details for an existing supplier, or to delete details for a supplier. You would need to check whether the maintenance of the supplier details falls within the scope of this system or not. If it does, the DFD should be amended accordingly. If it does not, a reference should be made to the procedures/subsystem where this takes place. It is assumed that the maintenance of the data store `Suppliers` falls beyond the scope of this subsystem.
8. The `Orders Placed` data store is subject to both reads and writes, but it might be the case that goods are not received for a particular order. This would need to be checked. There may be a number of reasons why this is the case. The supplier might not be able to fulfil the order. The order might have got lost. The Animal Park might subsequently have cancelled the order. It is not easy to make a reasonable assumption here, so we will leave the query.
   Query outstanding – what happens if orders placed are not delivered?

If this is an assessed piece of work, the assessor will know that you have given the matter some thought, and you can reasonably expect that your DFD will be assessed in the light of your assumptions (if plausible) rather than your assessor's assumptions.

At this stage of development the DFD remains unchanged.

9. The `Awaiting Payments` data store is subject to both reads and writes, but the scenario does not actually state what happens to the discrepancy responses if they are unresolved.
   Query outstanding – what happens to discrepancy responses if they are unresolved? No changes have been made to the DFD at this time.
10. `Back Orders` is a write-only data store – how long are back orders retained?
   Query outstanding – how long are back orders retained? No changes have been made to the DFD at this time.

For the required system, it might be that an additional process for archiving needs to be developed. Obviously, storage requirements and volumes would need to be considered when making this decision. This is discussed further in Worked example 3.1 in Chapter 3.

The queries and assumptions sheet in its entirety is included for completeness.

Queries and assumptions sheet:

1. The checks made of the delivery note against the goods received may be conducted by the keeper or the headkeeper depending on availability. Keeper has been entered in the process location to cover both eventualities.
2. The scenario states that the discrepancies are taken up with the supplier. It is assumed that this task is performed by a clerical assistant.
3. The scenario states that after the discrepancies have been checked the 'two documents are filed'. It is assumed that they are filed in the `Awaiting Payments` data store.
4. The scenario states that invoices are received and checked, and payments made but not by whom! It is assumed that the payments are made by a clerical assistant.
5. The scenario states that the 'office staff' coordinate requests for item ordering. It is assumed that this is performed by a clerical assistant.
6. It is assumed that 'items' and 'other necessary supplies' are ordered in the same way, so no extension to the DFD is necessary to deal with the latter category.
7. `Suppliers` is a read-only data store – that is there is no capability to add details for a new supplier, or to change details for an existing supplier, or to delete details for a supplier. You would need to check whether the maintenance of the supplier details falls within the scope of this system or not. If it does, the DFD should be amended accordingly. If it does not, a reference should be made to the procedures/subsystem where this takes place. It is assumed that the maintenance of the data store `Suppliers` falls beyond the scope of this subsystem.
8. The `Orders Placed` data store is subject to both reads and writes, but it might be the case that goods are not received for a particular order. This would need to be checked. There may be a number of reasons why this is the case. The supplier might not be able to fulfil the order. The order might have got lost. The Animal Park might subsequently have cancelled the order. It is not easy to make a reasonable assumption here, so we will leave the query.
   Query outstanding – what happens if orders placed are not delivered?

At this stage of development the DFD remains unchanged.

9. The `Awaiting Payments` data store is subject to both reads and writes, but the scenario does not actually state what happens to the discrepancy responses if they are unresolved.
   Query outstanding – what happens to discrepancy responses if they are unresolved? No changes have been made to the DFD at this time.
10. `Back Orders` is a write-only data store – how long are back orders retained?
   Query outstanding – how long are back orders retained? No changes have been made to the DFD at this time.

Reminder: In a real-life or role-play situation, facts would be checked/clarified with the user. The DFD would need to be amended accordingly.

Sometimes within assignments/examinations you are asked to prepare further questions for users. The query raised prior to the statement of an assumption in the example above can be rephrased into one or more fact finding questions. In other words, you imagine you are a systems analyst where the facts found have to be complete and consistent.

Processes on the DFD have been renumbered in Figure 2.6 for clarity, but remember they are for identification only and do not indicate the sequence of processing. A general 'rule' is to number from top to bottom of the diagram.

During the development of the DFD, as symbols are added, data flows will inevitably cross one another. This should be avoided where possible. One solution is to redraw the DFD, which is common practice during construction. Alternatively, duplicate symbols may be used for terminators and data stores. Figure 2.6 shows duplication of the terminator `Supplier` and of the data store `Suppliers`.

Note that there is a terminator `Supplier` and a data store `Suppliers` – do not get confused.

If, for simplicity, data flows still need to cross they should be drawn as shown in the margin.

A duplicate data store is indicated by an additional vertical bar.

Represent crossing data flows as:

Summary of DFD notation

- Process — transforms data
  active verb followed by object(s)
  process numbers do not indicate sequence
- Data Flow — data in motion
- Data Store — data at rest
- Terminator — originator or recipient of data *outside* the defined system.

See Figure 2.2 for diagrammatic symbols.

Exercise 2.1

Identify syntax errors on the current physical DFD given in Figure 2.7.

If you examine the DFD shown in Figure 2.7 closely, you will realize that, despite the syntax errors, the DFD still appears to be incomplete. How does an applicant know whether an offer is made, and how does UCAS know the decision made by the applicant on that offer? The answer is that the offer is communicated by UCAS directly to the applicant, who in turn informs UCAS of his/her decision. That is, data is flowing between terminators. Most DFD notations do not allow data flows between terminators stating that these communications are beyond the scope of the system. However, Structured Systems Analysis and Design Method (SSADM) Version 4 suggests that data flows of this kind may be included.

See the next section for a more detailed reference to SSADM.

The authors agree that the inclusion of these data flows aids the understanding of the system being modelled.

Data flows between terminators are shown by a dotted line, and are labelled in the usual way.

Exercise 2.2

Develop a current physical top level DFD from the following scenario.

Clients wishing to put their property on the market visit the estate agent, who will take details of their house, flat or bungalow and enter them on a card which is filed according to the area, price range and type of property.

Potential buyers complete a similar type of card which is filed by buyer name in an A4 binder.

Weekly, the estate agent matches the potential buyers' requirements with the available properties and sends them the details of selected properties.

When a sale is completed, the buyer confirms that contracts have been exchanged, client details are removed from the property file, and an invoice is sent to the client. The client receives the top copy of a three part set, with the other two copies being filed.

On receipt of the payment the invoice copies are stamped and archived. Invoices are checked on a monthly basis and for those accounts not settled within two months a reminder (the third copy of the invoice) is sent to the client.

Notation for other methods

SSADM is a set of procedural, technical and documentation standards for systems development. See the SSADM Reference Manuals (1990) for full details. SSADM is currently at Version 4, and any references in this text relate to that version.

For other texts on DFDs, see DeMarco (1979), Gane and Sarson (1977) and Yourdon (1989).

DFDs are central to most structured methodologies, and the notation is similar in concept in all of them. The notation used in this text is not used in any single methodology, though it is closest to that used in SSADM.

The symbols used in certain other methodologies are illustrated in Figure 2.8.

Figure 2.8 shows the additional symbols of a Resource Flow and a Resource Store. SSADM recognizes that it may be useful to document the movement of physical resources either as a basis for the definition of data

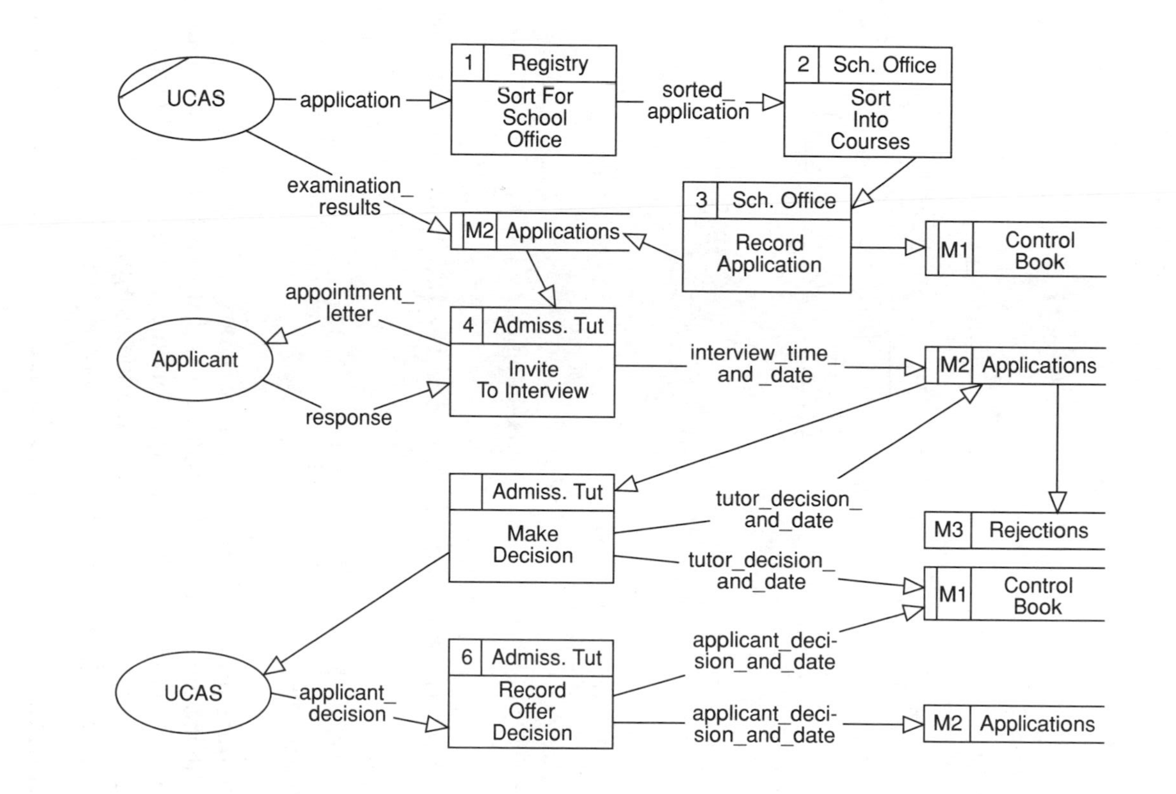

UCAS is the Universities' Central Admissions System.

Figure 2.7 Exercise 2.1.

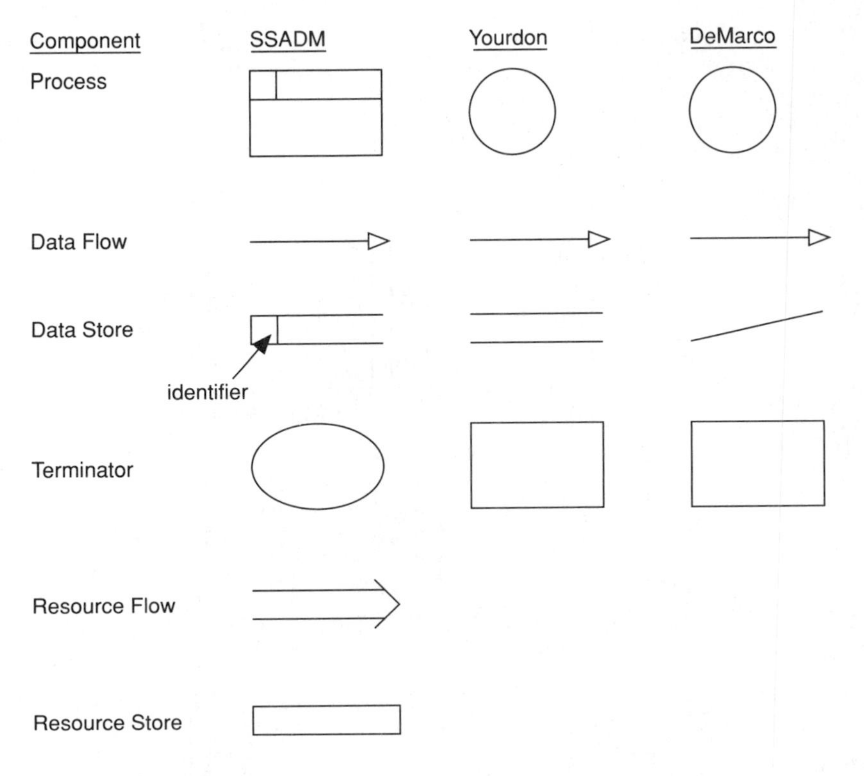

Figure 2.8 DFD symbols for other methodologies.

flows, or where data flows do not exist. See SSADM Reference Manuals (1990) for a full explanation.

Context diagrams

A **context diagram** comprises one process box for the entire system, together with the terminators and the data flows that pass between them and the system. The purpose of the context diagram is to identify and examine the interfaces between the terminators and the system. Once the top level DFD has been drawn, the context diagram may be derived from it.

Worked example 2.2

Develop a context diagram for the DFD developed in Worked example 2.1.

Solution: A single line is drawn on the top level DFD crossing all data flows passing to or from a terminator, as shown in Figure 2.9. The resulting line defines the system boundary and, indeed, the single process box required. The context diagram is shown in Figure 2.10.

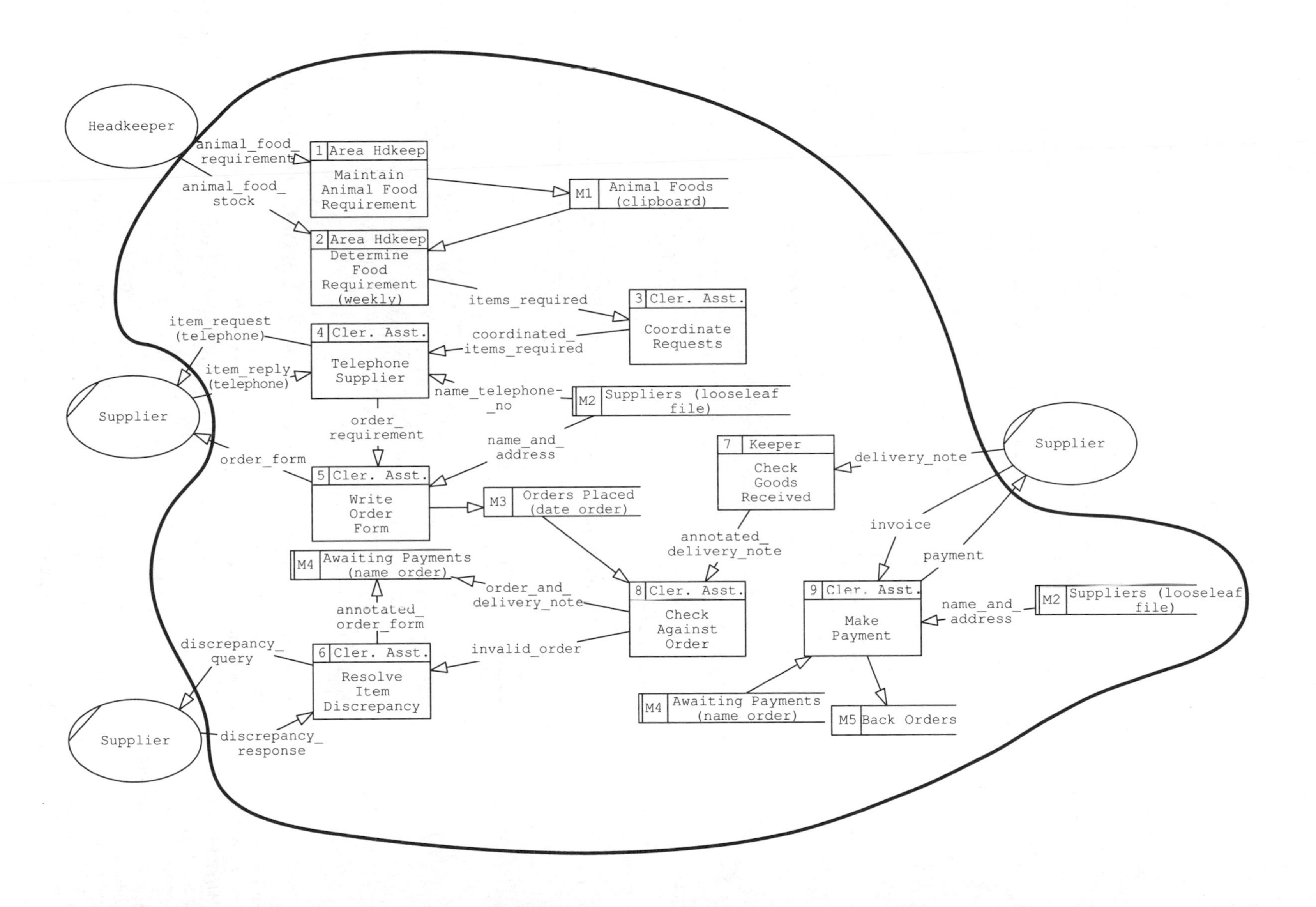

Figure 2.9 Worked example 2.1 with system boundary marked.

Figure 2.10 Context diagram for Worked example 2.2.

Note that the title in the process box is that of the system being modelled, not just the name of the organization.

Exercise 2.3

Develop a context diagram for the top level DFD given in Figure 2.1.

Logical Data Flow Diagrams

As stated earlier, the current physical DFD shows the 'what, how, who, when and where' of a system. However, to obtain a clearer picture and to avoid prejudging any design of the proposed system, it is important to focus on *what* happens. A DFD showing only *what* happens is known as a **logical DFD.** Guidelines for the conversion are as follows.

Guidelines for conversion of a current physical DFD to a current logical DFD

For example, references to physical locations, methods of communication and copies of documents, including their colours, should all be removed.

1. Retain 'what' happens and remove the 'how', 'who', 'when' and 'where'. Any references to the physical facilities or constraints of the system are removed, leaving the fundamental processing and the data required for that processing.
2. Processes should transform data as a business requirement. Remove those that do not, for example those that merely reorganize data.
3. If data is unaltered by a process, is the process needed?
4. Combine processes where: identical activities are performed; two or

more processes are always performed together or as a series; or more than one process exists only because the action is in a different location.
5. Remove from data stores any data elements that are not used by any processes.
6. Remove any data stores that exist only as an implementation dependent time delay between processes; consider renaming data stores that have different names but hold the same data.
7. Check the diagram for consistency and completeness.

Further development in the use of DFDs is explained in Chapter 3. More detailed consideration will be given to these and additional logicalization steps at that stage.

Worked example 2.3

Convert the current physical DFD for Worked example 2.1 (see Figure 2.6) to a current logical DFD.

Solution: See Figure 2.11.

Guideline step 1:
- Remove all process locations.
- Remove any physical resource flows and stores.
- Define the main data stores.
- Remove references to how data is held in data stores and how data is passed in data flows.
- Remove reference to sequence within data stores.
- Remove reference to time triggers within processes.
- Remove physical aspects from process titles, for example `Telephone Supplier` becomes `Select Supplier`.

Processes are triggered by events, usually the arrival of one or more data flows or a time trigger. This time trigger would be recorded in the data dictionary (see Chapter 4).

Figure 2.11 shows the DFD after the above actions have been performed.

Guideline step 2: The DFD has been checked and no changes are required.

Guideline step 3: The DFD has been checked and no changes are required.

Guideline step 4: Consideration is given to the possibility of combining processes (2) and (3). They are performed one after the other, and process (3) is accumulating data created by process (2). However, in this case further thought will need to be given to the method for the determination of food requirements, and whether this will be part of the automated system or not. No change is made at present.

See Chapter 3, Worked example 3.1.

Guideline step 5: It is not possible to consider this fully unless a data dictionary is being used (see Chapter 4).

See Chapter 4 ('Data stores').

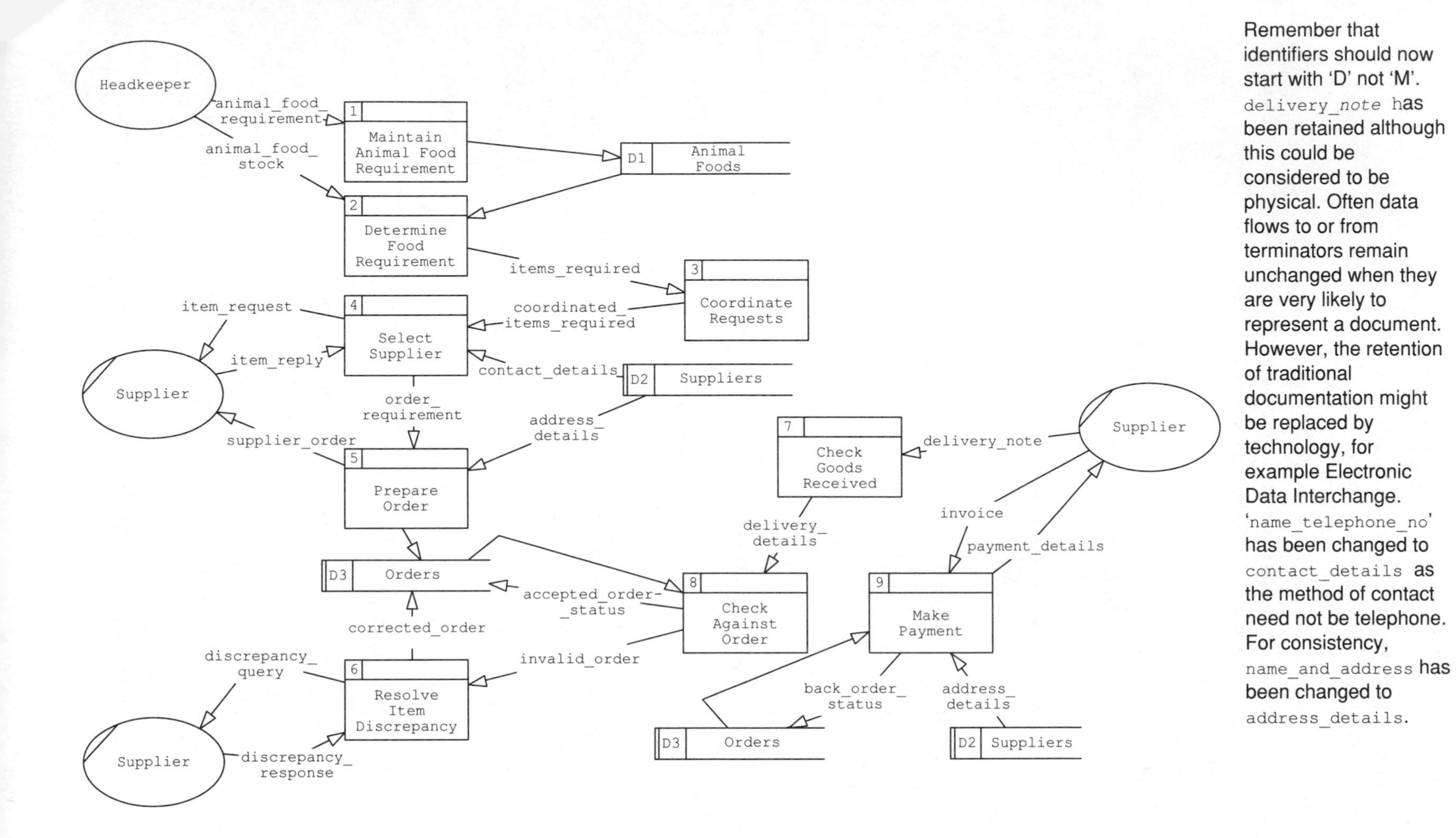

Figure 2.11 Worked example 2.3.

Remember that identifiers should now start with 'D' not 'M'. delivery_*note* has been retained although this could be considered to be physical. Often data flows to or from terminators remain unchanged when they are very likely to represent a document. However, the retention of traditional documentation might be replaced by technology, for example Electronic Data Interchange. 'name_telephone_no' has been changed to contact_details as the method of contact need not be telephone. For consistency, name_and_address has been changed to address_details.

Guideline step 6: This step is partially designed to remove time delays introduced unnecessarily within the system. For example, a clerk collects a pile of documents, on his/her desk and actions them as a batch. This action may involve calculation or the physical transportation of documents, making it more time-efficient to process the documents as a group rather than one at a time. However, logically the data store may be eliminated.

There are no examples of this type in the worked example.

You might think that the data stores `Orders Placed` and `Awaiting Payments` fall within this category. However, the existence of these data stores is necessitated by timing considerations in the environment. Orders placed will not be checked until deliveries are received. This might be some time after the order is placed, and is beyond our control. There will be a similar time delay concerning payments – we certainly do not want to pay suppliers until after the invoice has been received.

If there are multiple triggers for a process a data store is needed for delayed data.

This step is also designed to look at the data stores that have different names but hold the same data. It is often the case that the data elements are the same but that the status is different. For example, the data stores `Orders Placed`, `Awaiting Payments` and `Back Orders` hold many of the same data elements. One likely difference is that the data stores `Awaiting Payments` and `Back Orders` will hold the actual quantity delivered if different from the quantity ordered. If the two quantities are different, we need to find out if both figures are required or if the corrected quantity only is stored. The actual data is determined when creating the data dictionary and is considered in detail in Worked example 4.1 in Chapter 4. In principle, in this case the data held relates to the order which has one of three statuses – placed, awaiting payment, or back. Logically, the data stores could all be renamed as `Orders` with a data element for the status of the order. This change has been made.

Guideline step 7: The DFD has been checked and remains the same as that in Figure 2.11.

The context diagram should be checked and amended if necessary to reflect any changes made.

In this case, the following changes are needed, all to data flows: the word telephone is removed from `item_request` and `item_reply`; `order_form` becomes `supplier_order`; `payment` becomes `payment_details`.

Exercise 2.4

Convert the current physical DFD from Exercise 2.2 to a current logical DFD.

Either use your own physical DFD or use Figure E2.2.

Exercise 2.5

Convert the current physical DFD given in Figure 2.1 to a current logical DFD.

The action taken within guideline step (1) is to some extent mechanical. However, this is not so for the subsequent guideline steps, demonstrating that logicalization is not a trivial activity.

See Yourdon (1989) Chapters 7 and 17. He also refers to essential models (models showing the 'essence' of a system) instead of logical models, and implementation models instead of physical models.

Yourdon (1989) suggests that too much time might be spent considering the current system, making the user impatient to the point of possibly cancelling the project. He also notes that the terms 'physical' and 'logical' are confusing with the distinction between them poorly defined. As stated earlier, spending some time on the current models enables the systems analyst to become familiar with the business activity, and may be used to gain the user's confidence. The authors therefore suggest that a model of the current system is useful, but that the time spent on this should be monitored within a project management system. The terms physical and logical are retained.

Exercise 2.6

From the given case scenario for the Albany Hotel, develop for the guest booking and accounting system:

You should check your context diagram – it may need to be changed after logicalization.

(a) current physical top level DFD;
(b) context diagram;
(c) current logical top level DFD.

Chapter 3

Process Modelling 2

OBJECTIVES

In this chapter you will learn:

- □ using guidelines, how to develop a levelled set of DFDs;
- □ to appreciate alternative approaches to data flow modelling.

Introduction

Even in small systems, a top level DFD might contain many processes and look cluttered. Also, it can be difficult to see in sufficient detail exactly what a process does. To overcome this, the top level DFD is usually broken down, and a set of DFDs produced. This practice is known as **levelling** or **partitioning**, allowing us to view the system at different levels of detail.

See Chapter 2 ('Approach to drawing a current physical DFD') for an explanation of a top level DFD.

Levelled Data Flow Diagrams – basic concepts

Figure 3.1 shows part of a top level DFD with the process `Maintain Stock`, which may be decomposed as shown in Figure 3.2. At the lower level, the action taken by each individual data flow trigger is clearly shown. Also, it can be seen that the data flow `purchase_requisition` is only created as a result of the process `Issue Part`. Therefore, it is demonstrated that the lower level DFD depicts an additional level of detail.

Other names for decompose are expand or explode.

Notation

The notation used on levelled DFDs is the same as for the top level. The process identifier from the top level process is used as the first part of the identifier for the lower level processes. In Figure 3.1, the process identifier is 2. As shown in Figure 3.2, the decomposed processes have as their identifiers 2.1, 2.2, 2.3 and 2.4.

As with the top level DFD, the numbering sequence does not indicate the order in which the processes will be activated.

DFDs might have a number of levels, though as a student it is unlikely that you will need more than three (excluding the context diagram). The top level DFD is alternatively known as level 1, with successive levels being labelled as 2, then 3, and so on. The notation for the levels and for process identification is shown in Figure 3.3.

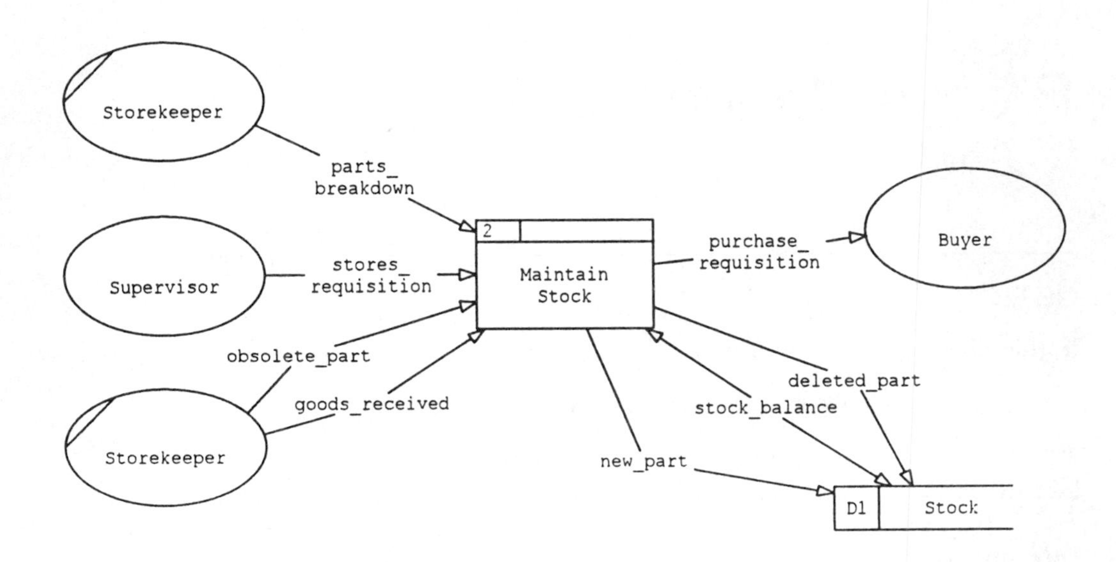

Figure 3.1 DFD notation – fragment of a parent diagram.

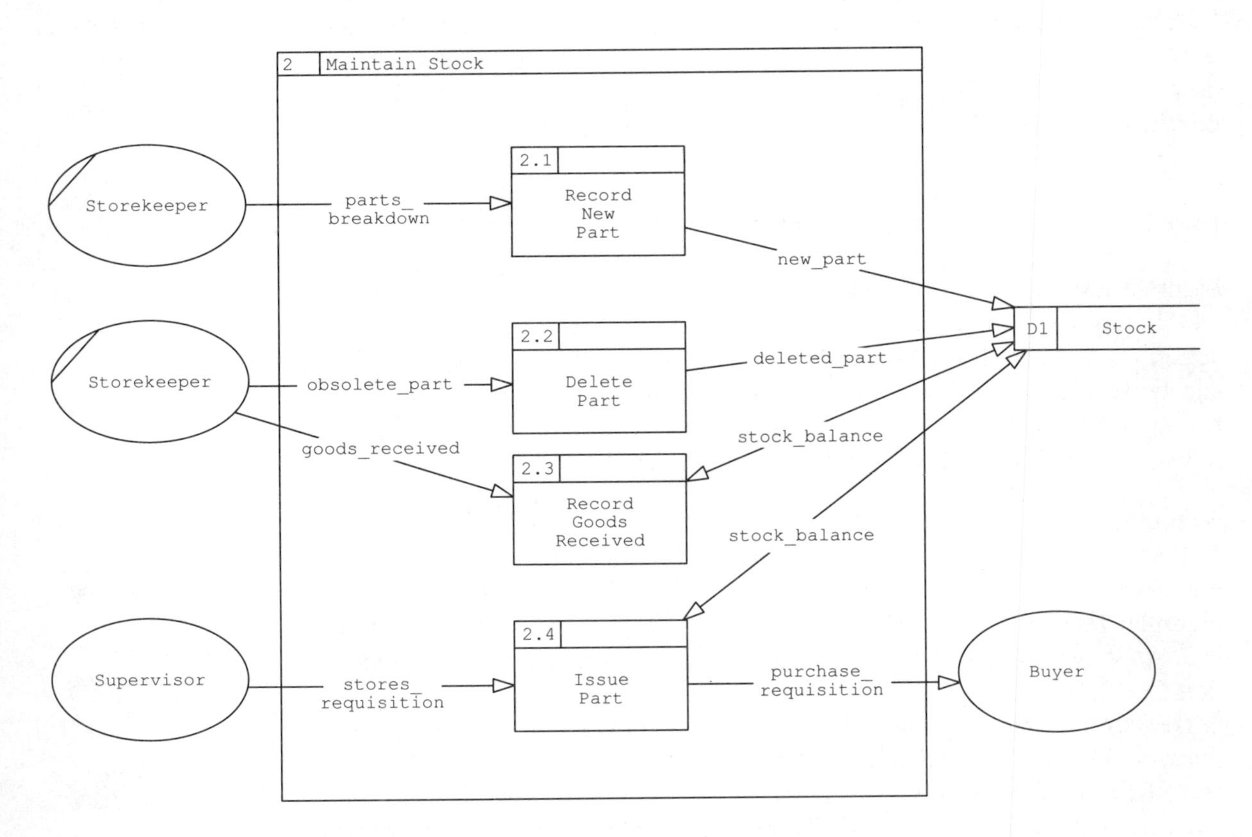

Figure 3.2 DFD notation – child diagram.

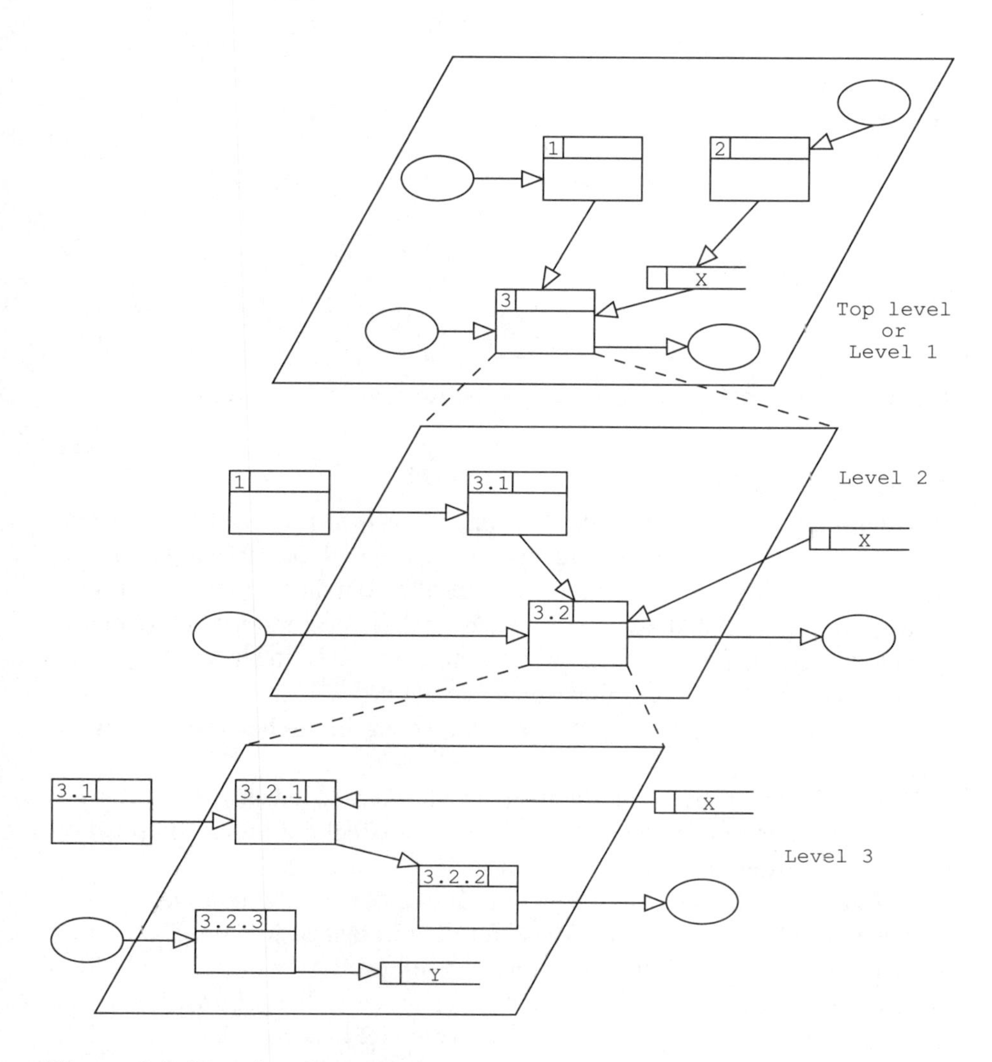

Figure 3.3 Notation for DFD levels.

In SSADM, for identification, bottom level processes are marked with an asterisk in the bottom right-hand corner of the box.

Functional primitives

Processes that will be decomposed are those that are complex, often having a number of data flows. This means that at any level of a DFD some processes will require further decomposition, whereas others will not. Processes that are not broken down any further are known as **bottom level** or **elementary processes,** or **functional primitives**.

In SSADM, a process specification is known as an Elementary Process Description. See Chapter 4 ('Processes') for references to the use of alternative methods of representation to Structured English.

Each functional primitive is supported by a process specification, which shows the detailed logic of the process in Structured English. The logic for one process is normally contained within one side of A4 paper. For anything longer, the process should be considered for further decomposition. Figure 3.4 shows the process specification for Figure 3.2 process 2.4.

The process specification will be part of the data dictionary, discussed in Chapter 4.

```
BEGIN
ACCEPT stores_requisition
RETRIEVE stock_balance FROM Stock
SUBTRACT quantity_issued FROM stock_balance
IF stock_balance LESS THAN reorder_level
  CREATE purchase_requisition
ENDIF
WRITE stock_balance TO Stock
END
```

Figure 3.4 Example process specification for Figure 3.2 process 2.4.

Balancing

It is important to check that the decomposition of a process has been performed correctly. This checking can be facilitated by drawing a frame around the lower level DFD, which denotes the boundary of the higher level process. In Figure 3.2, the frame drawn around the decomposed DFD makes it easier to count and then check the data flows entering and leaving the process. There are four data flows entering the process, three leaving it, and two both entering and leaving it, the same as on the higher level in Figure 3.1.

Note, however, that the data flow `stock_balance` appears once on the higher level and twice on the lower level.

The data flow names should then be checked to ensure that they are exactly the same at each level. This method of checking the decomposition is known as **balancing.**

In Figure 3.2, the data store `Stock` is drawn outside the boundary frame. This is normally the case if the data store is referenced by any of the other higher level processes. If the data store is referenced only by the level of the diagram in question then it is *local* to that process. It is shown inside the frame and then omitted from the higher level DFD making it less cluttered.

However, some systems analysts prefer to retain the data store at the higher level for completeness. In this case, the data store should be shown outside the frame on the lower level diagram.

On a more complete DFD the other elements (terminators or processes) from/to which a data flow passes are shown outside the frame. The DFDs are easier to read, which makes it easier to balance them. The remaining examples in this chapter use this technique.

Dictionary balancing

So far it has been suggested that the number of data flows entering and leaving a higher level or parent process should be the same as those on the associated lower level or child DFD. In fact, there is a case when this is not true, but the diagrams still balance! Figure 3.5 shows part of a parent DFD, and its associated child diagram.

The fact finding showed that the enquiry could be either a stock enquiry or an order status enquiry. On decomposition, the lower level or child diagram would be drawn as shown in Figure 3.5. To reduce the clutter on the parent diagram the more general terms ‘enquiry’ and ‘reply’ are used, with

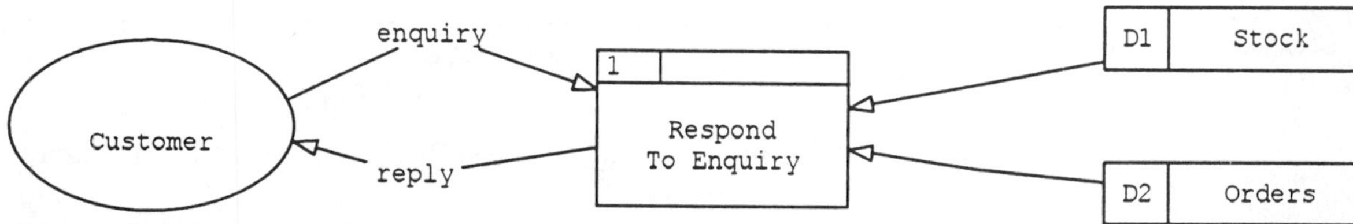

Figure 3.5 Example DFD process. (a) Parent.

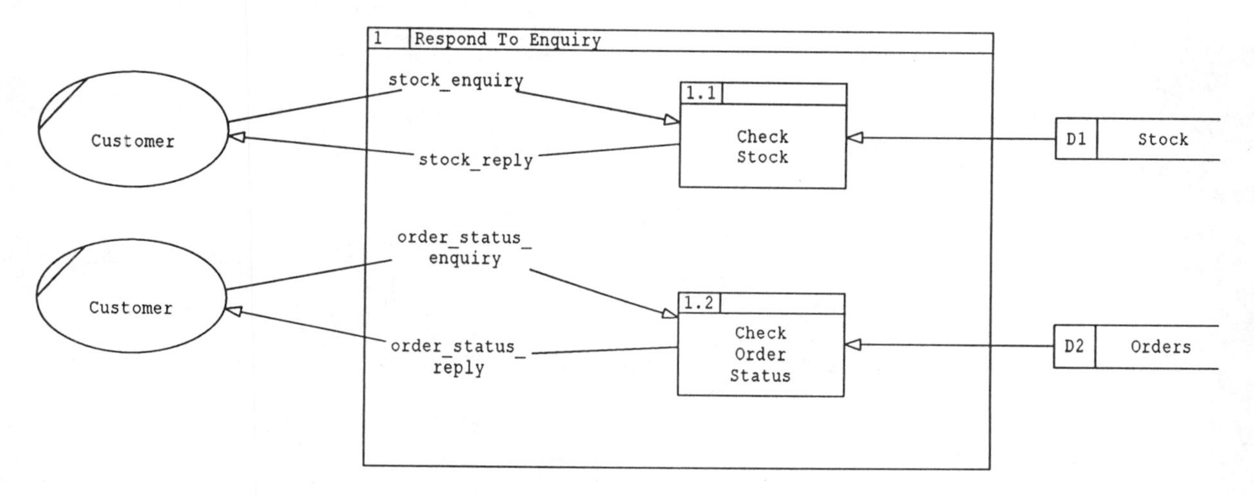

Figure 3.5 (b) Child.

more detail being included on the lower level diagram. It is not sufficient to *assume* that the diagram balances, so we need to make an entry in the data dictionary as follows:

```
enquiry = [stock_enquiry | order_status_enquiry]
reply = [stock_reply | order_status_reply]
```

`[] selection,`
`| separation of choices.`

The data dictionary is discussed in Chapter 4.

This practice may be used whenever it is possible to specify generic data flows at the higher level, and is known as **dictionary balancing.**

One of the dilemmas within data flow diagramming is whether or not to include error handling. Error handling should be included for completeness, but it clouds the 'big picture'. Diagrams become cluttered with many additional data flows. The common consensus of other authors is to exclude error handling from the diagrams completely or at most to confine it to the lower level diagrams. If it is the latter then obviously this does not facilitate balancing.

See below for Gane and Sarson's solution.

In this text, error handling is generally omitted where the error is a result of validation, but included where the data is incomplete. For example, in the Albany Hotel example (see Exercise 2.6c), the data flows have been included where further information is requested from the potential guest/guest if the booking form is incomplete.

Further discussion will take place in Chapter 4 in relation to the definition of the data dictionary.

Exercise 3.1

In Figure 3.6, check the balancing between the given extract of the parent diagram and the child diagram.

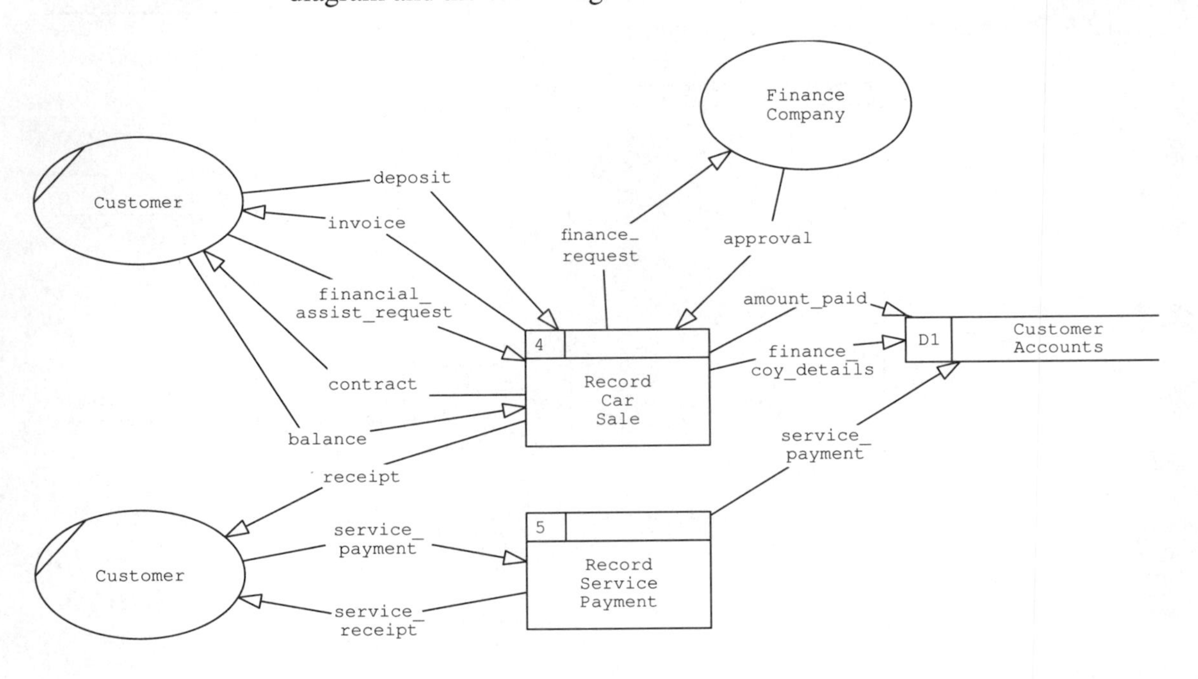

Figure 3.6 Fragment of DFD for Exercise 3.1. (a) Parent.

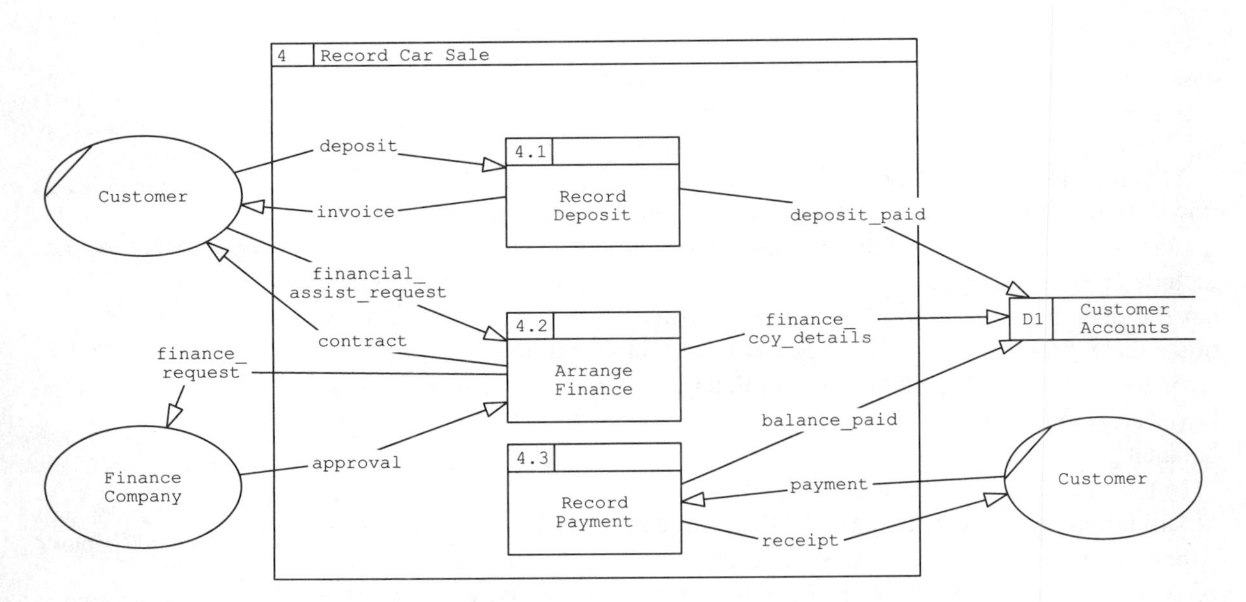

Figure 3.6 (b) Child.

The fact finding showed that on the sale of a car a deposit is received. An invoice is sent to the customer and acts as a receipt for the deposit paid and gives the balance outstanding. The garage will arrange finance for the balance on behalf of the customer if required. The customer pays the balance on delivery of the car and is given a receipt.

Exercise 3.2

For the solution to Exercise 2.5, decompose process 1, `Verify Order`. Orders are refused either due to credit check failure or if there is insufficient stock. The customer is informed of the reason.

Do not forget to check that your diagrams balance.

Combining processes

So far the levelling of a DFD has been considered only by decomposing processes. It is also possible to combine two or more processes at the higher level, and then to show the detail at the lower level. Figure 3.7 shows the combination of processes 3, 4 and 5 from the solution to Exercise 2.5. These processes are combined as they are all concerned with the recording of accounts and they access common data.

That is, from the user's viewpoint, they denote a logical functional area.

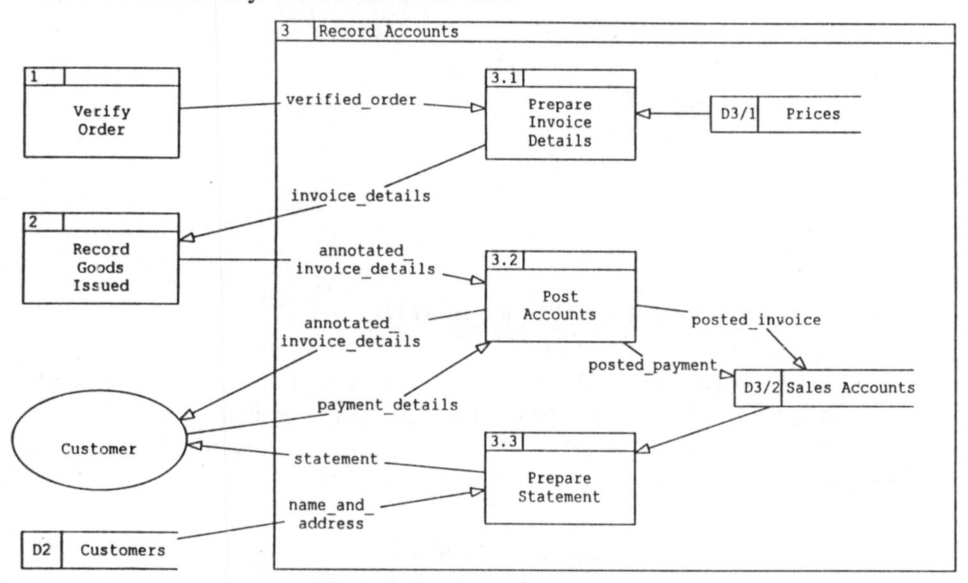

Figure 3.7 Lower level diagram for merged process.

The data stores `Prices` and `Sales Accounts` are local to this level, and are drawn inside the boundary frame. They may be omitted from the higher level diagram. The higher level diagram is redrawn, and balancing performed to check consistency between the two diagrams. Figure 3.8 shows the parent diagram now reflecting both the combination of processes 3, 4 and 5; and the decomposition of process 1, `Verify Order` from Exercise 3.2.

It is assumed that the maintenance of the data stores `Prices` and `Sales Accounts` is part of another subsystem, and is not included here.

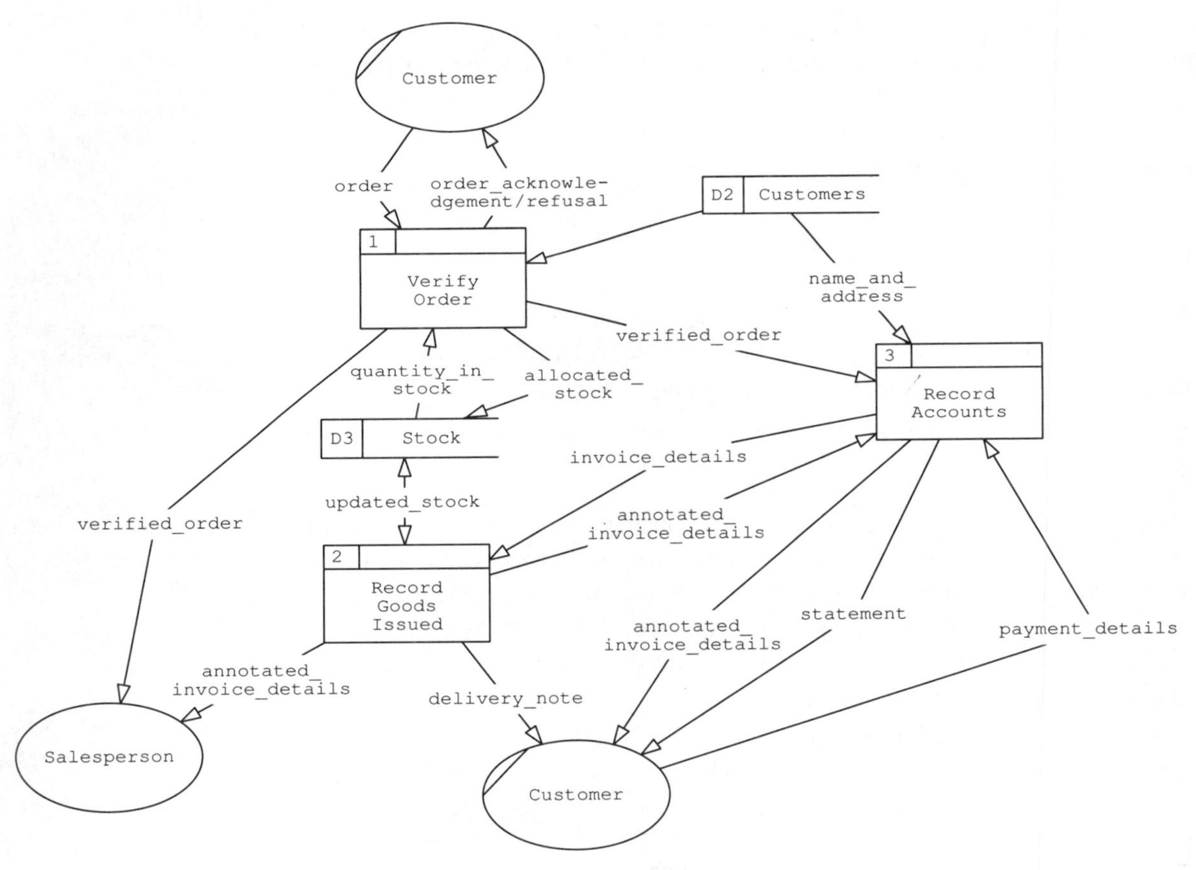

Figure 3.8 Revised top level diagram.

Approach to drawing a levelled set of DFDs

The points are labelled as guideline steps for consistency.

The current logical top level DFD may be used as a basis for the development of a levelled set of DFDs. The guidelines for levelling are less well-defined than those given in Chapter 2 for drawing top level DFDs; they are really points to keep in mind.

Guidelines for drawing a levelled set of DFDs

The number of processes on any DFD is commonly seven plus or minus two. This guideline is based on the work of George Miller (1956) who first observed that people found difficulty in being presented with too much information at

1. Review the top level DFD. If it does not contain more than nine processes then levelling may not be needed.
2. For DFDs where levelling is required, consider the two major techniques covered earlier: decomposition and merging.
3. Decompose complex processes, and check that both levels of the diagram balance.
4. Combine or group processes at the level you are considering that belongs to the same logical function from the user's viewpoint, or that

access the same data. Group them together in a single diagram at the next level down. Again, check that both levels of the diagram balance.

5. During steps (3) and (4), look for data stores that might be drawn on lower level DFDs and omitted from the top level.
6. During steps (3) and (4), look for opportunities to create generic data flows at the top level.
7. Repeat steps (2)–(6) for the lower level DFDs.
8. Review the set of DFDs. Whereas each process on the top level DFD will not be decomposed through the same number of levels, it is unlikely that one process will be a functional primitive and another will be decomposed through a number of levels. If this is the case, consider reorganizing the DFD to make it less skewed.
9. Prepare process specifications for all functional primitives. If the process specification covers more than one page of A4, the process might still be too complex and further levelling should be considered.
10. Finally, check that the levelled set of DFDs balances.

any one time. Yourdon (1989) concurs with this approach. However, SSADM specifies that typically 15 processes may be included at the top level, with four to eight at lower levels.

At this point, you would need to make the relevant entries in the data dictionary.

Step (9) is included for completeness, although you will not be able to do this until you have studied Chapter 4.

There is often a number of requirements for reports or enquiries that need to read but not update data stores. These are known as **retrievals**. SSADM advocates that a DFD will be less cluttered if many of the processes associated with retrievals are removed and entered in the requirements catalogue. Only the major retrievals are retained on the DFDs.

SSADM Reference Manual (1990).

Worked example 3.1

Figure 2.11 shows the current logical DFD for the Somerleyton Animal Park ordering supplies section. From the diagram, prepare a set of levelled DFDs.

Solution:

Guideline step 1: A review of the top level DFD shows that there are nine processes, so we would need to decide whether to produce a levelled set of DFDs or not. In this case, we have no choice – the worked example requests it! At the end of the example we hope you will see that the levelled set of DFDs with an increasing level of detail is easier to read.

Guideline step 2: Both decomposition and merging will be considered.

Guideline step 3: The only complex process on the top level DFD is the process `Make Payment`. This is decomposed as shown in Figure 3.9.

Note that the process identifiers are 9.1, 9.2 and 9.3. The data stores `Suppliers` and `Orders` are shown outside the frame as they are accessed by other processes.

In Worked example 2.3 guideline step 6, if separate data stores had been maintained for `Awaiting Payments` and `Back Orders`, then the data store `Back Orders` would be drawn inside the frame boundary as it is local to this level, and therefore removed from the top level.

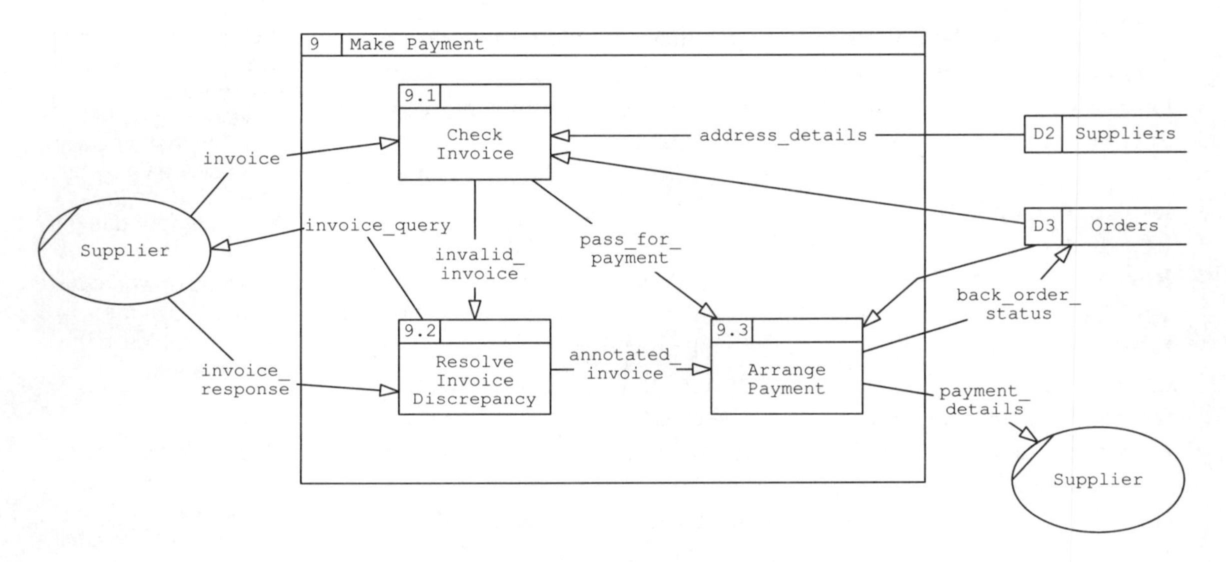

Figure 3.9 Worked example 3.1 after guideline step 3.

See Worked example 3.2. Within SSADM, the fact that this would need to be considered would be recorded in the requirements catalogue to ensure that it would not be forgotten.
The data flows `invoice_query` and `invoice_response` have been added to deal with these discrepancies.

As mentioned in Chapter 2 (solution to Worked example 2.1 queries and assumptions 9), there is no way of deleting the orders with back order status from the data store `Orders`. It will contain more and more records. The current action should be checked with the user and recorded. If no action is taken, then any proposal should be considered when the required system DFDs are developed.

It is possible that there could be a discrepancy over the invoice. This point would need to be checked with the user so that the appropriate action may be determined. It is assumed that any discrepancy will be resolved.

Guideline step 4: The top level DFD appears to have four functions from the user's viewpoint. Processes 1–3 all relate to the establishment of requirements. These processes are combined (or merged) to give one process `Establish Requirement` at the top level with the detail included at level 2. Similarly, processes 4 and 5 relate to the ordering of supplies; processes 6–8 relate to the recording of the receipt of supplies; and process 9 to making payments (as described in Guideline step 3). Figure 3.10 shows the levelled set of DFDs drawn to reflect these decisions, and includes for completeness the decomposition of the process `Make Payment`.

Note that the processes have been renumbered.

Guideline step 5: The data store `Animal Foods` is omitted from the top level diagram as it is local to the lower level. It is at this point, however, that we should resolve an earlier issue about food supplies. We need to be clear how the food requirements will be determined. The data store `Animal Foods` is used to maintain the requirements for each animal and the data flow

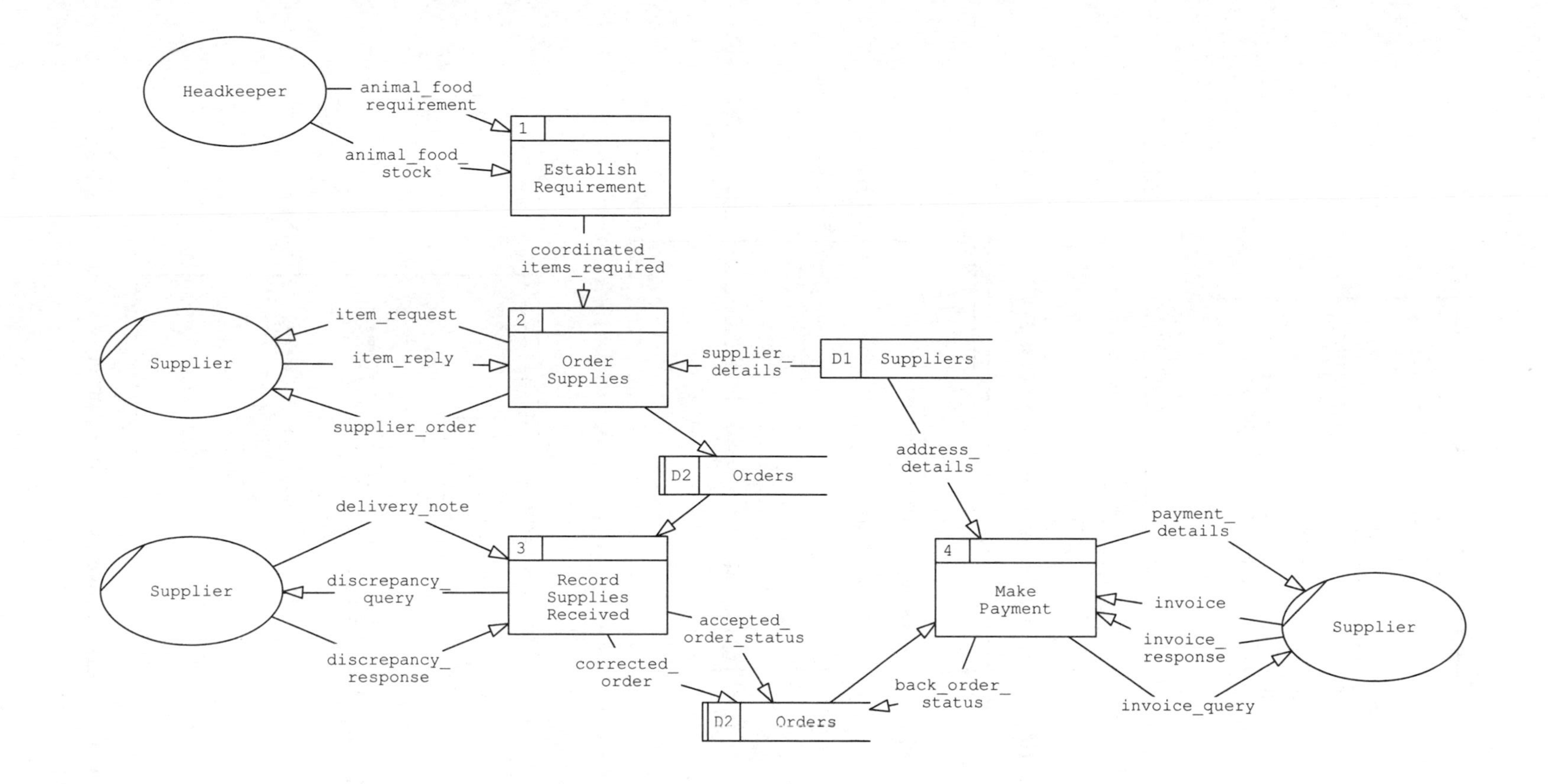

Figure 3.10 Worked example 3.1 after guideline step 4. (a) Top level diagram.

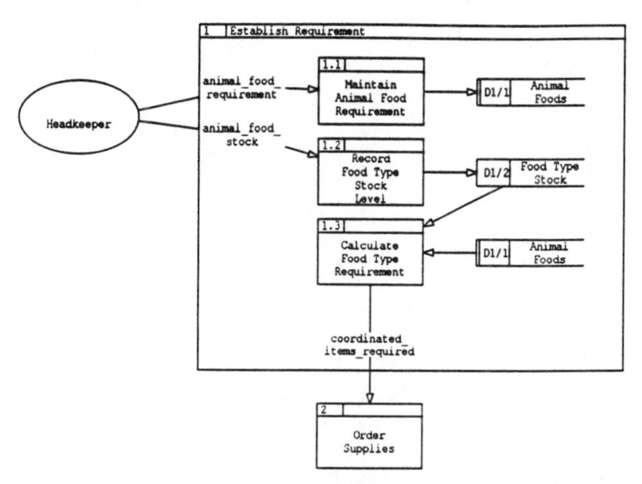

Figure 3.10 (b) Lower level diagram for process `Establish Requirement`.

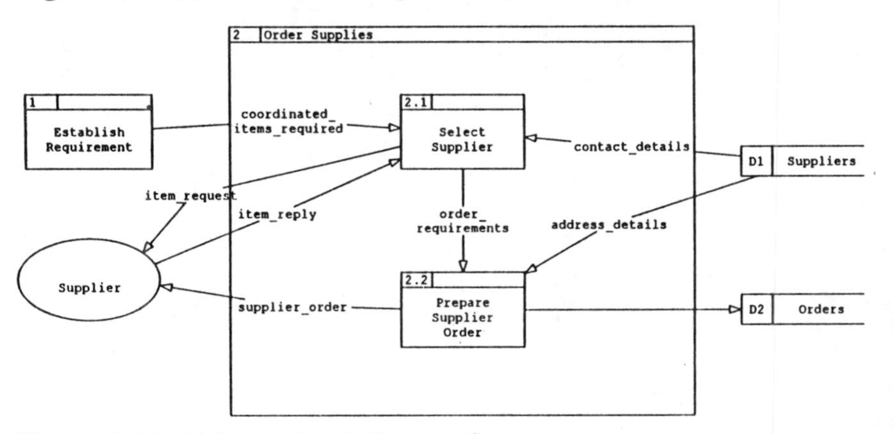

Figure 3.10 (c) Lower level diagram for process `Order Supplies`.

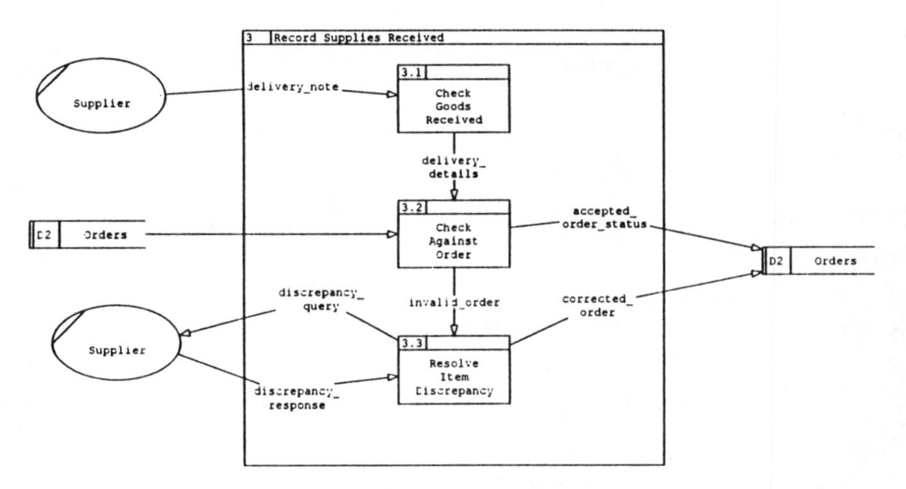

Figure 3.10 (d) Lower level diagram for process `Record Supplies Received`.

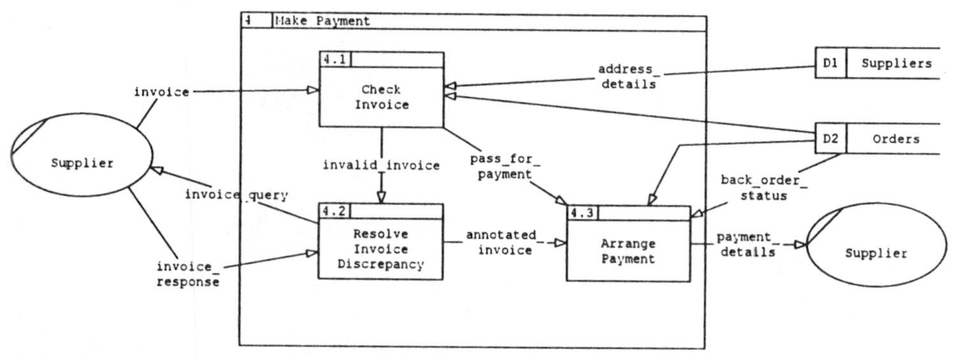

Figure 3.10 (e) Lower level diagram for process `Make Payment`.

`animal_food_stock` is used to show how much food we still have in stock. In the current system this is determined by the headkeepers physically checking the food supplies. This data will be required for the food requirement to be determined accurately. It is decided that the headkeepers actually check the food supplies and in fact record the animal food stock. So `animal_food_stock` remains as a data flow but with the data being recorded in the data store `Food Type Stock`. On a weekly basis, the quantity of each food type required is calculated taking into account requirements and the existing stock level for each area.

See Chapter 4, Worked example 4.1, for more details on the data dictionary entries for the proposed system.

This emphasizes the important fact that logicalization should not be just a mechanical process.

Guideline step 6: The data flow `supplier_details` on the top level DFD is a generic data flow for data flows `contact_details` and `address_details` on the level 2 diagram. The data dictionary entries for these are:

```
supplier_details = [contact_details | address_details]
```

As commented in Worked example 2.3 guideline step 6, the content of these data flows will be considered again in Chapter 4, Worked example 4.1.

Guideline step 7: The lower level diagrams have been checked and no further breakdown is needed.

Guideline step 8: The DFDs have been checked and the decomposition is even. No changes are necessary.

Guideline step 9: Process specifications will be discussed and prepared in Chapter 4.

Guideline step 10: The notation and balancing has been checked and no further changes are needed.

As is seen from this example, the preparation of the set of levelled DFDs with the added detail also highlights areas where further clarification is needed.

This worked example has been drawn for the ordering supplies subsystem. As can be seen from the case scenario, other subsystems exist. On

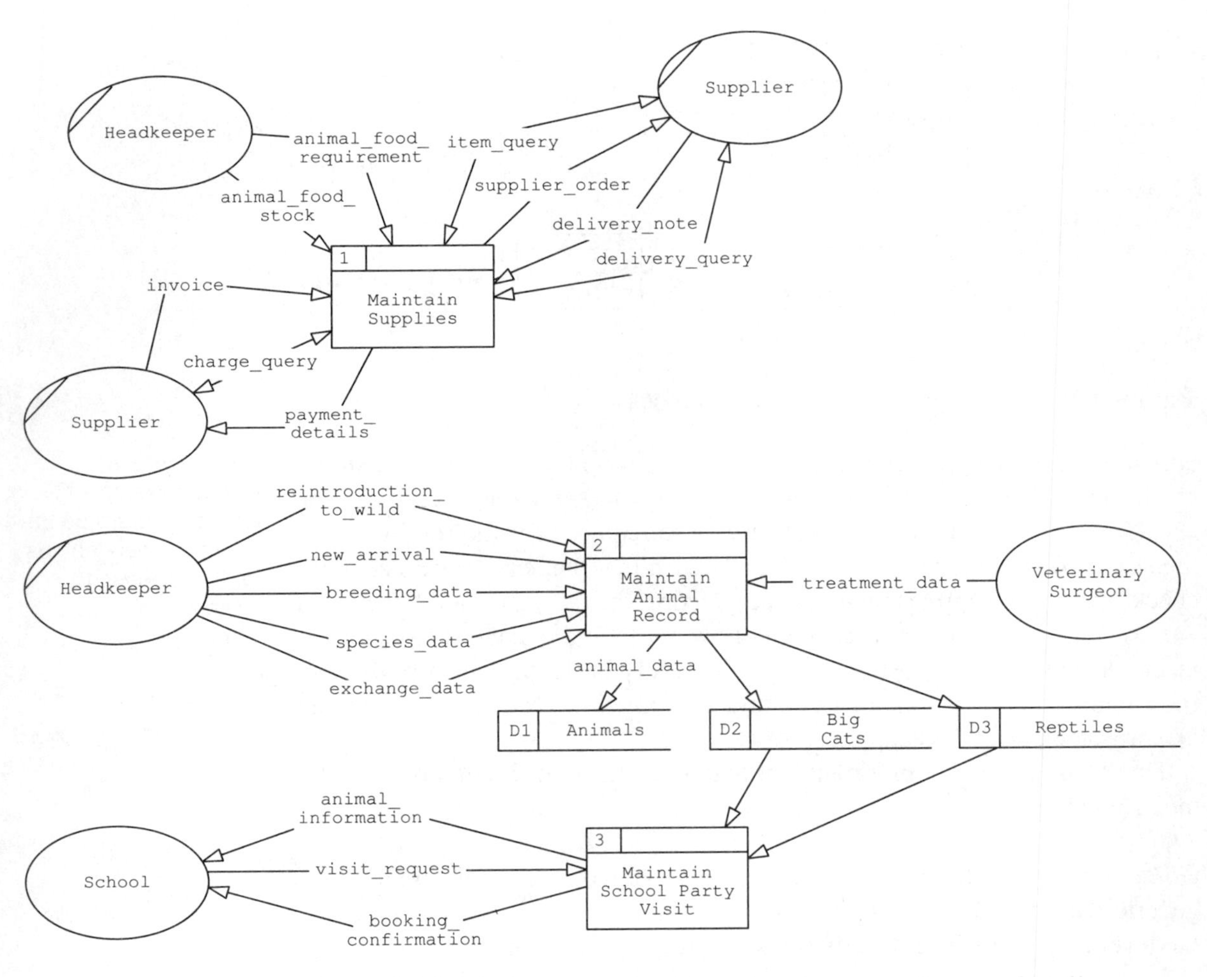

Figure 3.11 Worked example 3.1 top level current diagram for all subsystems.

investigation it may be seen that two further subsystems may be defined: maintain animal record and maintain school party visit.

It is common to depict the subsystems on a top level DFD as shown in Figure 3.11.

The lower level diagram for the process `Maintain Supplies` has not been redrawn here, and will be as shown in Figure 3.10a but with process identifiers 1.1, 1.2, 1.3 and 1.4. This diagram now becomes a level 2 diagram. Note that generic data flows have been introduced on the top level diagram. The data dictionary entries for these are:

```
item_query = [item_request | item_reply]
delivery_query = [discrepancy_query | discrepancy_response]
charge_query = [invoice_query | invoice_response]
```

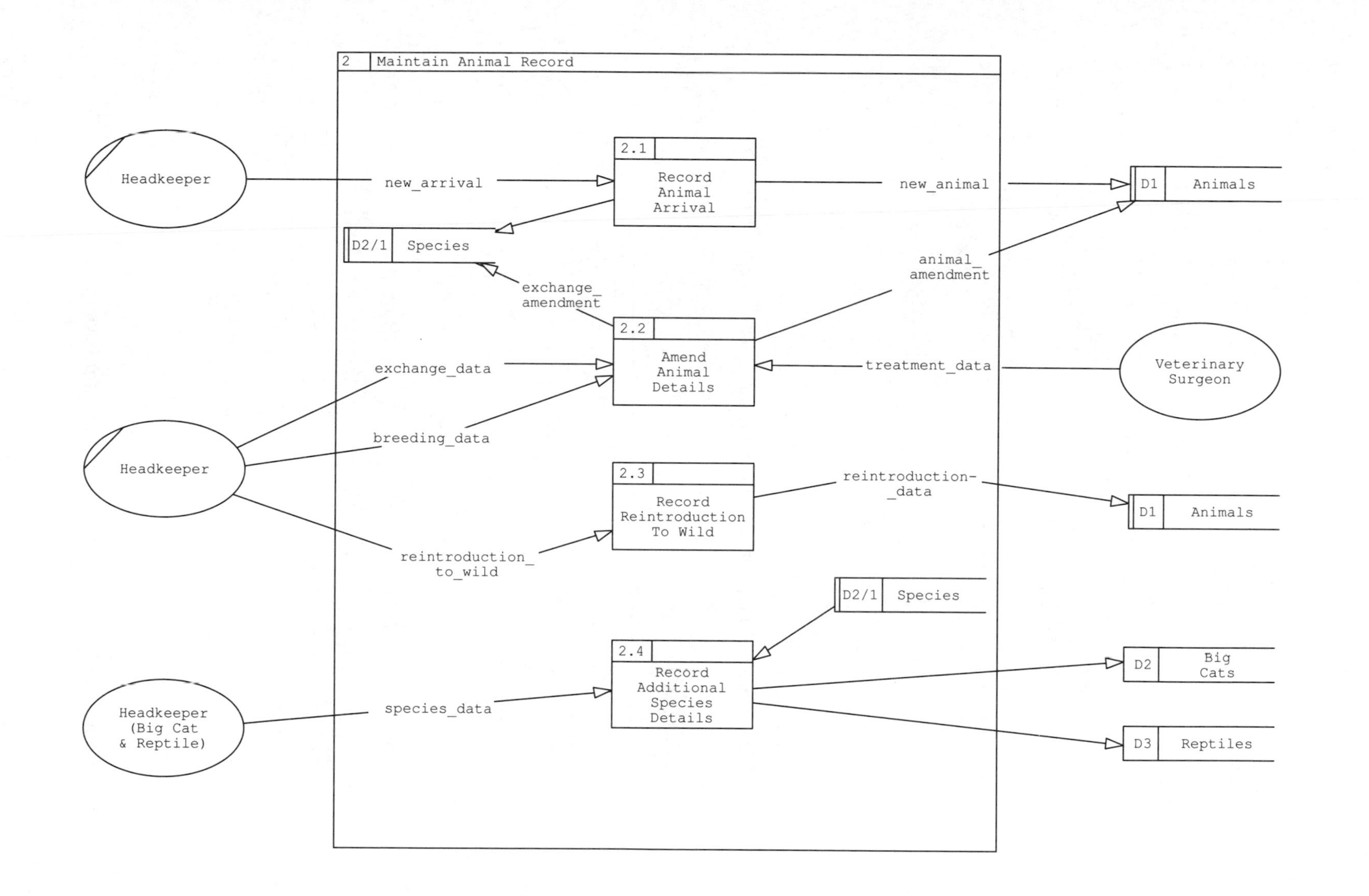

Figure 3.12 Worked example 3.1. (a) Lower level diagram for process `Maintain Animal Record`.

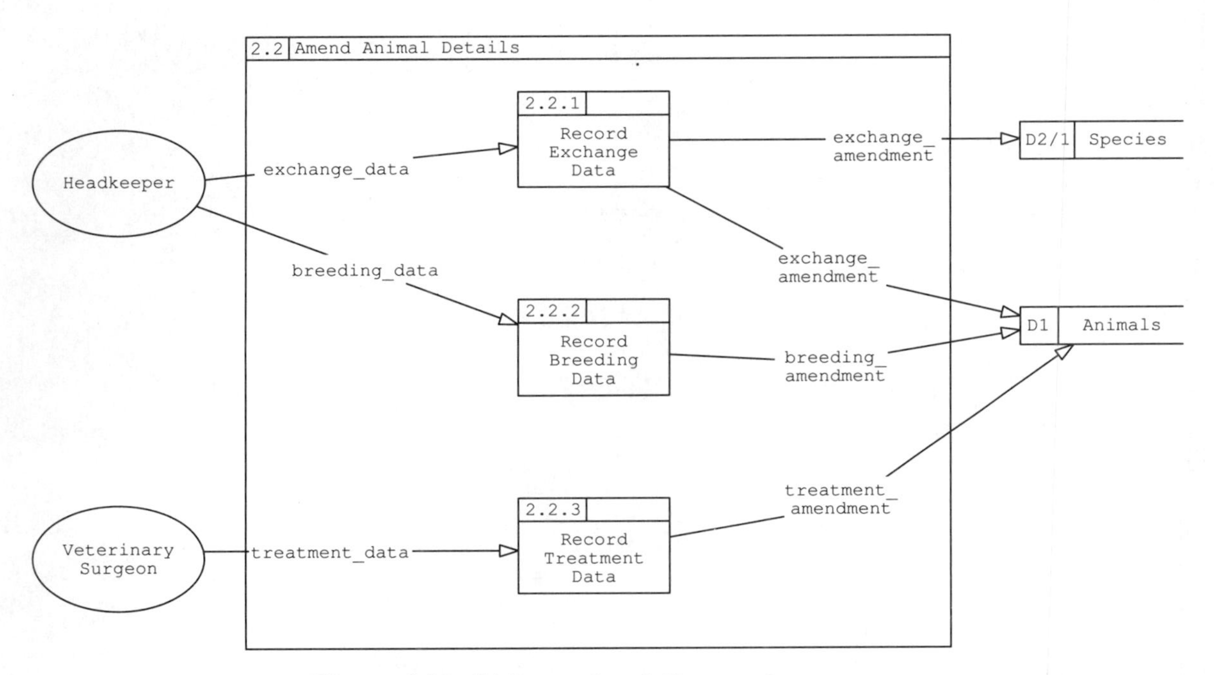

Figure 3.12 (b) Lower level diagram for process `Amend Animal Details`.

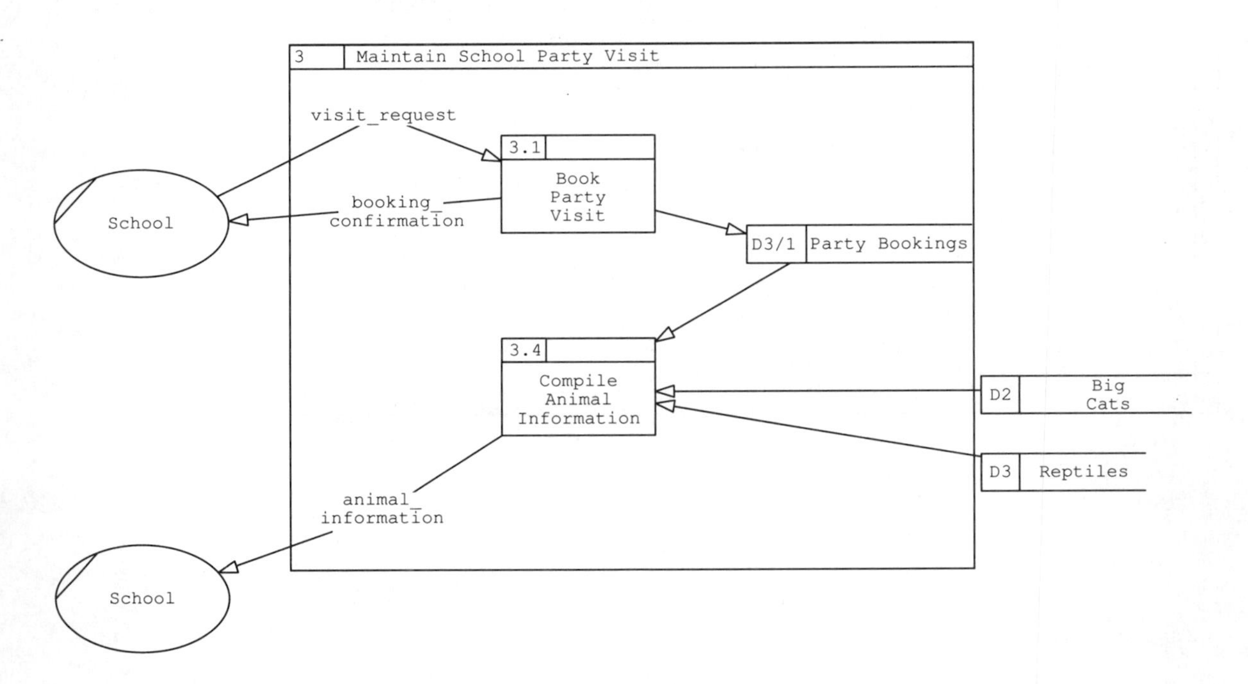

Figure 3.12 (c) Lower level diagram for process `Maintain School Party Visit`.

The data flows for `item_request` and `item_reply` flow in different directions, hence the bidirectional arrowheads on the generic data flow `item_query`. The same principle applies to generic data flows `deliver_query` and `charge_query`. Again, the diagram shown in Figure 3.10b has not been redrawn, but now becomes a level 3 diagram with the processes having process identifiers 1.1.1, 1.1.2 and 1.1.3. Similarly, the diagrams in Figures 3.10c, d and e also become level 3 with process identifiers as 1.2.1 and 1.2.2; 1.3.1, 1.3.2 and 1.3.3; and 1.4.1, 1.4.2 and 1.4.3, respectively.

The renumbered processes are all shown on the required levelled DFDs in Figure 3.13.

The lower level diagrams for the processes `Maintain Animal Record` and `Maintain School Party Visit` are shown in Figure 3.12.

In Figure 3.12a the data flows from the process `Amend Animal Details` are unidirectional towards the data stores `Animals` and `Species`. Although it is necessary to retrieve the details from the data stores this is *only* for the purpose of amendment, that is read and written back. In this case, the output data flow is dominant, and it is a convention to show the data flow as unidirectional rather than bidirectional. There are a number of other examples of this convention on the diagrams.

SSADM specifies that data flows on the lowest level diagrams should be unidirectional.

There is no provision to record that an animal has died. An experienced systems analyst would notice this at this stage (if not earlier) and check with the user. Any action taken in the current system should be recorded, otherwise this should be resolved when developing the required system.

See Worked example 3.2.

Figure 3.11 shows the generic data flow `animal_data`. The data dictionary entry for this is:

```
animal_data = [new_animal | animal_amendment | reintroduction_data]
```

Figures 3.12a and b show that the data flow `animal_amendment` is also generic. The data dictionary entry for this is:

```
animal_amendment = [breeding_amendment | exchange_amendment |
                                         treatment_amendment]
```

Exercise 3.3

Exercise 2.6c shows the current logical DFD for the Albany Hotel guest booking and accounting system. From the diagram, prepare a set of levelled DFDs.

Our examples have shown the development of levelled DFDs for the current logical model. It is possible to prepare a levelled set of DFDs for the current physical model using the same approach as for a current logical model.

However, if little emphasis is to be placed on the current physical model, then it would be inappropriate to spend a lot of time on levelling.

Logicalization of the current physical model will follow the guidelines specified in Chapter 2. However, the rationalization of the processes will start at the bottom level as these processes will be the most detailed. The guideline steps defined in Chapter 2 are applicable and should be followed. This approach will avoid the repetition of much of the analysis that would take place if we levelled in a top down manner. After the rationalization of the bottom level processes, it will be necessary to regroup these processes into higher level processes to revise the set of levels. The levelling here is in fact upwards rather than downwards. Again, the DFDs should be checked for balancing.

SSADM Reference Manual (1990).

Logical DFDs for the proposed (required) system

Coverage is given to this subject in a number of texts on systems analysis. Useful texts include: Blethyn and Parker (1990, Chapter 5), Flynn (1992, Chapters 6 and 7) and Senn (1989, Chapters 1 and 3).

When faced with the task of developing a system, whether automated, manual or a combination of the two, it is important to consider carefully the requirements of the proposed system. The specification of those requirements is not always an easy task.

The required logical DFDs are based on the current logical DFDs tailored to satisfy the requirements of the proposed system. There are no new guidelines for the preparation of the required logical DFDs. A top down approach is usually used with the concepts of levelling being applied.

Worked example 3.2

The proposed system will cover all the specified requirements for Somerleyton Animal Park, not just those for the ordering supplies subsystem.

Develop a set of levelled logical DFDs for the proposed system, incorporating the specified system requirements into the current logical DFDs. Figures 3.10–3.12 are referenced.

Solution: Reviewing the top level diagram, the specified requirements may be clarified as follows:

- Maintain animal record — Requirements 1, 2, 4
- Maintain school party visit — Requirements 3, 5
- Maintain supplies — Requirements 6, 7.

Hence, all three subsystems are required for the proposed system, and no additional subsystem need be created. It is therefore sensible to consider each subsystem in turn.

Maintain animal record

Requirements 1, 2 and 4 are satisfied by the DFD shown in Figures 3.13a, b and c. All data relating to the animals and species required for the educational notes and proposed booklets will be maintained by this subsystem.

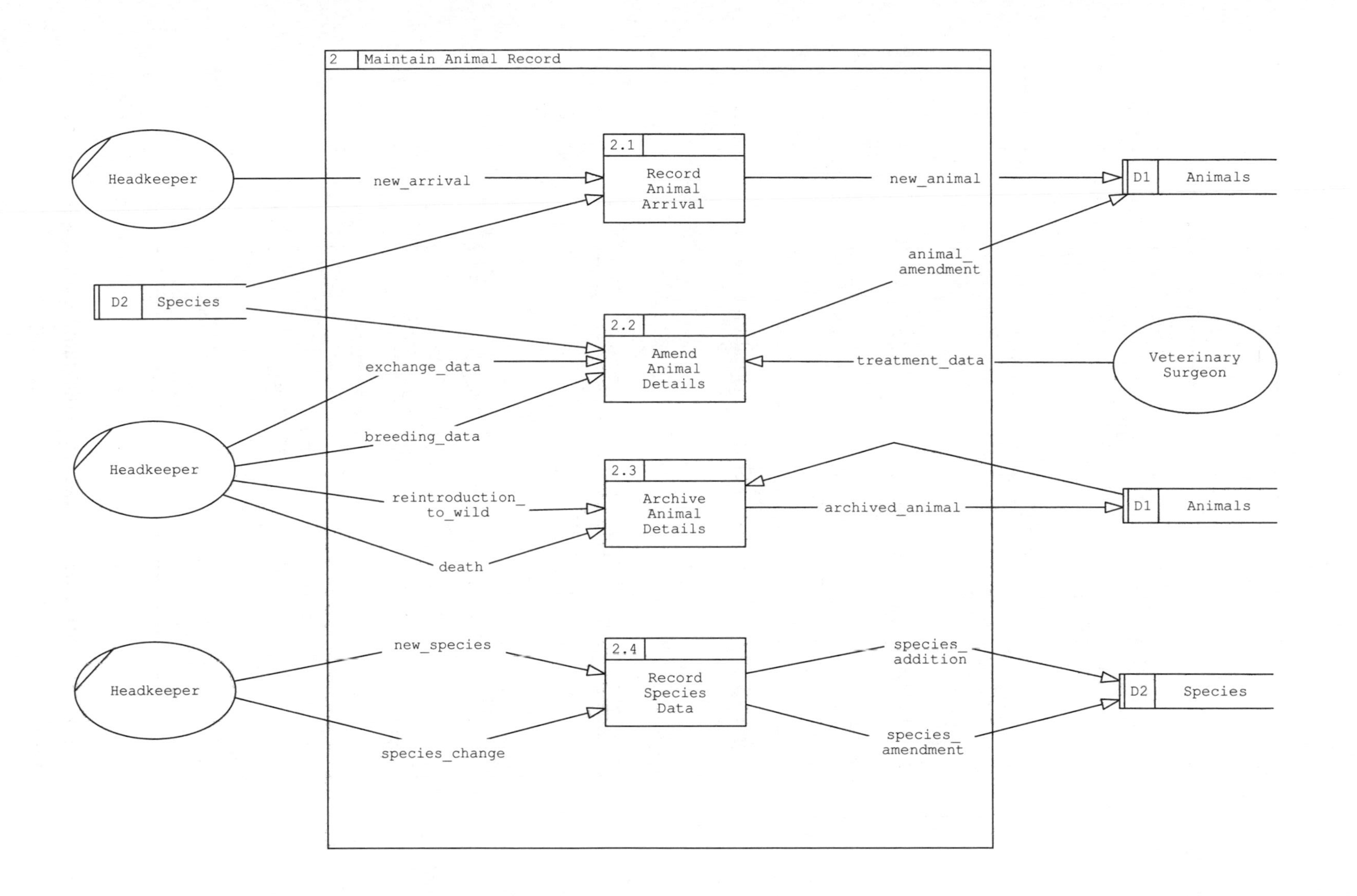

Figure 3.13 Worked example 3.2. (a) Required lower level diagram for process `Maintain Animal Record`.

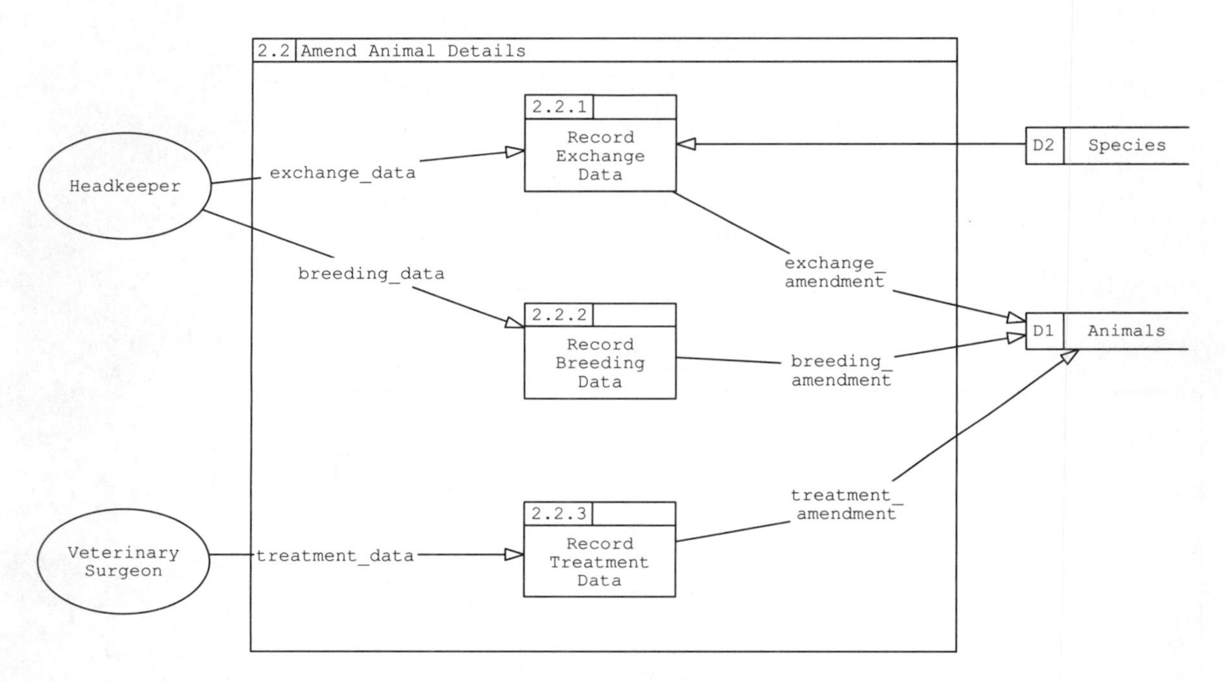

Figure 3.13 (b) Required lower level diagram for process `Amend Animal Details`.

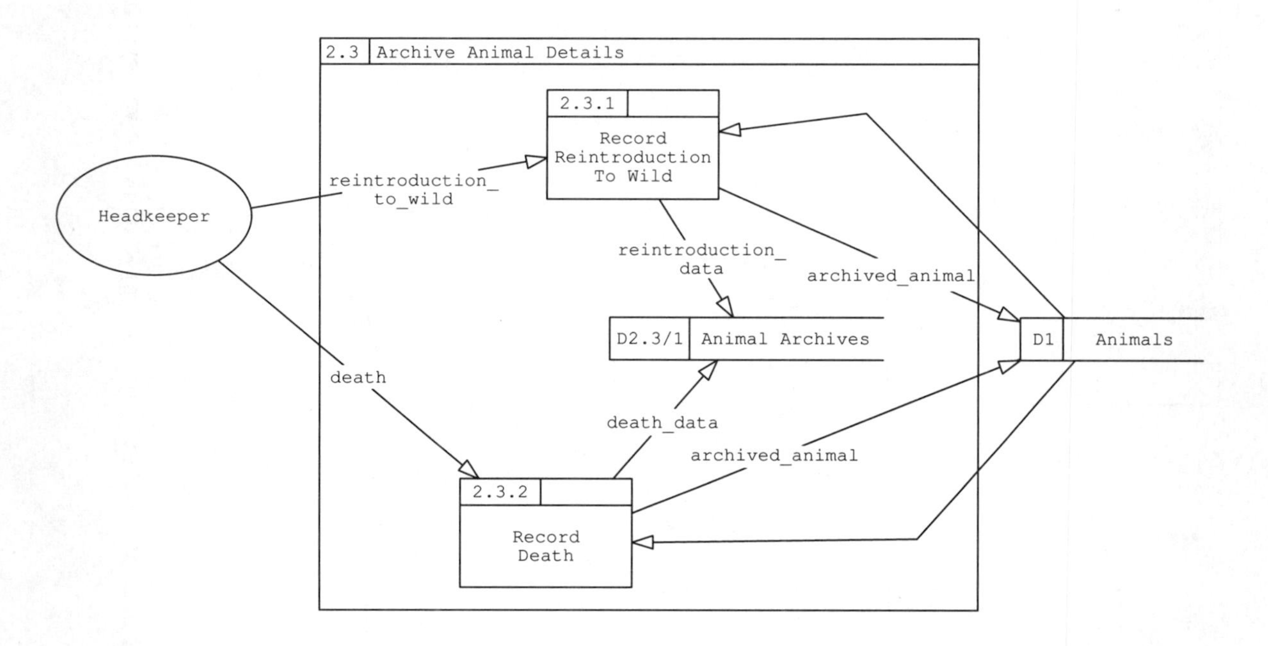

Figure 3.13 (c) Required lower level diagram for process `Archive Animal Details`.

In the current logical DFDs the data stores `Animals` and `Species` hold data for all animals. The data stores `Big Cats` and `Reptiles` hold additional data about these two species as well as duplicating some of the data held on the data stores `Animals` and `Species`. There is an argument for the data stores `Big Cats` and `Reptiles` to be subsumed into the data stores `Animals` and `Species`. The only reason that the duplicate data is held is because one of the systems is manual and one is computerized. This is a physical attribute and should be ignored during logicalization. However, additional data is only stored for big cats and reptiles. Should this additional data be maintained for all animals? This question needs to be resolved when developing the required system. For this reason, the data stores `Big Cats` and `Reptiles` were kept separately when modelling the current system. Requirement 3 states that educational notes are required for *all* species. Therefore, we have the answer to the question. Two data stores `Animals` and `Species` will be used to hold all the required data.

The requirements for the proposed system are stated at the end of the Somerleyton Animal Park case scenario.

The species to which an animal belongs will be recorded in the data store `Animals`.

This means that processes 2.1, `Record Animal Arrival`, and 2.2, `Amend Animal Details`, need only reference the data store `Species` for validation purposes.

Further discussion on the maintenance of the data store `Species` will be undertaken in Chapter 8.

The process `Record Additional Species Details` on the current logical DFDs is no longer required because of the subsumption of the data stores `Big Cats` and `Reptiles` described earlier. However, a process `Record Species Data` will be required to maintain the species' details for all the animals.

Maintain school party visit

Figure 3.13d shows the DFD for the maintenance of school party visits. The subsystem has been extended to allow for the booking, amendment and cancellation of school party visits, and retrieval by request of visit details. The data flows have been changed accordingly, and the data dictionary entry for the generic data flow `visit_booking` is:

```
visit_booking = [visit_request | visit_change]
```

The compilation of the educational notes will now reference the data stores `Animals` and `Species`. The data stores `Big Cats` and `Reptiles` are no longer required. In Figure 3.13a the data store `Species` is shown outside the frame boundary as it is no longer local to the process `Maintain Animal Record`. The process `Compile Animal Information` is triggered by a time trigger, mentioned in Chapter 2, Worked example 2.3, in this case a specified time prior to a school visit. Alternatively, the details could be produced on request which would mean the addition of a data flow to trigger the process.

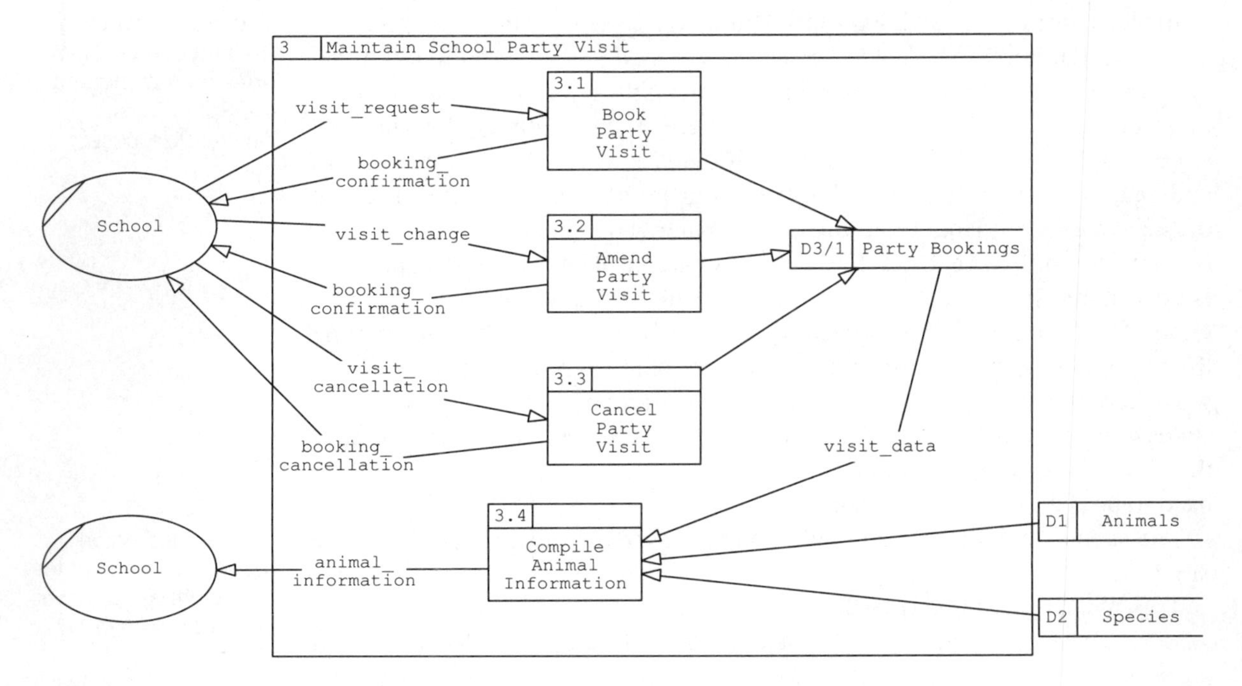

Figure 3.13 (d) Required lower level diagram for process `Maintain School Party Visit`.

Maintain supplies

Figure 3.13e–i shows the DFDs for the maintenance of supplies. Figure 3.13i shows the deletion of back orders triggered by a time trigger, the details of which will be recorded in the data dictionary.

Figure 3.13j shows the revised top level diagram to complete the set. Note that the data stores `Big Cats` and `Reptiles` are replaced by the data stores `Animals` and `Species`.

Additional data dictionary entries for the generic data flows are:

```
species_info = [new_species | species_change]
species_data = [species_addition | species_amendment]
visit_booking = [visit_request | visit_change]
```

Exercise 3.4

Develop a set of levelled logical DFDs, for the proposed system for the Albany Hotel, incorporating the specified system requirements into the current logical DFDs (from the given solution to Exercise 3.3).

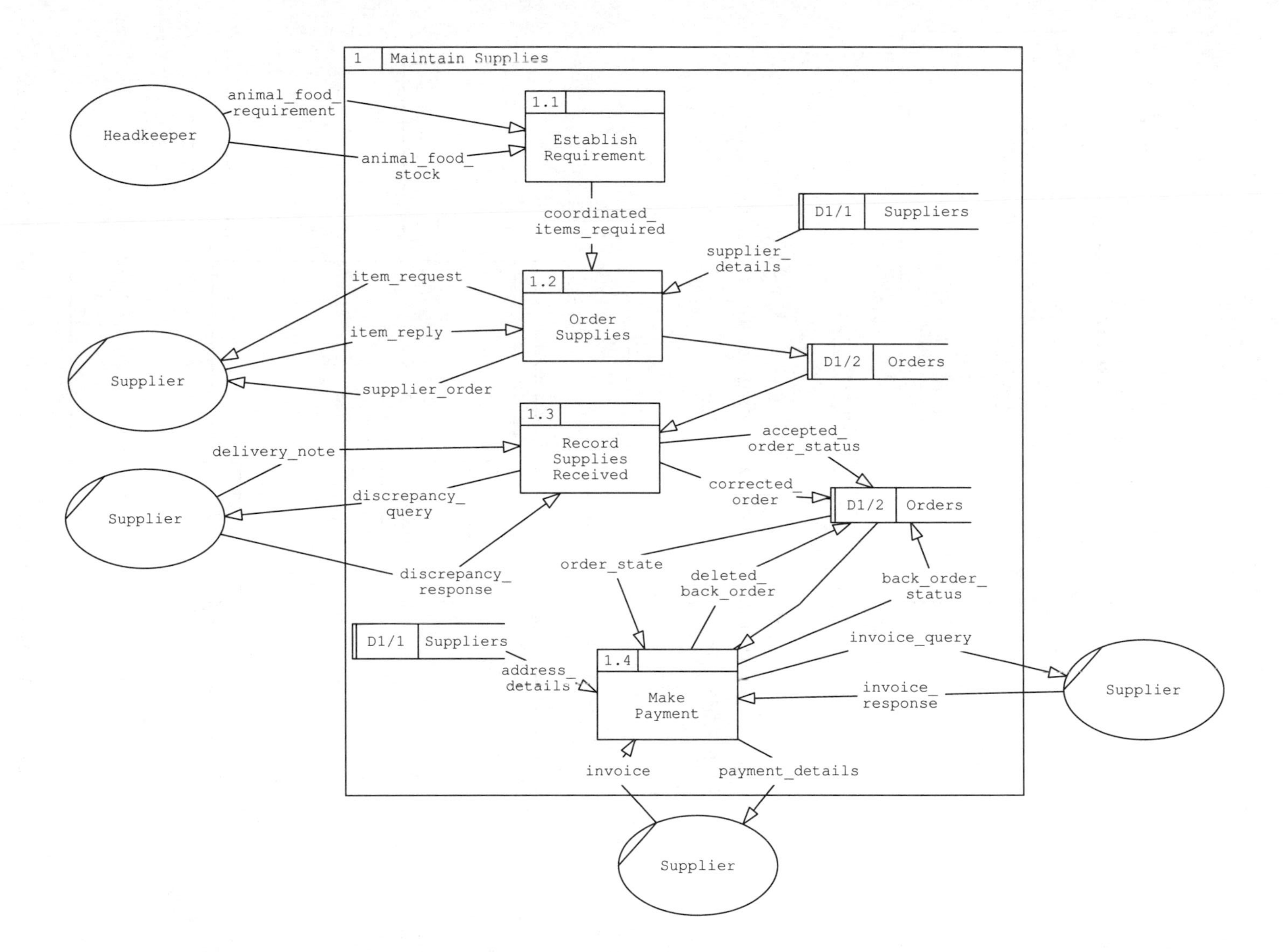

Figure 3.13 (e) Required lower level diagram for process `Maintain Supplies`.

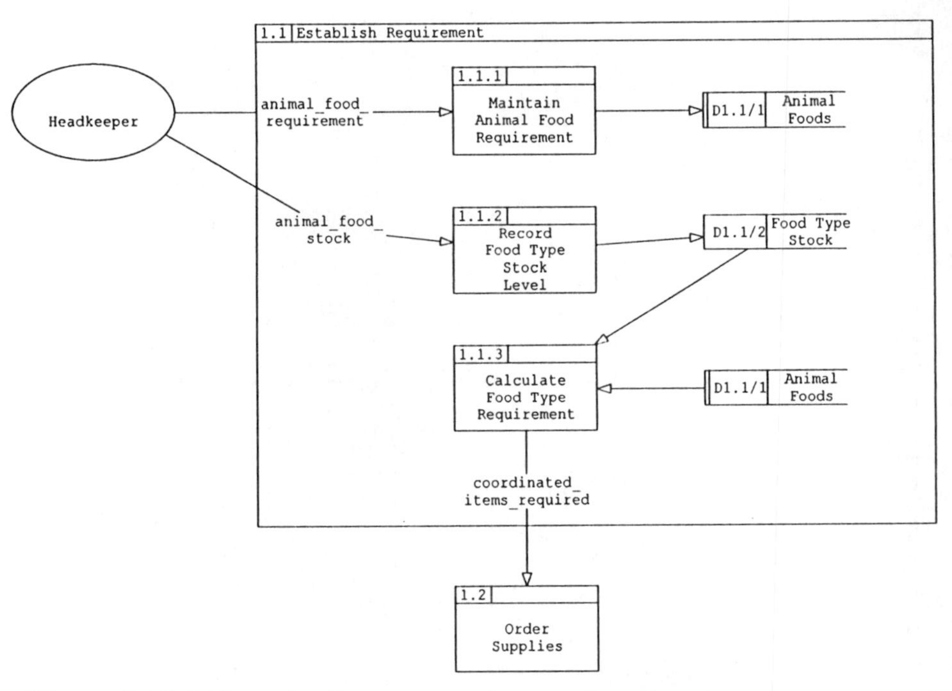

Figure 3.13 (f) Required lower level diagram for process `Establish Requirement`.

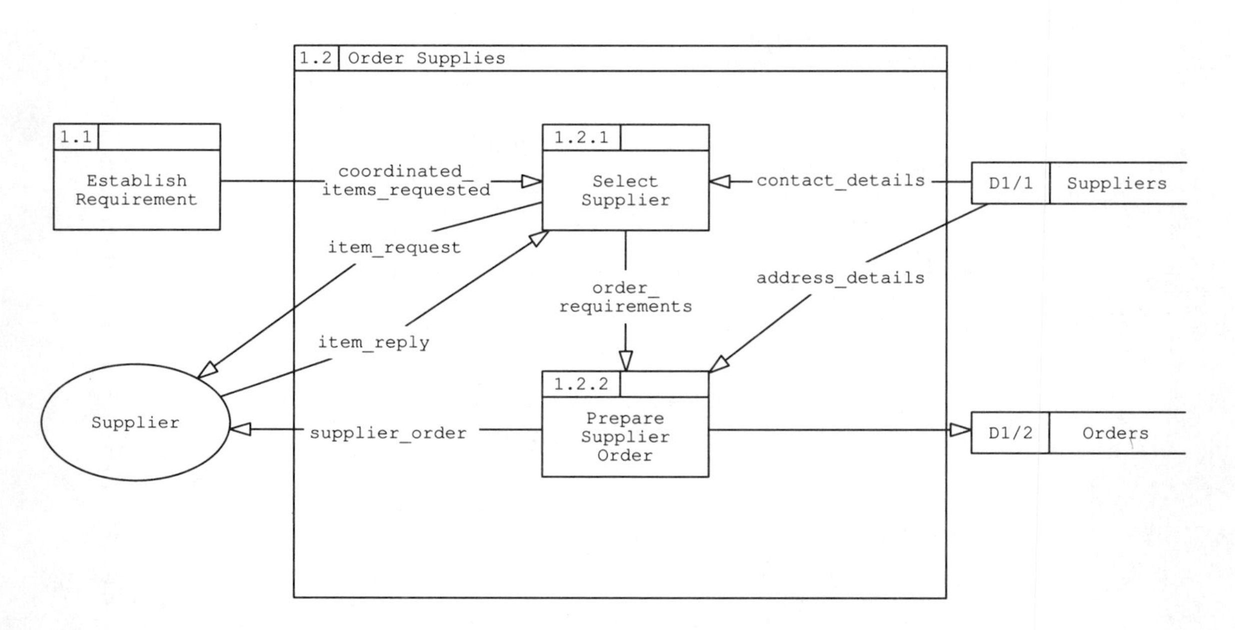

Figure 3.13 (g) Required lower level diagram for process `Order Supplies`.

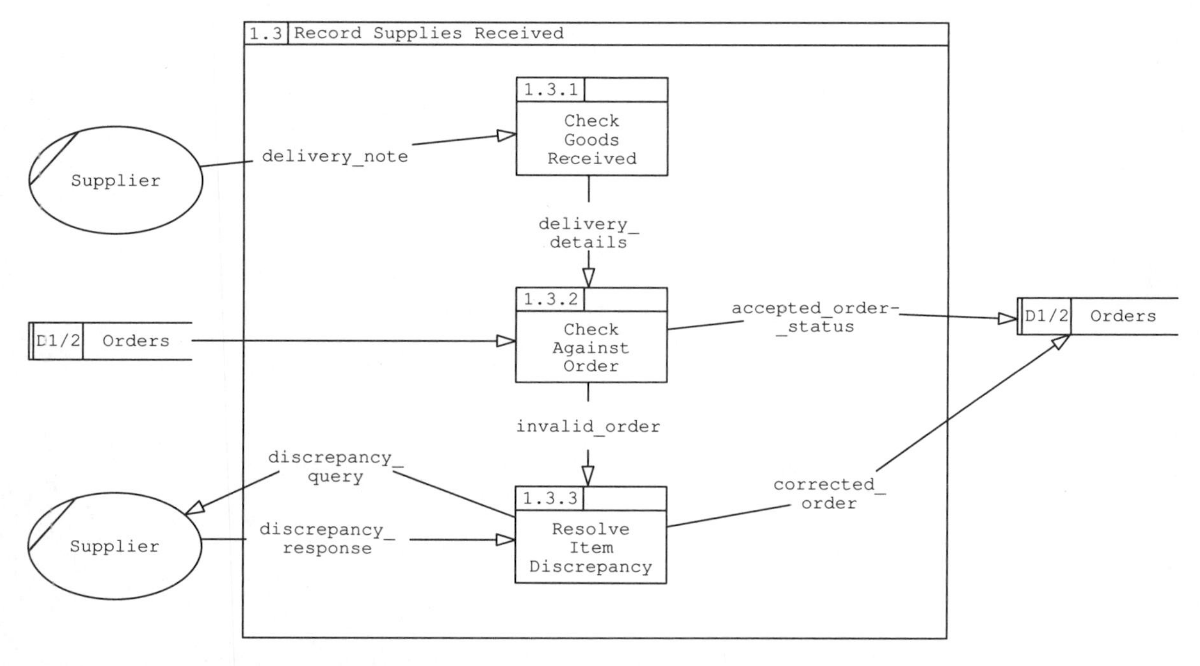

Figure 3.13 (h) Required lower level diagram for process `Record Supplies Received`.

Other approaches to process modelling

In the 1970s, there was much criticism of the traditional approaches to systems analysis and design. The arguments are well documented in texts on systems analysis and design, and will not be repeated here.

See Avison and Fitzgerald (1988), DeMarco (1979), Gane and Sarson (1977) and Yourdon (1989).

From this debate, the philosophy of structured analysis and design emerged. The idea of a structured approach to systems analysis and design was developed from structured programming, where the underlying model is functional decomposition. Functional decomposition takes a top down approach, where a problem is broken down into smaller units in a disciplined manner. Process modelling forms an important part of the concept of structured analysis and design.

DeMarco, and Gane and Sarson

DeMarco's (1979) view of structured analysis concentrates on the production of a structured specification dominated by the use of DFDs, and data dictionaries (DDs), which are supported by the use of Structured English, decision tables and decision trees.

The importance he attributes to the portrayal of the possible paths that data may take through the system is demonstrated by the use of the DFD. He starts with the current physical DFD which he believes could be verified and validated by the user. The model produced is then logicalized to

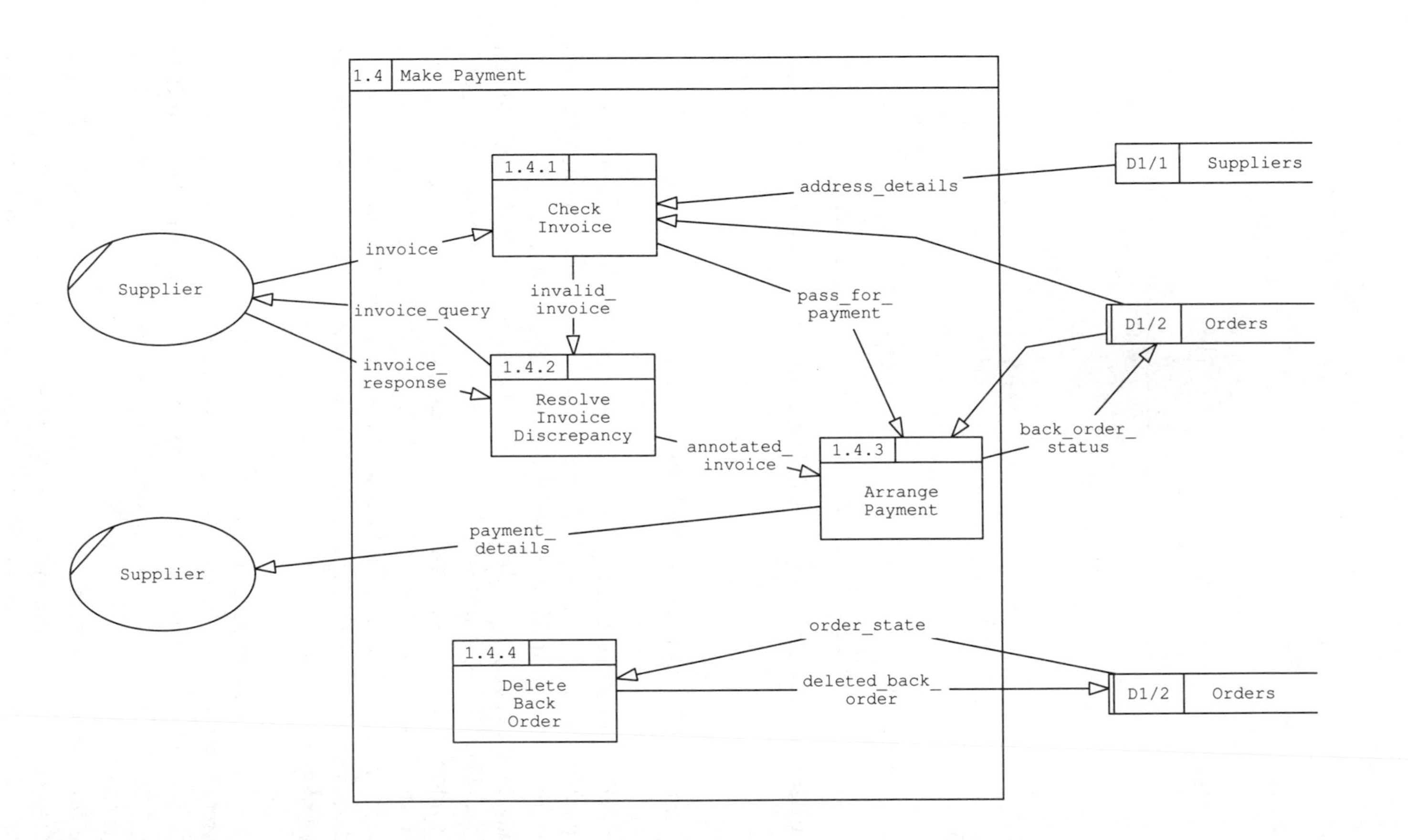

Figure 3.13 (i) Required lower level diagram for process `Make Payment`.

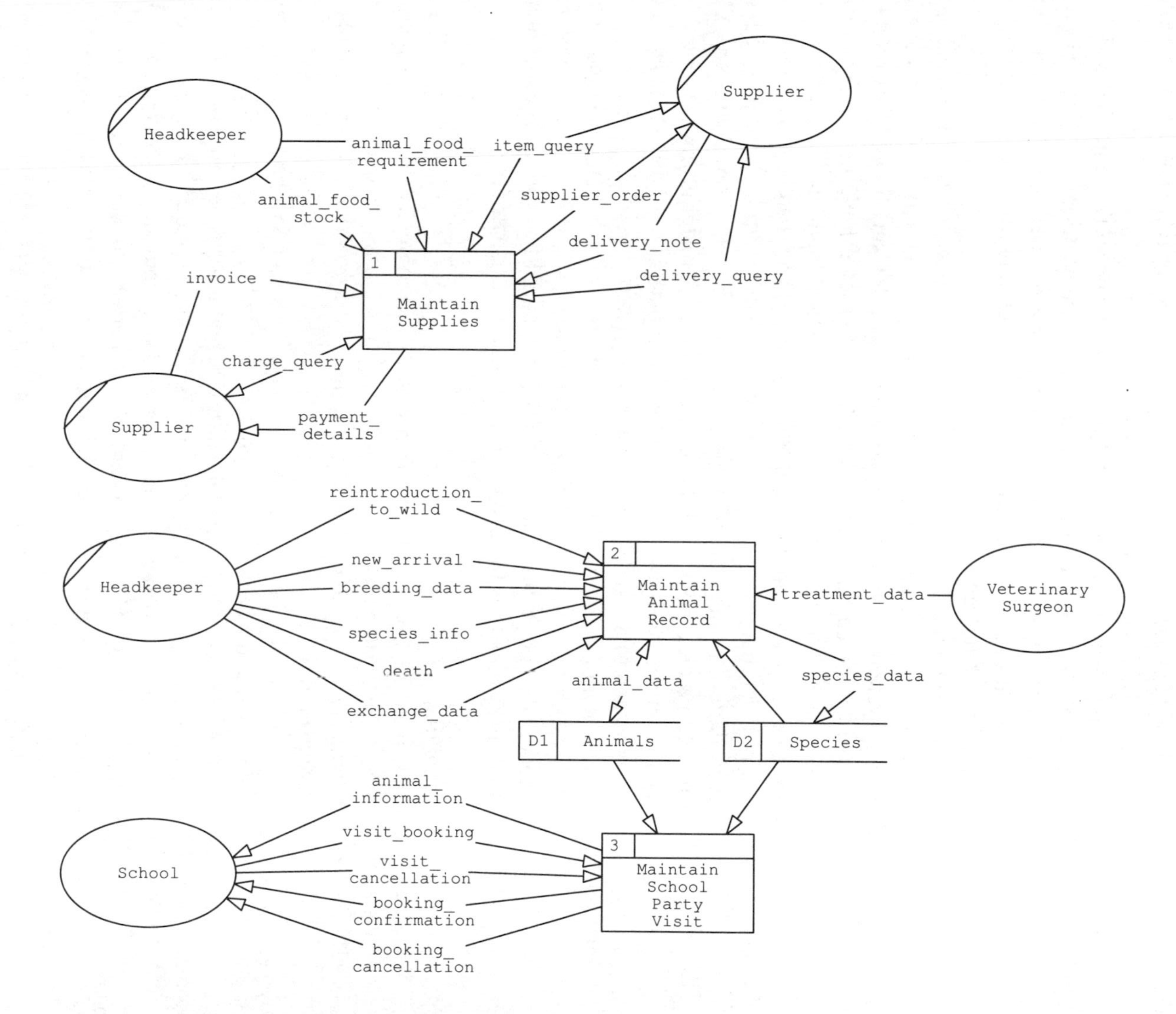

Figure 3.13 (j) Required top level diagram.

divorce the underlying objectives of the current operation from the methods of carrying out those objectives, that is separating the what from the how. The new logical DFD comprises a model of the system to be installed – what has to be done, but again not how it will be accomplished. The human–machine boundary is marked on the new logical DFD to make it into the new physical DFD, though this means that the scope only of the automated system has been identified, not the actual hardware/software implementation.

One of the main differences in the DFD conventions used by DeMarco (and Yourdon) is that terminators are shown on the context diagram only. Also, frames are not drawn on the lower level diagrams, as described above to facilitate balancing.

Gane and Sarson (1977) place the same emphasis on the functional specification as DeMarco, but their starting point is modelling the logical view of the system. However, they do suggest that too much emphasis might be placed on the current system, and that a top level DFD only be produced. Their use of frames on the lower level DFDs to facilitate balancing is similar to that advocated earlier in this chapter. Error handling is confined to lower level diagrams so that it does not interfere with the 'big picture'. The data flows showing errors at the lower level may be drawn with an 'X' at the exit point to assist with balance checking. A material flow diagram may be drawn and used as a basis for the development of a DFD. This is similar in approach to SSADM's resource flow diagram.

See Gane and Sarson (1977) for full details on material flow diagrams, and the SSADM Reference Manuals (1990) for resource flow diagrams.

One major difference in their approach from ours is that Gane and Sarson suggest that the definition of data store contents is driven from the process model, though normalization is used for simplification of data contents. Further comment is made on this at the end of Chapter 6.

Yourdon

As stated in Chapter 2, Yourdon (1989) recommends that whenever possible, the systems analyst should avoid modelling the current system and should develop a model of the required system, which Yourdon calls the **essential model.**

If it is necessary to model the current system, Yourdon suggests that you may draw the top level physical DFD with a breakdown only of the more critical processes. The DFDs would then be logicalized. See Yourdon (1989) for a more detailed explanation.

This essential model consists of two major components: an environmental model, and a behavioural model.

The environmental model starts to build a model to show the user requirements of a new system, and defines the boundary of the system, that is the interfaces between what is inside the system and what is outside. This is achieved by the development of a statement of purpose, a context diagram, and an event list containing stimuli to which the system must respond.

A process is drawn for each event, with the process being named as the response to the stimulus provided by that event. The other DFD elements are then added. The resulting DFD can then be refined and improved by levelling both upward and downward so that a levelled set of DFDs is produced. This is sometimes called the **middle-out** approach.

The techniques introduced for process modelling in Chapters 2 and 3 are suitable for business-oriented systems. When modelling real-time applications, where it is necessary to be able to show the synchronization and coordination of different processing tasks as well as the notation for showing interrupts and signals, additional notation is required. Control processes are available which monitor the state of the system by examining event flows which have no data content as such, and are either set to true or false. Figure 3.14 shows an example of part of a DFD with control flows and processes.

If you are interested in real-time applications, consult Yourdon (1989), Ward and Mellor (1985) or Goldsmith (1993).

Structured Systems And Design Method (SSADM)

SSADM is the most widely used methodology for systems analysis and design in the UK. This was originally because of its use within the government sector, but increasingly it is also being used in the private sector. It takes a top down approach, as mentioned earlier, and is more rigorous with its tightly defined structure than most of the other methods. Data flow modelling within SSADM comprises a context diagram (optional), a levelled set of DFDs and a set of associated textual descriptions. The DFDs follow through from current physical, to current logical and then required (new) logical. The context diagram is optional and is often drawn if the initial scope of the system is unclear. Additionally, a resource flow diagram showing the physical resources of the system may be developed and then used to prepare the DFDs.

On a levelled set of DFDs, requirements to retrieve and report data, where no data stores are updated, are specified in a requirements catalogue, leaving the DFDs less cluttered. The associated textual descriptions support the DFDs and comprise elementary process descriptions, external entity descriptions and input/output descriptions.

The elementary process descriptions are equivalent to the process specifications mentioned earlier. The external entity descriptions detail the responsibilities or functions of the external entities, together with any possible constraints on how they might interface with the system. Input/output descriptions describe each data flow crossing the system boundary, and list all data items contained in the data flow. In other methods of systems analysis and design, this information is usually held in the data dictionary.

For systems where the current system is predominantly clerical, SSADM suggests that a document flow diagram be drawn, which charts the progress of individual documents. This document flow diagram can then be developed into a DFD.

SSADM also advocates the use of quality criteria to enable the systems analyst to check the development of the product, in our case DFDs. Figure 3.15 shows the quality criteria for a lower level DFD.

See the SSADM Reference Manuals (1990) for a detailed explanation of data flow modelling and for product descriptions, including quality criteria.

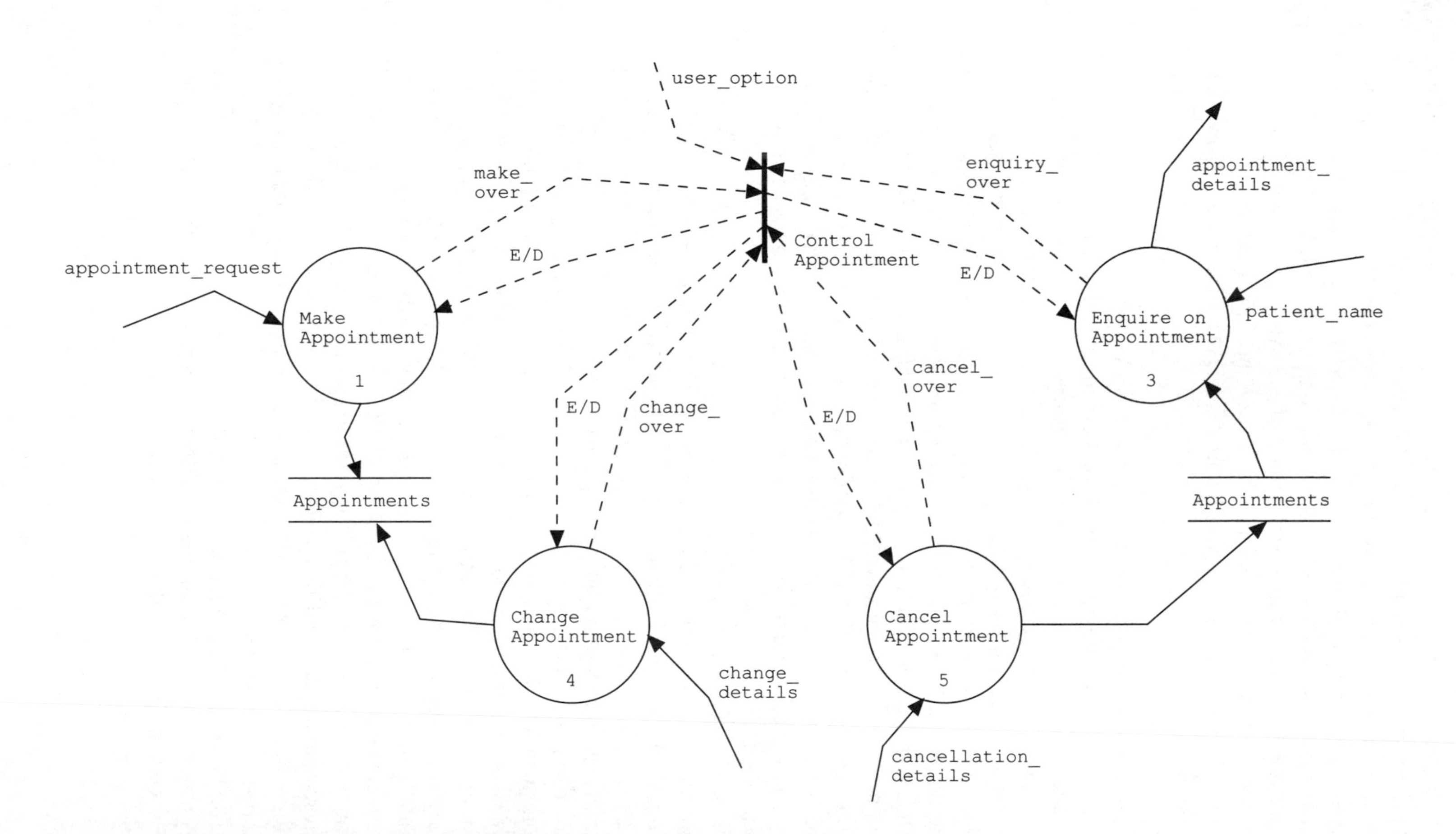

Figure 3.14 Example DFD notation with real-time extensions. E/D = Enable/Disable.

Quality criteria:

For each:

1. Is the variant correctly identified?
2. Are notational conventions correctly applied?
3. Is the boundary of the process clear?
4. Are meaningful names used for processes and data stores?
5. Do external entity names accurately reflect the environment outside the system?
6. Does the diagram avoid giving an inappropriate level of detail, such as sequencing or detailed processing logic?

For Logical Data Flow Diagram:

7. Are all physical aspects of the current system removed, unless constraints on the requirement?
8. Are any enquiries remaining after logicalization major ones?

For Required System Data Flow Diagram:

9. Are all and only the facilities defined by the Selected Business System Option modelled in the Required System Data Flow Diagrams?

For the set:

10. Are all identifiers unique?
11. Is the set of diagrams complete?

Figure 3.15 Example quality criteria for DFD Lower Level (SSADM Reference Manual (1990), Volume 4. Crown copyright. Reproduced with the permission of the Controller of Her Majesty's Stationery Office).

Summary of process modelling

The method we have adopted towards process modelling follows a traditional approach, starting with a top level current physical DFD, and logicalizing it into a current logical DFD. This logical diagram is then levelled to provide more detail, before incorporating the new systems requirements into a set of required logical DFDs. In the majority of cases, there is merit in finding out what happens in the existing system, yet Yourdon's point that it is counter-productive to spend too much time and effort on this is accepted. The conventions used for DFDs are closer to those advocated by SSADM, and provide an approach which allows us to check more easily and thoroughly the completeness and consistency of the DFDs. The DFDs are developed in parallel with the entity-relationship model (see Chapters 5 and 6), and are supported by the data dictionary (see Chapter 4). We have been following guidelines to ensure as far as possible that the DFDs within themselves provide accurate models of the requirements. In later chapters you will see how the various models may be cross-checked to enable us to feel more confident that our analysis and subsequent design is sound. So far no decisions have been made as to which processes in the proposed system will be computer-based and which will be undertaken manually. This will be a design issue, and is not considered further in this text.

Chapter 4

The Data Dictionary

OBJECTIVES

In this chapter you will learn:

- ☐ how a data dictionary (DD) may be used to support other system models;
- ☐ the components of a DD;
- ☐ using guidelines, how to develop a DD;
- ☐ to appreciate the use of an automated DD.

Introduction

Data flow diagrams (DFDs) give a diagrammatic representation of the process view of the system. However, we often ask ourselves what exactly is meant by a particular process? What are the contents of a data flow or data store? We really need the answers to these questions when developing DFDs; also, how do we know that the name given to one of the DFD elements is unique? The repository where all this information is kept is known as a **data dictionary**.

Data Dictionary – basic concepts

Martin (1976) defined a data dictionary as 'a repository of data about data'. DeMarco (1979), with his emphasis on the process view of the system, was more specific, saying that a DD is an ordered set of definitions of terms used in a DFD; in fact, a rigorous definition of all DFD elements.

DFD components or elements as previously defined are

- processes;
- data flows;
- data stores; and
- terminators.

Processes

A process is used to transform data from one form or state to another. What is not obvious from the DFD is the way in which this transformation is achieved. The DD records the logical steps within each functional primitive (bottom level or elementary process) within a process specification. A process specification should be unambiguous but concise.

It is inappropriate and redundant to define the logical steps for processes which are further broken down.

The logical steps of a process specification in this text are shown in the form of Structured English. Decision tables and decision trees are sometimes used as alternatives.

Useful texts explaining decision tables and decision trees include DeMarco (1976), Gane and Sarson (1977) and Martin and McClure (1988).

Structured English uses limited vocabulary and syntax. It comprises imperative verbs, components from the DD such as data flows, data structures or data elements, and specified reserved words used to formulate logic.

See below for a discussion on data flows, data structures and data elements.

The logical steps themselves are based on the well-understood structured programming constructs of sequence, selection and iteration.

This text includes an explanation of the fundamental components of the constructs.

Full details of programming constructs are not given in this text; see Jackson (1975) and Ingevaldsson (1979). For other explanations of Structured English used in this context, see DeMarco (1979), Gane and Sarson (1977), Martin and McClure (1988) and Yourdon (1979).

Sequence consists of one or more logical steps applied one after the other with no interruptions. This is expressed in Structured English by the use of sentences comprising an imperative verb followed by one or more objects. Figure 3.4 gave an example of the logical steps within a process specification. It is repeated here as Figure 4.1 for convenience.

```
BEGIN
ACCEPT stores_requisition
RETRIEVE stock_balance FROM Stock
SUBTRACT quantity_issued FROM stock_balance
IF stock_balance LESS THAN reorder_level
  CREATE purchase_requisition
ENDIF
WRITE stock_balance TO Stock
END
```

Figure 4.1 Example process specification.

ACCEPT, RETRIEVE, SUBTRACT and WRITE are examples of imperative verbs which are, by definition, action-oriented. Other examples include:

ADD	DETERMINE
MULTIPLY	ANNOTATE
DIVIDE	RECEIVE
MOVE	SELECT
SET	READ
RETRIEVE	CHECK

DELETE
CREATE
FIND
VALIDATE

Most organizations determine a set of verbs that will be used within process specifications.

The objects specified by the verbs should consist, where possible, of data flows, data structures or data elements specified elsewhere in the DD. In our example, this means that there should be DD entries for stores_requisition, quantity_issued, stock_balance, reorder_level and purchase_requisition. For more complex process specifications, there might be a requirement to show the results of an intermediate calculation that we do not need to store permanently. This is permissible, but there would be no reference to it in the DD. At this analysis stage, it may be more realistic on occasions to use more general terms to convey the meaning of the logical steps rather than using actual components from the DD. In this text, if the objects are not to be found in the DD they are shown in *italics*. These entries may then be clarified at the design stage.

Selection consists of a choice of one of two or more alternatives. The action taken on selection may be expressed by any one of the three specified constructs, though the advice is to *keep it simple*. Specified reserved words are used to express selection logic:

```
IF <condition>
  <action 1>
ENDIF

IF <condition>
  <action 1>
ELSE
  <action 2>
ENDIF
```

The specified reserved words are IF, ELSE and ENDIF. ENDIF indicates the end of the selection logic.

Figure 4.1 includes an example of a selection statement. In <condition> IS EQUAL TO or GREATER THAN may be used as alternatives to LESS THAN. Stock_balance and reorder_level must be defined elsewhere in the DD in the same way as the objects mentioned earlier. It is also permissible to use literals such as 27 or 'FINAL'. It is convention to include alphanumeric literals in quotation marks. The <action> may contain one or more sentences from any of the three constructs, though as mentioned earlier, the advice is to *keep it simple*. This leads to fewer misunderstandings between systems analysts and users, and between systems analysts and programmers.

OTHERWISE is often used as an alternative to ELSE.
THEN is sometimes used to aid understanding:

```
IF <condition>
THEN  <action 1>
ENDIF
```

Examples of use are shown later in this chapter.

The following data shows the discounts to be applied to given invoice values:

Invoice value (£)	Discount (%)
> 5000	20
1001–5000	10
501–1000	5
101–500	3

The logical steps that model the details given above are:

```
IF invoice_value GREATER THAN £5000
  CALCULATE 20% discount
ELSE
  IF invoice_value GREATER THAN £1000
    CALCULATE 10% discount
  ELSE
    IF invoice_value GREATER THAN £500
      CALCULATE 5% discount
    ELSE
      IF invoice_value GREATER THAN £100
        CALCULATE 3% discount
      ENDIF
    ENDIF
  ENDIF
ENDIF
```

The use of the selection construct here is clumsy. A neater solution is to use the CASE construct:

```
CASE invoice_value GREATER THAN £5000
   CALCULATE 20% discount
CASE invoice_value £(1001-5000)
   CALCULATE 10% discount
CASE invoice_value £(501-1000)
   CALCULATE 5% discount
CASE invoice_value £(101-500)
   CALCULATE 3% discount
ENDCASE
```

The CASE construct may be used when there are a number of possibilities, only one of which is applicable.

Iteration consists of the repetition of logical steps until a pre-defined condition is obtained. The iteration may take place zero times. Each construct has a single start and end point. Specified reserved words are used to express iteration logic. Three different constructs are described here.

1. A set of actions may be repeated:

```
FOR <scope>
  <action>
ENDFOR
```

For example,

```
FOR all stores_requisitions
  ACCEPT stores_requisition
  RETRIEVE stock_balance FROM Stock
  SUBTRACT quantity_issued FROM stock_balance
  IF stock_balance LESS THAN reorder_level
    CREATE purchase_requisition
  ENDIF
  WRITE stock_balance TO Stock
ENDFOR
```

In this case, all the logical steps between `FOR` and `ENDFOR` are repeated for each stores requisition. It may be necessary within `<scope>` to specify the repetition of action until a certain condition is met.

As you can see, this is a simple example. In a real system, we may have to check whether a purchase requisition has already been raised, and to determine what will happen when we run out of stock!

2. `REPEAT UNTIL` a certain condition is met:

```
REPEAT
   <action>
UNTIL <condition>
```

For example,

```
REPEAT
  ACCEPT stores_requisition
  RETRIEVE stock_balance FROM Stock
  SUBTRACT quantity_issued FROM stock_balance
  IF stock_balance LESS THAN reorder_level
    CREATE purchase_requisition
  ENDIF
  WRITE stock_balance TO Stock
UNTIL no more stores_requisitions
```

This means that the logical steps between `REPEAT` and `UNTIL` will be repeated until the specified condition is fulfilled. This means that the logical steps will be actioned at least once. In our example, it might be the case that the `stock_balance` is already negative and we do not want the logical steps to take place even once. If it is the case that the repetition could take place zero times, then the following construct should be used.

3. `DO WHILE` a specified condition is true:

```
DO WHILE <condition>
     <action>
ENDDO
```

In this case, the condition is tested first and the action only takes place if the condition is true.

For example,

```
DO WHILE more stores_requisitions
  ACCEPT stores_requisition
  RETRIEVE stock_balance FROM Stock
  SUBTRACT quantity_issued FROM stock_balance
  IF stock_balance LESS THAN reorder_level
    CREATE purchase_requisition
  ENDIF
  WRITE stock_balance TO Stock
ENDDO
```

It is also possible to use a negative condition (`NOT`).

Each process specification should have one start (entry) point and one end (exit) point specified by `BEGIN` and `END`, respectively.

See below for a discussion on data flows, data structures and data elements.

In addition to the specification of the logical steps within a process specification, it is also possible to record the inputs to and outputs from a process. These inputs and outputs may comprise data flows, data structures and/or data elements.

Further discussion on the cross-checking of different views of the system will take place in Chapter 8.

If the inputs and outputs are explicitly specified, the cross-checking of the different views of the system is easier.

The inputs and outputs for the logical steps expressed in Figure 4.1 are:

Inputs	*Outputs*
`stores_requisition`	`purchase_requisition`
`stock_balance`	`stock_balance`

This example shows the `stock_balance` as an input and an output as specified in the logical steps. The specification of inputs and outputs is not included in all examples, but will be discussed again later in this chapter, and again in Chapter 8.

See Worked example 4.1a process specification of the process `Calculate Food Type Requirement` for an example of a comment for a time trigger.

As stated in Worked example 2.3 in Chapter 2, a process may take place at specified intervals, and is therefore triggered by time. For clarification, a comment line indicating the time trigger is included as the first logical step within the process specification.

A comment line may be used in this way within any DD entry.

Data flows

In DD terms, a **data flow** may be defined as a packet of data which comprises data elements or a combination of the contents of other data flows and data elements.

Useful information about a data flow includes:

- data flow name – as on DFD;
- data flow description – if needed;
- contents;
- source of the data flow – as on DFD;
- destination of the data flow – as on DFD; and
- volumetric information – if known, otherwise added prior to systems design.

Volumetric information will indicate the number of occurrences of the data flow within a specified time span; for example, 100 stores requisitions per day.

Most DD entries are made up of a set of related components. For this purpose a set of relational operators are used to show these 'relationships'. The principle of the three constructs sequence, selection and iteration can be used for data as well as process. Sequence is the concatenation of two or more elements separated by a comma. For example,

```
order_item = foodtype_name, foodtype_requirement, price
```

= means 'is composed of'

+ or the word AND are sometimes used as an alternative to a comma.

Selection is the choice of one of two or more alternatives shown within square parentheses. For example,

```
item_query = [item_request | item_reply]
```

Iteration is the repetition of a component zero or more times shown by curly parentheses. For example,

```
animal_food_stock = area_no, {foodtype_name, stock_level}
```

An optional entry may be shown by parentheses ().

An example of a DD entry for a data flow is:

Data flow name: `animal_food_requirement`
Data flow description: Comprises the quantities of different food types required by an animal.
Contents:

```
animal_food_requirement = animal_no, area_no,
              {foodtype_name, daily_quantity, feeding_notes}
```

Source of data flow: `Headkeeper`
Destination of data flow: `Process 1.1 Maintain Animal Food Requirement`
Volumetric information: To be determined.

Data stores

In DD terms, a **data store** may be defined as a collection of packets of data. These packets of data comprise a number of individual data elements.

Useful information about a data store includes:

- data store name – as on DFD;

Chapters 5 and 6 look at the data view of a system rather than the process view. It should become obvious, and will be clearly stated in Chapter 8, that there is a relationship between data stores and entities. This relationship will be described in these later chapters.

- data store description;
- contents;
- reference to 'incoming' data flow(s);
- reference to 'outgoing' data flow(s); and
- physical organization to be added at system design stage.

During the development of DFDs through logicalization and levelling, and the progression from the current to the required system, data store entries should be checked to ensure that all the data elements are used.

An example of a DD entry for a data store is:

Data store name: `Suppliers`
Data store description: Details are held for each supplier.
Contents:
`suppliers = {supplier_name, supplier_struct, supplier_tel_no, supplier_fax_no, {foodtype_name}, standard_discount}`
Data flows in: None
Data flows out: `supplier_details`
`address_details`
`contact_details`
Physical organization: To be determined.

Note the definition of `supplier_struct` in Worked example 4.1a. The specification of the fact that there are no data flows into the data store should be investigated, and will be discussed further in Chapter 8.

An underlined data element(s) indicates the identifier of the data store, in the above case the `supplier_name` uniquely identifies each supplier.

Terminators

Terminators are often not defined with the DD as there are far fewer of them. This omission is true of DeMarco, Gane and Sarson, and Yourdon. Within SSADM, however, external entity descriptions are defined and included as part of the documentation for the data flow model.

Useful information on terminators includes terminator name and function (description).

An example of a DD entry for a terminator is:

Terminator name: `Headkeeper`
Function: Each headkeeper is responsible for an area of the Animal Park. Additional keepers are assigned to the larger areas.

So far, the contents of the DD have been discussed in terms of DFD components. However, data flows and data stores have specified the requirement for data elements.

Data elements

A **data element** is a meaningful item of data that will not be decomposed further. Some useful information includes:

- data element name;
- data element description; and
- aliases.

A data element may be referred to by different names in different parts of an organization. For example, a customer may be identified by a customer number in the Sales Department, and that same customer number may be referred to as the account number in the Accounts Department.

The following information may be added during systems design:

- type – usually character, numeric or alphanumeric;
- format – make up of element for example, XX9999 means two characters followed by four numeric digits; and
- security – authority to retrieve/update data element.

An example of a DD entry for a data element is:

Data element name: `animal_no`
Data element description: A unique number given to each animal.
Aliases: None.

Data structures

Often when creating a DD, packets of data or data elements are repeated within a data flow or data store. To avoid this repetition of definition these data elements may be grouped and defined as a **data structure**.

Some useful information includes:

- data structure name;
- data structure description;
- contents; and
- volumetric information, if appropriate.

An example of a DD entry for a data structure is:

Data structure name: `order_item`
Data structure description: a description of each line on the order
Contents:
`order_item = foodtype_name, foodtype_requirement, price`
Volumetric information: To be determined.

Worked example 4.1a

From the levelled set of logical DFDs for the proposed system for the Somerleyton Animal Park:

Prepare DD entries for the subsystem 'Maintain Supplies'.

Solution: For this example, DD entries are identified and defined starting at the top level DFD and working down through the levels as explained:

1. For the top level DFD;
   - prepare data flow, terminator and data store entries;
   - prepare suitable data structures, as appropriate;
   - prepare data element entries for data flows, data stores and data structures; and
   - prepare process specifications for functional primitives.
2. For each level 1 DFD, prepare further DD entries for data flows, data stores, data elements and process specifications. Further data structures may also be needed.
3. For each level 2 DFD, continue as for level 1.

DD entries for the top level DFD are prepared.

Generic data flows are used to reduce the 'clutter' on parent DFDs, where a more general data flow is defined and broken down into two or more data flows on a child diagram. See Chapter 3 ('Dictionary balancing').

Figure 3.13j shows the top level DFD for the proposed system for the Somerleyton Animal Park.

Considering the maintenance of supplies subsystem only, data flows are defined, first checking the lower level diagrams for the identification of generic data flows (shown in Figures 3.13e–i).

It is assumed that fact finding is complete and copies of sample documentation with any additional requirements are available. Data structures are defined for repeating data as necessary.

In this example, the complete DD entry is prepared for the data flow `animal_food_requirement` as described earlier. For all other data flows, abbreviated versions of the DD entries are shown.

Data flows

A data flow shows where data is being passed between components on a DFD, and is known as 'data in motion'. In DD terms, a data flow may comprise a combination of data flows and data elements. See Chapter 2 ('Data Flow') and above ('Data flows').

Data flows for the top level DFD:

Data flow name: `animal_food_requirement`
Data flow description: Comprises the quantities of different food types required by an animal.
Contents:

```
animal_food_requirement = animal_no, area_no,
                {foodtype_name, daily_quantity, feeding_notes}
```

Source of data flow: `Headkeeper`
Destination of data flow: `Process 1.1 Maintain Animal Food Requirement`
Volumetric information: To be determined.

The contents of the generic data flows are identified as data flows at a lower level.

Data structures used here are defined below.

```
animal_food_stock = area_no, {foodtype_name, stock_level}
charge_query = [invoice_query | invoice_response]
delivery_note = delivery_no, supplier_name,
                {foodtype_name, quantity_delivered, {order_no}}
delivery_query = [discrepancy_query | discrepancy_response]
invoice = invoice_struct
```

```
item_query = [item_request | item_reply]
payment_details = payment_no, supplier_name, supplier_struct,
            {invoice_no}, amount_paid
supplier_order = order_struct, supplier_struct
```

Terminators are defined for headkeeper and supplier.

The complete DD entry is prepared for the terminator `Headkeeper` as described earlier. For the terminator `Supplier`, an abbreviated version of the DD entry is shown.

Terminators show the origin or recipient of data outside the system boundary. See Chapter 2 ('Terminator') and above ('Terminators').

Terminators for the top level DFD.

Terminators

Terminator name: `Headkeeper`

Function: Each headkeeper is responsible for an area of the Animal Park. Additional keepers are assigned to the larger areas.

Supplier

There are no data stores at this level on the DFD so no DD entries are defined.

It is recognized that details relating to suppliers, orders including order items, and invoices will be repeated. Therefore, data structures are defined.

It is usual for the identification of data structures to be an iterative process necessitating the redefinition of previous DD entries. The complete DD entry is prepared for the data structure `order_item` as described earlier. For all other data structures abbreviated versions of the DD entries are shown.

A data store shows where data is held either temporarily or permanently, and is known as 'data at rest'. In DD terms, a data store may be defined as a collection of packets of data. See Chapter 2 ('Data Store') and above ('Data stores').

Data stores for the top level DFD – none in this case.

Data structures

Data structures identified from other entries.

```
invoice_struct = invoice_no, invoice_date,
          {order_struct}, discount_value, invoice_value
```

A data structure defines groups of repeated data to avoid repetition within the DD. See above ('Data structures').

Data structure name: `order_item`

Data structure description: a description of each line on the order

Contents:

```
order_item = foodtype_name, foodtype_requirement, price
```

Volumetric information: To be determined.

```
order_struct = order_no, supplier_name, order_date, {order_item},
                              total_discount_offered
supplier_struct = supplier_address, main_contact
```

The complete DD entry is prepared for the data element `animal_no` as

A data element is a meaningful item of data that will not be decomposed further. See above ('Data elements').

described above. For all other data elements, abbreviated versions of the DD entries are shown.

Data elements

amount_paid

Data elements identified from the top level DFD.

Data element name:	animal_no
Data element description:	A unique number given to each animal.
Aliases:	None.

area_no
daily_quantity
delivery_no
discount_value
feeding_notes
foodtype_name
foodtype_requirement
invoice_date
invoice_no
invoice_value
main_contact
order_date
order_no
payment_no
price
quantity_delivered
stock_level
supplier_address
supplier_name
total_discount_offered

A process specification is prepared for each functional primitive and shows the operations performed on the data by the process.

No process specifications are defined as the process Maintain Supplies is not a functional primitive.

Now move down one level to identify further DD entries relating to Maintain Supplies.

DD entries are defined from the level 1 DFD Maintain Supplies (see Figure 3.13e). Abbreviated DD entries are shown throughout, apart from the first entry of a data store.

Most data flow entries are already defined where they cross the frame boundary, although we need to ensure that entries which relate to generic flows at the higher level are prepared. Additional DD entries are required for data flows within the frame boundary.

Level 1 data flows either from within the frame boundary or from generic data flows at the higher level.

Data flows

accepted_order_status = order_status
back_order_status = order_status
coordinated_items_required = {foodtype_name, foodtype_requirement}

```
corrected_order = {order_struct}, order_status
deleted_back_order = order_no
discrepancy_query = delivery_no,
 {foodtype_name, quantity_ordered, quantity_delivered, {order_no},
                                   query_text}
discrepancy_response = delivery_no, {foodtype_name, query_response}
invoice_query = invoice_struct, query_text
item_reply = {foodtype_name, foodtype_requirement, price},
                              total_discount_offered
invoice_response = invoice_struct, query_response
item_request = {foodtype_name, foodtype_requirement}, terms_required
order = * hidden on DFD *
      * Process 'Prepare Supplier Order' to data store 'Orders' *
      * Data store 'Orders' to process 'Check Against Order' *
      * Data store 'Orders' to process 'Check Invoice' *
      * Data store 'Orders' to process 'Arrange Payment' *
order_state = order_no, order_status
supplier_details = [contact_details | address_details]
```

The process specification will include a logical step for the deletion of a packet of data from a data store. In this text, the data flow for a deleted packet of data is represented by the data element, which acts as the identifier for that packet of data within the data store.

It has been stated in earlier chapters that a DFD is less cluttered if a data flow is left unlabelled when a complete packet of data is to be transferred to or from a data store. For completeness, an entry is made in the DD to show this omission, showing the process title at the lowest level where the data flow occurs.

Data stores

```
Orders = {order_struct, order_status}
         * the identifier is order_no *
```

Level 1 data stores within the frame boundary.

Data store name: `Suppliers`
Data store description: Details are held for each supplier.
Contents:

```
Suppliers = {supplier_name, supplier_struct, supplier_tel_no,
         supplier_fax_no, {foodtype_name}, standard_discount}
```

Data flows in: None
Data flows out: `supplier_details`
`address_details`
`contact_details`
Physical organization: To be determined.

Additional data elements need to be defined.

Data elements

Level 1 data elements not previously defined.

```
order_status
quantity_ordered is an alias for foodtype_requirement
query_text
query_response
standard_discount
supplier_fax_no
supplier_tel_no
terms_required
```

Level 1 process specifications.

No process specifications are defined as none of the processes are functional primitives.

Now move down one more level to identify further DD entries from Level 2 DFDs. Firstly, the process `Establish Requirement`.

DD entries are defined from the level 2 DFD `Establish Requirement` (see Figure 3.13f).

Level 2 data flows for the process `Establish Requirement`.

Data flows

```
animal_food = * hidden on DFD *
     * Process 'Maintain Animal Food Requirement'
                                to data store 'Animal Foods' *
     * Data store 'Animal Foods'
                 to process 'Calculate Food Type Requirement' *
food_type_stock = * hidden on DFD *
                   * Process 'Record Food Type Stock Level'
                       to data store 'Food Type Stock' *
                   * Data store 'Food Type Stock'
                  to process 'Calculate Food Type Requirement' *
```

Level 2 data stores within the process `Establish Requirement`.

Data stores

```
Animal Foods = {animal_no, area_no,
               {foodtype_name, daily_quantity, feeding_notes}}
Food Type Stock = {area_no, foodtype_name, stock_level}
```

Level 2 data elements within the process `Establish Requirement`.

There are no additional data elements required at this level.

Process specifications are required for the functional primitives.

Level 2 process specifications within the process `Establish Requirement`.

Process specifications

```
Process 1.1.1: Maintain Animal Food Requirement
BEGIN
ACCEPT animal_food_requirement
WRITE animal_food_requirement TO Animal Foods
END

Process 1.1.2: Record Food Type Stock Level
BEGIN
ACCEPT animal_food_stock
DO WHILE more foodtype_name
  WRITE area_no, foodtype_name, stock_level TO Food Type Stock
ENDDO
END

Process 1.1.3: Calculate Food Type Requirement
BEGIN
* Time trigger, each week *
FOR each foodtype_name
  foodtype_requirement = 0
```

```
  FOR each area_no
    total_daily_quantity = 0
    DO WHILE more details for foodtype_name, area_no
      RETRIEVE daily_quantity FROM Animal Foods
      total_daily_quantity = total_daily_quantity + daily_quantity
    ENDDO
    RETRIEVE stock_level FROM Food Type Stock
    SUBTRACT stock_level FROM total_daily_quantity
    foodtype_requirement
      = foodtype_requirement + total_daily_quantity
  ENDFOR
  CREATE an iteration of coordinated_items_required
ENDFOR
END
```

DD entries are defined from the level 2 DFD `Order Supplies` (see Figure 3.13g).

Now move to the next level 2 process `Order Supplies.`

Data flows

```
address_details = supplier_struct
contact_details = supplier_name, main_contact,
[supplier_tel_no | supplier_fax_no], foodtype_name, standard_discount
order_requirements = {order_item, supplier_name}
```

Although these data flows cross the frame boundary, they still need to be defined as they were identified as part of the generic data flow `supplier_details` at the higher level.

There are no additional data stores or data elements.

Process specifications

```
Process 1.2.1: Select Supplier
Process 1.2.2: Prepare Supplier Order
```

DD entries are defined from the level 2 DFD `Record Supplies Received` (see Figure 3.13h).

Now move to the next level 2 process `Record Supplies Received.`

Data flows

```
delivery_details = delivery_struct
invalid_order = delivery_struct, query_text
```

There are no additional data stores.

Data structure

```
delivery_struct = delivery_no, supplier_name,
                  {foodtype_name, quantity_delivered,
                       quantity_accepted}
```

Data element

quantity_accepted

Process specifications

Process 1.3.1: Check Goods Received
Process 1.3.2: Check Against Order
Process 1.3.3: Resolve Item Discrepancy

Now move to the next level 2 process Make Payment.

DD entries are defined from the level 2 DFD Make Payment (see Figure 3.13i).

Data flows

```
annotated_invoice = invoice_struct
invalid_invoice = invoice_struct, query_text
pass_for_payment = invoice_struct
```

There are no additional data stores or data elements.

Process specifications

Process 1.4.1: Check Invoice
Process 1.4.2: Resolve Invoice Discrepancy
Process 1.4.3: Arrange Payment
Process 1.4.4: Delete Back Order

Abbreviated DD entries are shown for the data stores to assist understanding, but associated data element entries are omitted.

Worked example 4.1b

Identify the DD entries required for the subsystems 'Maintain Animal Record' and 'Maintain School Party Visit'.

Solution: For the top level DFD shown in Figure 3.13j, the DD entries identified for the subsystems 'Maintain Animal Record' and 'Maintain School Party Visit' are shown.

Top level data flow entries.

Data flows

```
animal = * hidden on DFD *
         * Data store 'Animals' to process 'Compile Animal
                                              Information' *
animal_data = [new_animal | animal_amendment | archived_animal |
                 animal]
animal_information
booking_cancellation
booking_confirmation
breeding_data
death
exchange_data
```

An exchange may be to or from the Animal Park.

```
new_arrival
reintroduction_to_wild
species = * hidden on DFD *
          * Data store 'Species' to process 'Compile Animal
                                            Information' *
    * Data store 'Species' to process 'Record Animal Arrival' *
    * Data store 'Species' to process 'Record Exchange Data' *
species_data = [species_addition | species_amendment]
species_info = [new_species | species_change]
treatment_data
visit_booking = [visit_request | visit_change]
visit_cancellation
```

The data flow `new_arrival` will record the species of the animal as a data element. Further discussion on species' information will take place in Chapter 8.

Data stores

Top level data stores within the frame boundary.

The contents of these two important data stores are included for clarity, and will be needed for further discussion in Chapter 8.

```
Animals = animal_no, species_no, animal_name, date_of_birth,
  area_no, enclosure_no,
  {treatment_no, treatment_date, vet_name, treatment_details},
  {breeding_no, mate_animal_no, date_gave_birth, birth_notes},
  {loan_no, start_date_of_loan, end_date_of_loan,
  park_loaned_to/from, loan_purpose, comments}
Species = species_no, species_name, species_details
```

Data elements

Top level data elements.

```
animal_name
birth_notes
breeding_no
comments
date_gave_birth
date_of_birth
enclosure_no
end_date_of_loan
loan_no
loan_purpose
mate_animal_no
park_loaned_to/from
species_details
species_name
species_no
start_date_of_loan
treatment_date
treatment_details
treatment_no
vet_name
```

Terminators (further)

Top level terminators.

```
School
Veterinary Surgeon
```

Data elements are identified for those items entered in the data stores above, but other data structures and data elements would be identified on closer inspection; that is when the DD entries are prepared rather than just identified. There are no process specifications needed at this level, as neither of the processes being considered are functional primitives.

DD entries for the level 1 process `Maintain Animal Record` and the level 2 processes `Amend Animal Details` and `Archive Animal Details`.

The DD entries for lower level diagrams for the subsystem 'Maintenance of Animal Record' are identified from Figures 3.13a–c. Remember that some of the data flows have been defined from the top level diagram.

Data flows

```
animal = * hidden on DFD *
   * Data store 'Animals' to process 'Archive Animal Details' *
   * Data store 'Animals' to
     process 'Record Reintroduction To Wild' *
   * Data store 'Animals' to process 'Record Death' *
animal_amendment = [breeding_amendment | exchange_amendment |
                                          treatment_amendment]

archived_animal
breeding_amendment
death_data
exchange_amendment
new_animal
new_species
reintroduction_data
species_addition
species_amendment
species_change
treatment_amendment
```

Data store

```
Animal Archives
```

Process specifications

```
Process 2.1: Record Animal Arrival
Process 2.2.1: Record Exchange Data
Process 2.2.2: Record Breeding Data
Process 2.2.3: Record Treatment Data
Process 2.3.1: Record Reintroduction To Wild
Process 2.3.2: Record Death
Process 2.4: Record Species Data
```

DD entries for the level 1 process `Maintain School Party Visit`.

The DD entries for lower level diagrams for the subsystem 'Maintenance of School Party Visit' are identified in Figure 3.13d.

Data flows

```
party_booking = * hidden on DFD *
  * Process 'Book Party Visit' to data store 'Party Bookings' *
  * Process 'Amend Party Visit' to data store 'Party Bookings' *
  * Process 'Cancel Party Visit' to data store 'Party Bookings' *
visit_data
visit_request
visit_change
```

Data store

```
Party Bookings
```

Process specifications

```
Process 3.1: Book Party Visit
Process 3.2: Amend Party Visit
Process 3.3: Cancel Party Visit
Process 3.4: Compile Animal Information
```

The DD entries created for this example, both parts (a) and (b) together, are repeated for completeness.

Data flows

```
accepted_order_status = order_status
address_details = supplier_struct
animal = * hidden on DFD *
   * Data store 'Animals' to process 'Compile Animal Information' *
   * Data store 'Animals' to process 'Archive Animal Details' *
   * Data store 'Animals' to
        process 'Record Reintroduction To Wild' *
   * Data store 'Animals' to process 'Record Death' *
animal_amendment = [breeding_amendment | exchange_amendment |
                                           treatment_amendment]
animal_data = [new_animal | animal_amendment | archived_animal |
                                                        animal]
animal_food = * hidden on DFD *
   * Process 'Maintain Animal Food Requirement'
                                 to data store 'Animal Foods' *
   * Data store 'Animal Foods'
                to process 'Calculate Food Type Requirement' *
```

Data flow name: `animal_food_requirement`
Data flow description: Comprises the quantities of different food types required by an animal.
Contents:

```
animal_food_requirement = animal_no, area_no,
                {foodtype_name, daily_quantity, feeding_notes}
```

Source of data flow: Headkeeper
Destination of data flow: Process 1.1 Maintain Animal Food Requirement
Volumetric information: To be determined.

```
animal_food_stock = area_no, {foodtype_name, stock_level}
animal_information
annotated_invoice = invoice_struct
archived_animal
back_order_status = order_status
booking_cancellation
booking_confirmation
breeding_amendment
breeding_data
charge_query = [invoice_query | invoice_response]
contact_details = supplier_name, main_contact,
  [supplier_tel_no | supplier_fax_no], foodtype_name,
                                         standard_discount
coordinated_items_required = {foodtype_name, foodtype_requirement}
corrected_order = {order_struct}, order_status
death
death_data
deleted_back_order = order_no
delivery_details = delivery_struct
delivery_note = delivery_no, supplier_name,
                {foodtype_name, quantity_delivered}, {order_no}
delivery_query = [discrepancy_query | discrepancy_response]
discrepancy_query = delivery_no,
 {foodtype_name, quantity_ordered, quantity_delivered,
                                      {order_no}, query_text}
discrepancy_response = delivery_no, {foodtype_name, query_response}
exchange_amendment
exchange_data
food_type_stock = * hidden on DFD *
                  * Process 'Record Food Type Stock Level'
                    to data store 'Food Type Stock' *
                  * Data store 'Food Type Stock'
                    to process 'Calculate Food Type Requirement' *
invalid_invoice = invoice_struct, query_text
invalid_order = delivery_struct, query_text
invoice = invoice_struct
invoice_query = invoice_struct, query_text
invoice_response = invoice_struct, query_response
item_query = [item_request | item_reply]
item_reply = {foodtype_name, foodtype_requirement, price},
                                         total_discount_offered
```

```
item_request = {foodtype_name, foodtype_requirement}, terms_required
new_animal
new_arrival
new_species
order = * hidden on DFD *
   * Process 'Prepare Supplier Order' to data store 'Orders' *
   * Data store 'Orders' to process 'Check Against Order' *
   * Data store 'Orders' to process 'Check Invoice' *
   * Data store 'Orders' to process 'Arrange Payment' *
order_requirements = {order_item, supplier_name}
order_state = order_no, order_status
party_booking = * hidden on DFD *
   * Process 'Book Party Visit' to data store 'Party Bookings' *
   * Process 'Amend Party Visit' to data store 'Party Bookings' *
   * Process 'Cancel Party Visit' to data store 'Party Bookings' *
pass_for_payment = invoice_struct
payment_details = payment_no, supplier_name, supplier_struct,
                              {invoice_no}, amount_paid
reintroduction_data
reintroduction_to_wild
species = * hidden on DFD *
   * Data store 'Species' to process 'Compile Animal Information' *
   * Data store 'Species' to process 'Record Animal Arrival' *
   * Data store 'Species' to process 'Record Exchange Data' *
species_addition
species_amendment
species_change
species_data = [species_addition | species_amendment]
species_info = [new_species | species_change]
supplier_details = [contact_details | address_details]
supplier_order = order_struct, supplier_struct
treatment_amendment
treatment_data
visit_booking = [visit_request | visit_change]
visit_cancellation
visit_change
visit_data
visit_request
```

Data structures

```
delivery_struct = delivery_no, supplier_name,
        {foodtype_name, quantity_delivered, quantity_accepted}
invoice_struct = invoice_no, invoice_date,
               {order_struct}, discount_value, invoice_value
```

Data structure name: `order_item`
Data structure description: a description of each line on the order
Contents:
```
order_item = foodtype_name, foodtype_requirement, price
```
Volumetric information: To be determined.

```
order_struct = order_no, supplier_name, order_date, {order_item},
                       total_discount_offered
supplier_struct = supplier_address, main_contact
```

Data stores

```
Animal Archives
Animal Foods = {animal_no, area_no,
                  {foodtype_name, daily_quantity, feeding_notes}}
Animals = animal_no, species_no, animal_name, date_of_birth,
  area_no, enclosure_no,
  {treatment_no, treatment_date, vet_name, treatment_details},
  {breeding_no, mate_animal_no, date_gave_birth, birth_notes},
  {loan_no, start_date_of_loan, end_date_of_loan,
   park_loaned_to/from, loan_purpose, comments}
Food Type Stock = {area_no, foodtype_name, stock_level}
Orders = {order_struct, order_status}
     * the identifier is order_no *
Party Bookings
Species = species_no, species_name, species_details
```

Data store name: `Suppliers`
Data store description: Details are held for each supplier.
Contents:
```
Suppliers = {supplier_name, supplier_struct, supplier_tel_no,
          supplier_fax_no, {foodtype_name}, standard_discount}
```
Data flows in: None
Data flows out: `supplier_details`
`address_details`
`contact_details`
Physical organization: To be determined.

Data elements

```
amount_paid
animal_name
```

Data element name: `animal_no`
Data element description: A unique number given to each animal.
Aliases: None.

```
area_no
birth_notes
breeding_no
comments
daily_quantity
date_gave_birth
date_of_birth
delivery_no
discount_value
enclosure_no
end_date_of_loan
feeding_notes
foodtype_name
foodtype_requirement
invoice_date
invoice_no
invoice_value
loan_no
loan_purpose
main_contact
mate_animal_no
order_date
order_no
order_status
park_loaned_to/from
payment_no
price
quantity_accepted
quantity_delivered
quantity_ordered is an alias for foodtype_requirement
query_response
query_text
species_details
species_name
species_no
standard_discount
start_date_of_loan
stock_level
supplier_address
supplier_fax_no
supplier_name
```

```
supplier_tel_no
terms_required
total_discount_offered
treatment_date
treatment_details
treatment_no
vet_name
```

Terminators

Terminator name: `Headkeeper`

Function: Each headkeeper is responsible for an area of the Animal Park. Additional keepers are assigned to the larger areas.

```
School
Supplier
Veterinary Surgeon
```

Process specifications

```
Process 1.1.1: Maintain Animal Food Requirement
BEGIN
ACCEPT animal_food_requirement
WRITE animal_food_requirement TO Animal Foods
END

Process 1.1.2: Record Food Type Stock Level
BEGIN
ACCEPT animal_food_stock
DO WHILE more foodtype_name
  WRITE area_no, foodtype_name, stock_level TO Food Type Stock
ENDDO
END

Process 1.1.3: Calculate Food Type Requirement
BEGIN
* Time trigger, each week *
FOR each foodtype_name
  foodtype_requirement = 0
  FOR each area_no
    total_daily_quantity = 0
    DO WHILE more details for foodtype_name, area_no
      RETRIEVE daily_quantity FROM Animal Foods
      total_daily_quantity = total_daily_quantity + daily_quantity
    ENDDO
    RETRIEVE stock_level FROM Food Type Stock
    SUBTRACT stock_level FROM total_daily_quantity
```

```
    foodtype_requirement = foodtype_requirement +
                                              total_daily_quantity
  ENDFOR
  CREATE an iteration of coordinated_items_required
ENDFOR
END

Process 1.2.1: Select Supplier
Process 1.2.2: Prepare Supplier Order
Process 1.3.1: Check Goods Received
Process 1.3.2: Check Against Order
Process 1.3.3: Resolve Item Discrepancy
Process 1.4.1: Check Invoice
Process 1.4.2: Resolve Invoice Discrepancy
Process 1.4.3: Arrange Payment
Process 1.4.4: Delete Back Order
Process 2.1: Record Animal Arrival
Process 2.2.1: Record Exchange Data
Process 2.2.2: Record Breeding Data
Process 2.2.3: Record Treatment Data
Process 2.3.1: Record Reintroduction To Wild
Process 2.3.2: Record Death
Process 2.4: Record Species Data
Process 3.1: Book Party Visit
Process 3.2: Amend Party Visit
Process 3.3: Cancel Party Visit
Process 3.4: Compile Animal Information
```

Exercise 4.1

From the levelled set of logical DFDs for the proposed system for the Albany Hotel:

(a) *Identify* the DD entries required. Include terminators, data flows with generic entries where applicable, data stores and process specifications.
(b) *Prepare* the DD entries in abbreviated form for the section of the DFD comprising the processes 1.1 Respond To Enquiry and 1.2 Make Provisional Booking. Include terminators, data flows, data stores, data structures, data elements and the Structured English for the process specifications.

Use the figures in Exercise solution 3.4a–c for the top level DFDs, and those in Exercise solution 3.3d–h for the lower level diagrams. (In this example, the lower level diagrams were almost the same for the current and proposed system and were not redrawn. The two exceptions are that the solution for

Exercise 3.4 requires the addition of the data flow `archived_booking` to move between the process `Receive Payment` and the data store `Guest Archives`; and the data flow `deleted_booking` between the process `Receive Payment` and the data store `Guests` should be bidirectional.)

Approach to the development of a DD

Ideally, the DD should be developed in parallel with the other models. For example, as a DFD component is drawn onto the diagram, its corresponding DD entry should be made. This would mean that the DD would be created in parallel with the current physical DFDs, modified at the time of logicalization, and again on the development of required logical DFDs. This may be possible if the DD is automated, but impossible if it is manual. A compromise would be to enter data flows and data stores with their content, where it is easy to establish and where it is likely to form part of the required system. Outline logic for processes should be included.

Guideline steps used for worked examples

There appear to be a number of 'sub steps' within a step here! This is true, but the 'sub steps' are often undertaken in parallel.

Guideline step 1: Identify inputs and outputs, sources and recipients for the DFD. Enter inputs and outputs as data flow names, and sources and recipients as terminators. Define data flow contents, complete their entries and also enter as data elements. Identify suitable data structures and make appropriate modifications to the current entries.

Guideline step 2: Draw a top level DFD. Define additional data flows. Define data stores. Identify and enter data structures and data elements for data flows and data stores as in guideline step 1. Define process specifications for functional primitives, that is those processes not to be decomposed further.

Guideline step 3: Decompose a process, and draw lower level DFDs. Check and amend any generic data flows that cross the frame boundary. Define data flows internal to this DFD. Define data stores local to this level. Identify and enter data structures and data elements for data flows and data stores. Define process specifications for functional primitives.

As levelling takes place and processes are merged or decomposed, review and revise the process specifications so that they are accurate and defined only for functional primitives.

Guideline step 4: Repeat guideline step 3 as appropriate.

Guideline step 5: Check balancing of DFD and total DD entries.

Worked example 4.2

For the given scenario, develop a set of levelled logical DFDs and prepare associated DD entries.

> Existing customers telephone or post their orders to the Sales Office, which transfers them to a standard order form. The order is then checked for customer validity and credit, and may be refused at this time. Availability of stock is checked for accepted orders, with the order being acknowledged or refused accordingly. Those order items that cannot be fulfilled are rejected. Stock is allocated for accepted order items prior to issue. One copy of the verified order is passed to the Area Salesperson, and one to Accounts who use a price list to calculate the total invoice value, and then print a four-part invoice set which is passed to the Warehouse. The goods are issued and despatched together with the delivery note, that is the green copy of the invoice. The blue copy of the invoice is sent to the Area Salesperson. The blue and green copies do not include the accounting information. The pink and white invoice copies are passed to Accounts, who post the details to the sales ledger, file the pink invoice copy and send the white invoice copy to the customer. On a monthly basis, Accounts prepare a statement and send it to the customer. On receipt of payment the amount paid is posted to the sales ledger and matched against the invoices on the basis of the oldest first. All paid invoice details and a record of payments are archived.

This example affords the opportunity to practise the development of levelled logical DFDs rather than physical DFDs followed by logicalization and levelling.

Solution:

Guideline step 1:

Inputs and outputs	*Sources*	*Recipients*
`order`	`Customer`	
`order_acknowledgement/refusal`		`Customer`
`verified_order`		`Salesperson`
`delivery_note`		`Customer`
`annotated_invoice_details`		`Salesperson`
`annotated_invoice_details`		`Customer`
`statement`		`Customer`
`payment_details`	`Customer`	

The inputs and outputs are defined as data flow entries in the DD and the sources and recipients as terminators. The content of the data flows is determined in the fact finding process mentioned in Chapter 2. Samples of any documents, for example the order, are obtained. The entries below assume that this work is complete. Abbreviated DD entries are used throughout.

The data flow entries for the first and second inputs and outputs are:

Iteration of possible eight items ordered.

```
order = order_no, customer_name, customer_address, order_date,
                1{catalogue_no, description, qty_ordered}8
```

```
order_ackowledgement/refusal = order_no, customer_name,
  customer_address, customer_no, order_date,
                  1{catalogue_no, description, qty_ordered}8, text.
```

Already there is some repetition of the data elements appearing within the data flows. This means that the common data elements may be defined as a data structure. Experienced systems analysts will define the data structures in parallel with the definition of the data flows. As stated earlier, the definition of the DD is an iterative process.

On examination of the other data requirements for the inputs and outputs, it is obvious that there are examples of the order details being used on a number of occasions. A data structure is defined:

```
order_details = order_no, order_date, 1{catalogue_no,
                                   description, qty_ordered}8
```

Customer details are also used many times, the majority of which include the customer_no, customer name and address so a data structure is defined:

```
customer_struct = customer_no, customer_name, customer_address
```

However, the two data structures are often used together, so it is possible to define a third data structure:

```
order_struct = customer_struct, order_details
```

Additionally, customer name and customer address are used together, so a further data structure may be defined:

```
customer_details = customer_name, customer_address
```

Therefore,

```
customer_struct = customer_no, customer_details
```

As you will see later in the example, the data elements making up an order item will be specified on a number of occasions, so a data structure is defined:

```
item = catalogue_no, description, qty_ordered
```

This means that the data structure `order_details` needs redefining again, and now contains a data structure within a data structure:

```
order_details = order_no, order_date, 1{item}8
```

The iteration in the data structure `order_details` is retained as there is no iteration defined in the data structure `item`. If the data structure is always used with the same number of iterations, then the two data structures could be defined as follows:

```
item = 1{catalogue_no, description, qty_ordered}8
order_details = order_no, order_date, item
```

In our example, the order and invoice have eight lines and the statement twenty. The first two data flows are now redefined using the data structures:

```
order = customer_details, order_details
order_acknowledgement/refusal = order_struct, text
```

The associated data structures are:

```
customer_struct = customer_no, customer_details
customer_details = customer_name, customer_address
order_details = order_no, order_date, 1{item}8
item = catalogue_no, description, qty_ordered
order_struct = customer_struct, order_details
```

The data flow entries for the other inputs and outputs are:

```
verified_order = order_struct
delivery_note = order_struct
annotated_invoice_details = customer_struct, invoice_struct,
                                                despatch_date
statement = customer_struct, statement_date, balance_b_f,
        statement_line, balance_c_f
payment_details = customer_struct, payment, payment_date
```

As you will see further data structures are defined:

```
invoice_struct = invoice_date, 1{invoice_line}8, vat, invoice_value
invoice_line = item, price, item_value
statement_line = 1{[(invoice_date, invoice_value) |
                                 (payment_date, payment)]}20
```

Terminators are defined for the sources and recipients:

```
Customer
Salesperson
```

Data elements are defined for:

```
balance_b_f
balance_c_f
catalogue_no
customer_address
customer_name
customer_no
description
despatch_date
invoice_date
invoice_value
item_value
order_date
order_no
payment
payment_date
price
qty_ordered
statement_date
text
vat
```

Some systems analysts often use a code as the first part of the DD entries, for example cust or c1 for items relating to customer, which would mean easier identification and understanding; also, the items would be listed together on a sorted list.

Guideline step 2: See Figure 4.2 for the top level DFD.

Additional data flows, data elements and possible data structures are now defined; also, the data stores and their associated data elements and data structures:

Data flows

Process reference for the lower level DFD shown here.

```
customer = * hidden on DFD *
        * Data store 'Customers' to process 'Check Credit Status' *

name_and_address = customer_details
invoice_details = customer_struct, invoice_struct
quantity_in_stock = * Data element 'quantity_in_stock' *
allocated_stock = * Data element 'allocated_stock' *
updated_stock = quantity_in_stock, allocated_stock
```

Most of the additional data flows are those passing between processes, or between processes and data stores.

The data stores on the top level DFD are now defined:

```
Customers = {customer_struct}
            * identifier is customer_no *
```

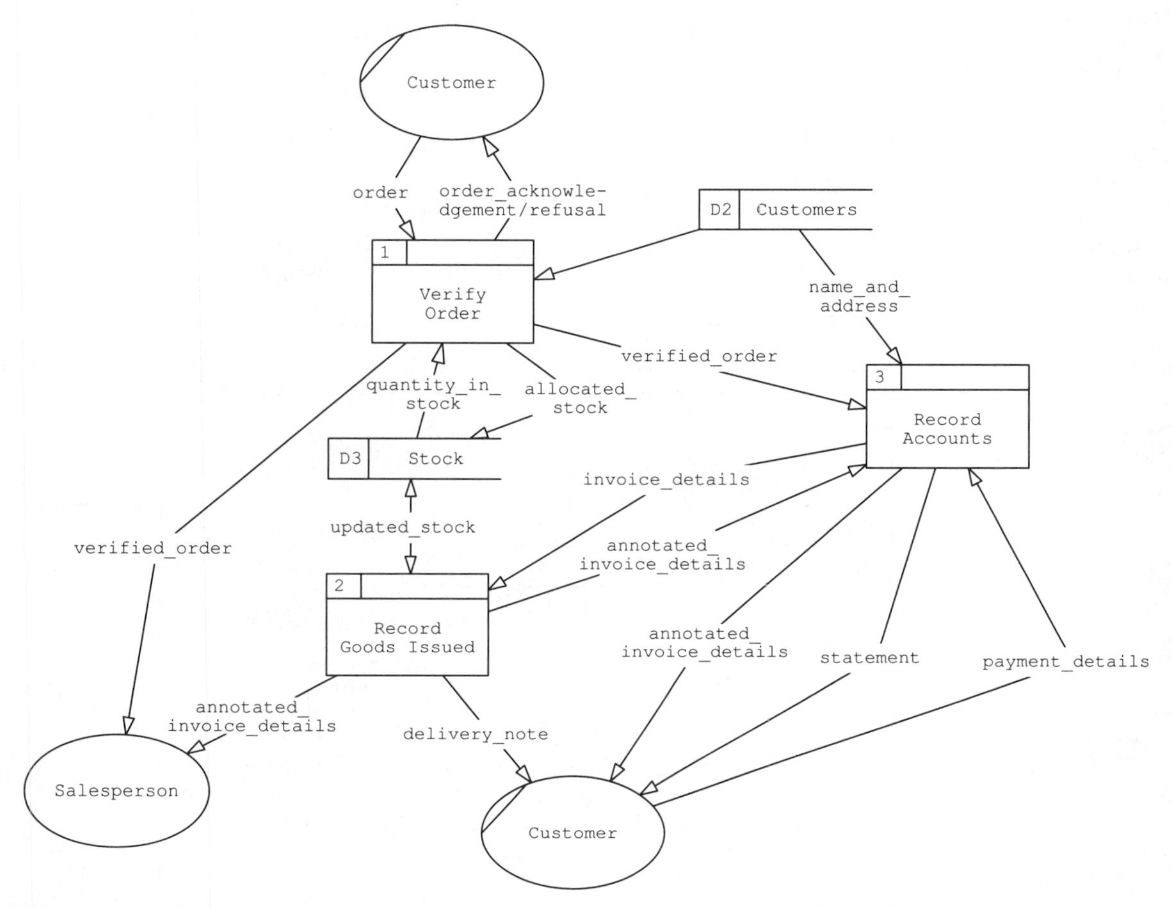

Figure 4.2 Worked example 4.2 after guideline step 2 top level diagram.

```
Stock = {catalogue_no, description, quantity_in_stock,
                                        allocated_stock}
```

Data elements

```
allocated_stock
quantity_in_stock
```

In this example there is no merging of processes. Process 2 `Record Goods Issued` is a functional primitive, but process 1 `Verify Order` and process 3 `Record Accounts` are considered complex enough to decompose.

A process specification is defined for process 2:

It is assumed that today's date will be available, and is not shown as a separate data element entering the system.

```
Process 2: Record Goods Issued
BEGIN
ACCEPT invoice_details
DO WHILE more invoice_lines
  RETRIEVE quantity_in_stock, allocated_stock FROM Stock
  SUBTRACT qty_ordered FROM quantity_in_stock, allocated_stock
```

```
  WRITE quantity_in_stock, allocated_stock TO Stock
ENDDO
CREATE delivery_note
despatch_date = today's date
CREATE annotated_invoice_details
END
```

Guideline step 3: Figure 4.3 shows the decomposition of process 1 `Verify Order`.

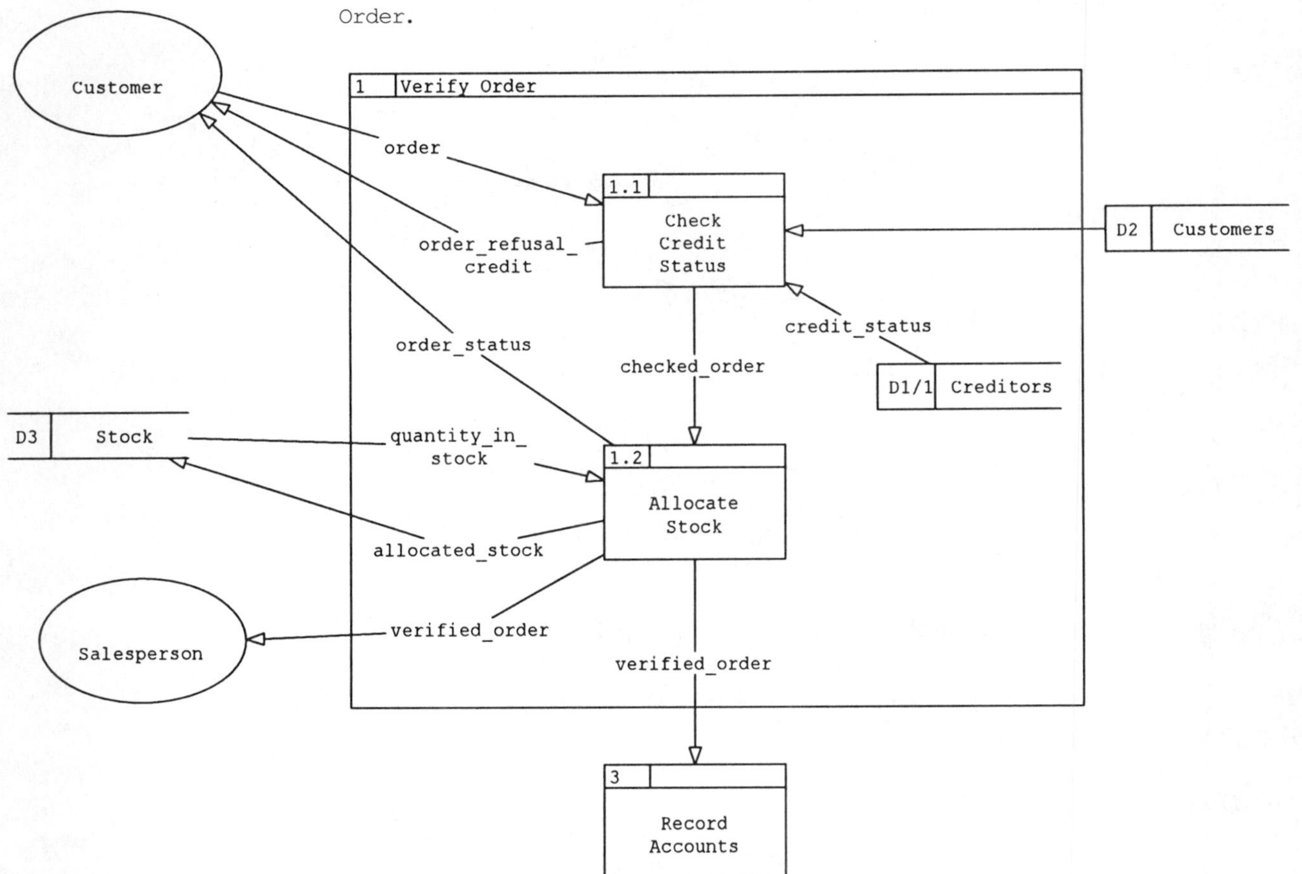

Figure 4.3 Worked example 4.2 after guideline step 3 – lower level diagram for process `Verify Order`.

The data flow `order_acknowledgement/refusal` is an example of a generic data flow and shows the different actions that may be taken when the order is verified. It is revised as follows:

```
order_acknowledgement/refusal = [order_refusal_credit |
                                      order_status]
order_status = customer_struct, order_no, order_date,
                                       1{item,text}8
```

The additional DD entries required are:

Data flows

```
order_refusal_credit = customer_details, order_details, text
credit_status = * Data element 'credit_status' *
checked_order = order_struct
```

Data store

```
Creditors = {customer_struct, credit_status}
            * identifier is customer_no *
```

Data element

```
credit_status
```

Process specifications are defined for:

```
Process 1.1: Check Credit Status
BEGIN
ACCEPT order
RETRIEVE name_and_address FROM Customer
IF name_and_address not found
  text = 'not existing customer'
  DISPLAY order_refusal_credit
ELSE
  RETRIEVE credit_status FROM Creditors
  IF credit_status NOT OK
    text = 'order refusal credit'
    DISPLAY order_refusal_credit
  ELSE
    CREATE checked_order
  ENDIF
ENDIF
END
```

Note the simplicity of this particular example. If any item of stock cannot be supplied in total, then that item is rejected. A more realistic system would allow for partial deliveries, as in the Somerleyton Animal Park example.

```
Process 1.2: Allocate Stock
BEGIN
ACCEPT checked_order
DISPLAY customer_struct, order_no, order_date
DO WHILE more order items
  RETRIEVE quantity_in_stock, allocated_stock FROM Stock

  IF (quantity_in_stock - allocated_stock - qty_ordered)
                                            LESS THAN ZERO
     text = 'order item rejected'
     DISPLAY item, text
     qty_ordered = 0
```

Note the output dominant data flow for `allocated_stock` on the DFD.

`Qty_ordered` set to zero to indicate insufficient stock.

```
  ELSE
     text = 'order item accepted'
     ADD qty_ordered TO allocated_stock
     WRITE allocated_stock TO Stock
     DISPLAY item, text
  ENDIF
ENDDO
CREATE verified_order for accepted order details
END
```

Figure 4.4 shows the decomposition of process 3 `Record Accounts`.

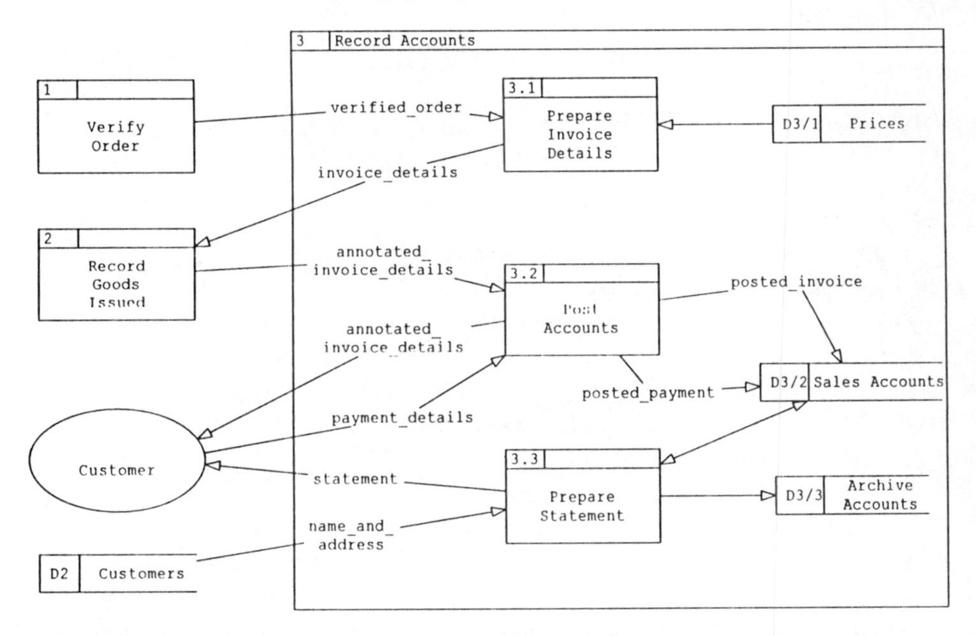

Figure 4.4 Worked example 4.2 after guideline step 3 – lower level diagram for process `Record Accounts`.

The additional DD entries required are:

Data flows

```
archive_account = * hidden on DFD *
                  * Process 'Prepare Statement' to
                        data store 'Archive Accounts' *
posted_invoice = customer_no, transaction_date,
                                    invoice_transaction
posted_payment = customer_no, transaction_date,
                                    payment_transaction
price = * hidden on DFD *
        * Data store 'Prices' to process 'Prepare Invoice Details' *
```

```
sales_account = * hidden on DFD *
                * Data store 'Sales Accounts' to and from
                              process 'Prepare Statement' *
```

Data structures

```
invoice_transaction = invoice_details
payment_transaction = payment, payment_date
```

Data stores

```
Prices = {catalogue_no, price}
Sales Accounts = {customer_no, transaction_date,
    [invoice_transaction | payment_transaction | statement_balance]}
Archived Accounts = {customer_no, transaction_date,
    [invoice_transaction | payment_transaction]}
```

Data elements

```
statement_balance
transaction_date
```

Process specifications are defined for:

```
Process 3.1: Prepare Invoice Details
BEGIN
ACCEPT verified_order
invoice_value = 0
DO WHILE more order items
  RETRIEVE price FROM Prices
  item_value = price * qty_ordered
  CREATE an iteration of invoice_line
  invoice_value = invoice_value + item_value
ENDDO
CALCULATE vat
invoice_value = invoice_value + vat
CREATE invoice_details
END
```

This sets up part of invoice_details shown below.

```
Process 3.2: Post Accounts
BEGIN
transaction_date = today's date
ACCEPT annotated_invoice_details or payment_details
IF annotated_invoice_details
  WRITE posted_invoice TO Sales Accounts
  CREATE annotated_invoice_details
ELSE
  WRITE posted_payment TO Sales Accounts
ENDIF
END
```

```
Process 3.3: Prepare Statement
BEGIN
* Time trigger, in last week of month *
DO WHILE more customers within Sales Accounts
  RETRIEVE all details for a customer FROM Sales Accounts
  RETRIEVE name_and_address FROM Customers
  statement_date = today's date
  balance_b_f = statement_balance
  SET statement_line for each invoice for current month and
                      each payment received in current month
  balance_c_f = balance_b_f +
         {invoice_value for current month} -
         {payment for current month}
  statement_balance = balance_c_f
  CREATE statement
  WRITE details for payments received this month TO Archive
                                                 Accounts and
  DELETE the same FROM Sales Accounts
  WRITE invoice details TO Archive Accounts
  for all totally or partially paid invoices on the basis of oldest
         first to the sum of payments received this month and
  DELETE those FROM Sales Accounts
  WRITE new statement balance details TO Sales Accounts
ENDDO
END
```

The entire DD for this example is repeated for completeness.

Data flows

```
allocated_stock = * Data element 'allocated_stock' *
annotated_invoice_details = customer_struct, invoice_struct,
                            despatch_date
archive_account = * hidden on DFD *
                  * Process 'Prepare Statement' to
                       data store 'Archive Accounts' *
checked_order = order_struct
credit_status = * Data element 'credit_status' *
customer = * hidden on DFD *
   * Data store 'Customers' to process 'Check Credit Status' *
delivery_note = order_struct
invoice_details = customer_struct, invoice_struct
name_and_address = customer_details
order = customer_details, order_details
order_acknowledgement/refusal = [order_refusal_credit | order_status]
order_refusal_credit = customer_details, order_details, text
```

```
order_status = customer_struct, order_no, order_date, 1{item,text}8
payment_details = customer_struct, payment, payment_date
posted_invoice = customer_no, transaction_date, invoice_transaction
posted_payment = customer_no, transaction_date, payment_transaction
price = * hidden on DFD *
  * Data store 'Prices' to process 'Prepare Invoice Details' *
quantity_in_stock = * Data element 'quantity_in_stock' *
sales_account = * hidden on DFD *
                * Data store 'Sales Accounts' to and from
                     process 'Prepare Statement' *
statement = customer_struct, statement_date, balance_b_f,
                statement_line, balance_c_f
updated_stock = quantity_in_stock, allocated_stock
verified_order = order_struct
```

Data structures

```
customer_details = customer_name, customer_address
customer_struct = customer_no, customer_details
invoice_line = item, price, item_value
invoice_struct = invoice_date, 1{invoice_line}8, vat, invoice_value
invoice_transaction = invoice_details
item = catalogue_no, description, qty_ordered
order_details = order_no, order_date, 1{item}8
order_struct = customer_struct, order_details
payment_transaction = payment, payment_date
statement_line = 1{[(invoice_date, invoice_value) |
                                 (payment_date, payment)]}20
```

Data elements

```
allocated_stock
balance_b_f
balance_c_f
catalogue_no
credit_status
customer_address
customer_name
customer_no
description
despatch_date
invoice_date
invoice_value
item_value
order_date
order_no
payment
```

```
payment_date
price
qty_ordered
quantity_in_stock
statement_balance
statement_date
text
transaction_date
vat
```

Data stores

```
Archived Accounts = {customer_no, transaction_date,
    [invoice_transaction | payment_transaction]}
Creditors = {customer_struct, credit_status}
      * identifier is customer_no *
Customers = {customer_struct}
      * identifier is customer_no *
Prices = {catalogue_no, price}
Sales Accounts = {customer_no, transaction_date, [invoice_transaction
                | payment_transaction | statement_balance]}
Stock = {catalogue_no, description, quantity_in_stock,
                                  allocated_stock}
```

Process specifications

```
Process 1.1: Check Credit Status
BEGIN
ACCEPT order
RETRIEVE name_and_address FROM Customer
IF name_and_address not found
  text = 'not existing customer'
  DISPLAY order_refusal_credit
ELSE
  RETRIEVE credit_status FROM Creditors
  IF credit_status NOT OK
    text = 'order refusal credit'
    DISPLAY order_refusal_credit
  ELSE
    CREATE checked_order
  ENDIF
ENDIF
END

Process 1.2: Allocate Stock
BEGIN
ACCEPT checked_order
```

```
DISPLAY customer_struct, order_no, order_date
DO WHILE more order items
  RETRIEVE quantity_in_stock, allocated_stock FROM Stock
  IF (quantity_in_stock - allocated_stock - qty_ordered) LESS THAN
        ZERO
     text = 'order item rejected'
     DISPLAY item, text
     qty_ordered = 0
  ELSE
     text = 'order item accepted'
     ADD qty_ordered TO allocated_stock
     WRITE allocated_stock TO Stock
     DISPLAY item, text
  ENDIF
ENDDO
CREATE verified_order for accepted order details
END

Process 2: Record Goods Issued
BEGIN
ACCEPT invoice_details
DO WHILE more invoice_lines
  RETRIEVE quantity_in_stock, allocated_stock FROM Stock
  SUBTRACT qty_ordered FROM quantity_in_stock, allocated_stock
  WRITE quantity_in_stock, allocated_stock TO Stock
ENDDO
CREATE delivery_note
despatch_date = today's date
CREATE annotated_invoice_details
END

Process 3.1: Prepare Invoice Details
BEGIN
ACCEPT verified_order
invoice_value = 0
DO WHILE more order items
  RETRIEVE price FROM Prices
  item_value = price * qty_ordered
  CREATE an iteration of invoice_line
  invoice_value = invoice_value + item_value
ENDDO
CALCULATE vat
invoice_value = invoice_value + vat
CREATE invoice_details
END
```

```
Process 3.2: Post Accounts
BEGIN
transaction_date = today's date
ACCEPT annotated_invoice_details or payment_details
IF annotated_invoice_details
  WRITE posted_invoice TO Sales Accounts
  CREATE annotated_invoice_details
ELSE
  WRITE posted_payment TO Sales Accounts
ENDIF
END

Process 3.3: Prepare Statement
BEGIN
* Time trigger, in last week of month *
DO WHILE more customers within Sales Accounts
  RETRIEVE all details for a customer FROM Sales Accounts
  RETRIEVE name_and_address FROM Customers
  statement_date = today's date
  balance_b_f = statement_balance
  SET statement_line for each invoice for current month and each
                    payment received in current month
  balance_c_f = balance_b_f +
         {invoice_value for current month} -
         {payment for current month}
  statement_balance = balance_c_f
  CREATE statement
  WRITE details for payments received this month TO Archive
                                 Accounts and
  DELETE the same FROM Sales Accounts
  WRITE invoice details TO Archive Accounts
  for all totally or partially paid invoices on the basis of oldest
  first to the sum of payments received this month and
  DELETE those FROM Sales Accounts
  WRITE new statement balance details TO Sales Accounts
ENDDO
END
```

Terminators

```
Customer
Salesperson
```

Use of a CASE tool

As you will have noticed by now, the manual development of a DD is a tedious, almost impossible, task. Abbreviated DD entries are used for most examples and exercises, and indeed, only part of the DD for the Somerleyton Animal Park and the Albany Hotel are included in this chapter. Even for the small Worked example 4.2, the DD is quite substantial. Obviously, therefore, most users of DDs use an automated DD of some kind. It is possible to use a word processing package, but this does not provide much support. By far the best solution is to use a CASE (Computer-Aided System/Software Engineering) tool. Front-end CASE tools support the activities within the systems analysis and design phases of systems development. Single function tools aim to support one aspect of development, for example DFDs. These tools can be used to develop DFDs and to perform the balancing and consistency checking discussed in Chapters 2 and 3. An integrated set of tools would typically support one or more aspects of the development process; some support a particular methodology, for example SSADM, whereas others do not. There are many different CASE tools, but to generalize, it is possible to use a CASE tool to set up a DD in parallel with the development of the DFDs. As data flows and data stores are entered on the DFD, DD entries will be created for them. It is then possible for the CASE tool user to add data structures and data elements to them. Process specifications may be completed for the functional primitives. Some CASE tools then automatically include the inputs to and outputs from the process specification mentioned earlier in the chapter. The degree to which the Structured English needs to specify DD entries rather than to convey the general understanding of the logical steps is dependent on the CASE tool used. Using a CASE tool increases the accuracy of DD entries and it is possible within most CASE tools to check the completeness and consistency of the DD entries, for example to identify data elements that are not used anywhere in data flows or stores.

Summary of data dictionaries

One of the steps identified when developing DFDs is to check the DFD for consistency and completeness. Some of this checking is performed 'automatically' or as a by-product of DD creation. Processes are carefully inspected when process specifications are prepared. It is easy to check for processes that have inputs and no outputs, or *vice versa*. Labelling is checked both for its presence and uniqueness. Read-only and write-only data stores should be identified and checked. The use of a CASE tool can help with the completeness and consistency checking both of the DFDs and the DD. When you have studied all three views of a system, that is process, data and time, it will become apparent that the DD underpins the models depicting these three views. Further discussion will take place on the use of DDs and CASE tools in later chapters, when entity relationship modelling and entity life histories are developed, and the models are cross-checked.

Chapter 5

Data Modelling 1

OBJECTIVES

In this chapter you will learn:

- □ the components of an entity-relationship (E-R) diagram;
- □ using guidelines, how to develop a simple E-R diagram;
- □ how to identify and correct connection traps.

Introduction

In Chapters 2 and 3 you have learned how to use data flow diagrams to model the process view of a system. These models necessarily include some consideration of the data used by the system, and the content of data stores and data flows is defined in the data dictionary (see Chapter 4). However, the construction of DFDs does not involve a close examination of the structure of the data. To determine the file structures that will be necessary to support the new computer system's procedures, it is necessary to examine the data more closely, and to model it more rigorously than DFDs allow.

E-R modelling is based on the original work of Chen (1976). A detailed examination of E-R modelling is provided in Howe (1989).

This chapter introduces a top-down approach to data modelling, using entity-relationship (E-R) diagrams. Chapter 6 describes a bottom-up approach, by developing a set of normalized tables. These two techniques are complementary, and together produce an entity-relationship model to represent the data view of the system.

A logical view

The aim of data analysis is to produce a logical data model that accurately represents the structure of the data needed to support the information system under investigation, without any reference to the implementation environment. This logical data model will almost certainly need to be adapted during the systems design phase to tailor it to a particular hardware/software platform. However, the effort involved in producing an independent, logical data model during the analysis phase is justified by the following advantages:

- Data tends to be more stable than processes; thus over time, the fundamental structure of an organization's data is unlikely to undergo radical change. An accurate data model can remain relevant, while the transactions that use it and the implementation environment may change.
- A poor understanding of the fundamental structure of the data can lead to badly designed and inflexible systems.
- The benefits of adopting a database approach can be realized.

A database approach implies the sharing of data across applications, instead of supplying each application with its own data. The latter approach often leads to the same piece of data being held in more than one place; for example, the Accounts Department of a company might hold a customer's name and address, and so might the Area Sales Office. This redundant (unnecessary) duplication can lead to inconsistencies in the data when one copy is updated and the other is not, and is a waste of storage space. Sharing data across applications introduces issues such as how access to sensitive data can be controlled; Database Management Systems provide a layer of software to address security and related problems.

See Chapter 6 for more about redundant and duplicated data. See Howe (1989), Date (1990) and Korth and Silbershatz (1991) for more information about Database Management Systems.

Entity-Relationship modelling – basic concepts

E-R modelling uses three basic components; entities, attributes and relationships.

Entity

An **entity** is something about which the organization wishes to collect and store data. It is capable of being uniquely identified. For example, in the Somerleyton Animal Park case scenario, `Animal`, `Order` and `Supplier` are likely entities.

Attribute

An **attribute** is a data element that is associated with an entity. For example, the entity `Animal` may have attributes such as `animal_name` and `date_of_birth` while the entity `Supplier` is likely to have attributes such as `supplier_name`, `supplier_address` and `supplier_telephone_number`.

Relationship

Entities are associated with each other via **relationships**. For example, the entity `Animal` is associated with the entity `Enclosure` by the relationship `Lives_in`.

Entity-Relationship diagrams

Figure 5.1 illustrates how entities and the relationships between them may be represented in an **E-R diagram**. The attributes associated with each entity are held in a table, or relation; this aspect of the E-R model will be considered in detail in Chapter 6.

Figure 5.1 An entity-relationship type diagram.

Types and occurrences

Note that there is an important distinction between an entity type (for example, `Animal`) and an entity occurrence (for example, Fred the chimpanzee). Similarly, a relationship type (for example, `Lives_in`) may be distinguished from a relationship occurrence (for example, the specific relationship that Fred lives in enclosure 39). The diagram in Figure 5.1 is therefore an E-R type diagram.

An E-R occurrence diagram is illustrated in Figure 5.2. In this diagram, there are two entity types, `Animal` and `Enclosure`, and one relationship type, `Lives_in`. Five occurrences of the entity type `Animal` (Fred, Jane, Bonzo, Stripe and Alfred) and two occurrences of the entity type `Enclosure` (enclosure number 39 and enclosure number 42) are specified. The lines between the entity occurrences represent occurrences of the relationship type `Lives_in`.

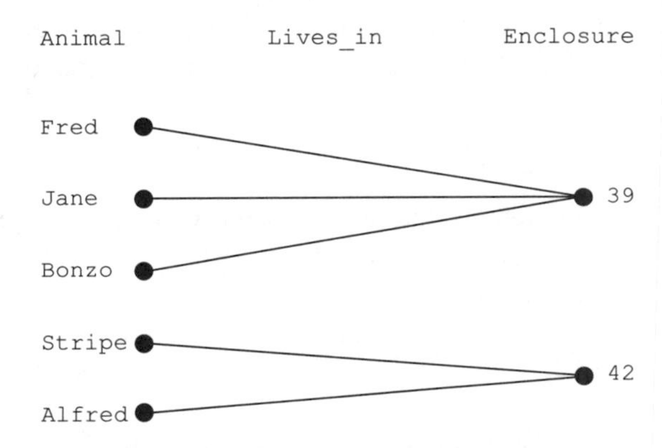

Figure 5.2 An entity-relationship occurrence diagram.

The term **E-R diagram** will be used in this text to refer to an E-R type diagram, and will be qualified with 'occurrence' where this meaning is intended. The terms **entity** and **relationship** will be used to refer to entity type and relationship type, respectively.

Identifiers

It was stated earlier that an entity is capable of being uniquely identified. That is, a particular entity occurrence must be capable of being distinguished from every other entity occurrence by an attribute or a combination of attributes. If an animal may be uniquely identified by its name, then `animal_name` is said to be the **identifier** of the entity type `Animal`. Note that if it is possible for more than one animal to have the same name, then some other identifier must be chosen; for example, a unique number may be allocated to each animal.

An identifier is sometimes called a **primary key**.

Degree of the relationship

A relationship has an important property; its **degree**. This may be 1:1 (one-to-one), 1:M (one-to-many) or M:M (many-to-many). Figure 5.3 illustrates the three types of degree. Note that 'M:1' is also a 'one-to-many' relationship.

Sometimes written 1:N and M:N.

A headkeeper may supervise just one enclosure.
An enclosure may be supervised by just one headkeeper.

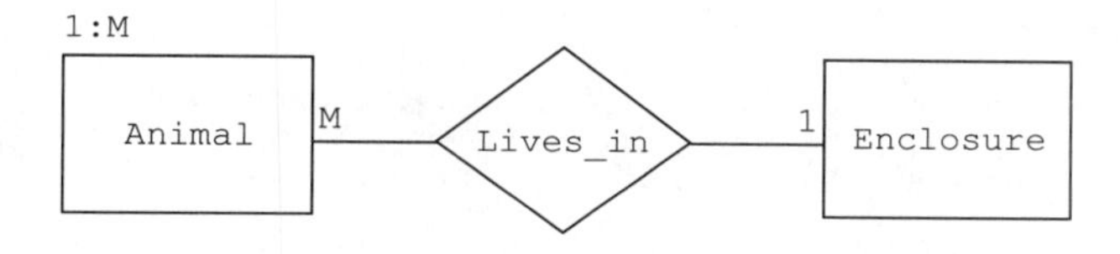

An animal may live in just one enclosure.
An enclosure may contain more than one animal.

A supplier may supply more than one type of food.
A type of food may be supplied by more than one supplier.

Figure 5.3 The degree of a relationship.

In general, 'many' includes zero, one and more than one. A type of food might be supplied by many suppliers, one supplier or none at all (for example, food grown by the Animal Park).

Note carefully from Figure 5.3 how a relationship is 'read' in each direction. The entity type that is named first is *always* read in the singular; for example in the M:M relationship that is illustrated, '*A supplier* may supply more than one type of food' and '*A type of food* may be supplied by more than one supplier'.

Membership class

Finally, precision can be added to the E-R diagram by the inclusion of the **membership class** for each relationship. That is, consideration of whether *every* entity occurrence takes part in a particular relationship (when membership class is said to be 'obligatory'), or whether only some entity occurrences will take part in the relationship (when the membership class is said to be 'non-obligatory').

Membership class is indicated by the inclusion of a dot inside the entity box for obligatory, and outside the box for non-obligatory membership. All possible cases are illustrated for the relationship between `Animal` and `Enclosure` in Figure 5.4.

An animal must live in just one enclosure.
An enclosure must contain at least one, and possible many, animals.

An animal must live in just one enclosure.
An enclosure may contain zero, one or many animals.

An animal may live in an enclosure.
An enclosure must contain at least one, and possible many, animals.

An animal may live in an enclosure.
An enclosure may contain zero, one or many animals.

Figure 5.4 Membership class.

Enterprise rules

The question as to which one of the possibilities illustrated in Figure 5.4 is an accurate model of the data held by Somerleyton Animal Park, will depend upon the Park's **enterprise rules**. If every animal *must* be assigned to an enclosure, then the entity `Animal` will be obligatory in the relationship with the entity `Enclosure`. However, if it is possible for details of an animal to be held by the Park but for it to be not currently assigned to an enclosure (because, for example, it has been loaned to another Animal Park), then it will be non-obligatory. Similarly, the systems analyst must determine whether an enclosure *must* at all times house at least one animal or whether from time to time an enclosure may be out of use. An important task for the systems analyst, therefore, is to determine the enterprise rules that govern the data. The process of preparing the data model will be an iterative one, as the systems analyst checks the draft models with the user and refines them as his/her understanding of the data, its structure and the enterprise rules grows.

An enterprise rule can be defined as a rule that relates to the enterprise's data model. It can therefore apply not only to membership class but to the degree of a relationship (a head-keeper supervises only one enclosure), and to issues such as validation (an enclosure number must be exactly three numeric digits).

Approach to drawing an E-R diagram

The systems analyst will gather information about the data used by the system during the fact finding phase. There might already be a computer system in operation, in which case the contents of existing data files will need to be carefully examined. Documents in use will be a valuable source of information, as they contain data that is being obtained, transmitted and stored within the organization. If work has begun on process modelling, the data dictionary entries for data elements, data structures, data flows and data stores will contain clues about possible entities and attributes.

Guidelines for drawing an E-R diagram

1. Select likely entities.
2. Select an identifier for each of the entities.
3. Identify relationships between the selected entities.
4. Sketch an E-R diagram, adding the degree of each relationship.
5. Add membership classes.

Note that guideline 2 aids the selection of likely entities, as an entity must be capable of being uniquely identified.

Worked example 5.1

Using the specified guidelines, develop an E-R diagram for the ordering supplies section (subsystem) of the Somerleyton Animal Park case scenario.

Solution:

Guideline steps 1 and 2:

It is unlikely that every possible entity will be identified at this stage; others will emerge as the data model is refined.

Note that a form is not usually an entity; the order form contains attributes which belong to several entities, namely `Order`, `Supplier` and `Foodtype`. The justification for selecting these entities will be explained further in Chapter 6.

As the user is not available for consultation, assumptions will need to be made for this and subsequent worked examples.

Likely entities	*Identifiers*
Order	order_no
Supplier	supplier_name
Foodtype	foodtype_name
Delivery	delivery_no

Assumptions

1. Each order is assigned a unique `order_no`.
2. Supplier names are unique.
3. Each delivery is assigned a unique `delivery_no`.

Note that an entity is given a singular name (`Order`, not `Orders`); it may help to remember that it is the name of a single entity type.

It has been assumed that each supplier has a unique name, as it is unlikely that two companies have exactly the same name. However, it would be equally acceptable to assign a unique number to each supplier, which would often be done in practice because a number would be more concise and less subject to transcription error.

Guideline step 3:
Relationships

An order is sent to a supplier.
A foodtype is ordered on an order.
A supplier supplies foodtypes.
A supplier makes a delivery.
A delivery corresponds to an order.

Not every relationship will be necessarily identified at this stage; as with entities, others may emerge later as the data model is refined.

A relationship represents an association between two entities that is of relevance to the system. Thus, it is important to know: to which supplier an order has been sent; which types of food were requested on that order; which supplier can supply a given type of food; from which supplier a delivery came; to which order a delivery corresponds.

Guideline step 4: See Figure 5.5.

Assumptions (continued)

4. A supplier can supply more than one type of food.
5. A foodtype can be supplied by more than one supplier.
6. A delivery could correspond to more than one order, and an order could be split and sent in more than one delivery (for example, if there was insufficient stock to satisfy the whole order immediately).

Assumptions 4, 5 and 6 relate to the degree of one of the relationships. Some of the degrees are made explicit by the case scenario, for example, the sample order form indicates that an order goes to just one supplier, and that

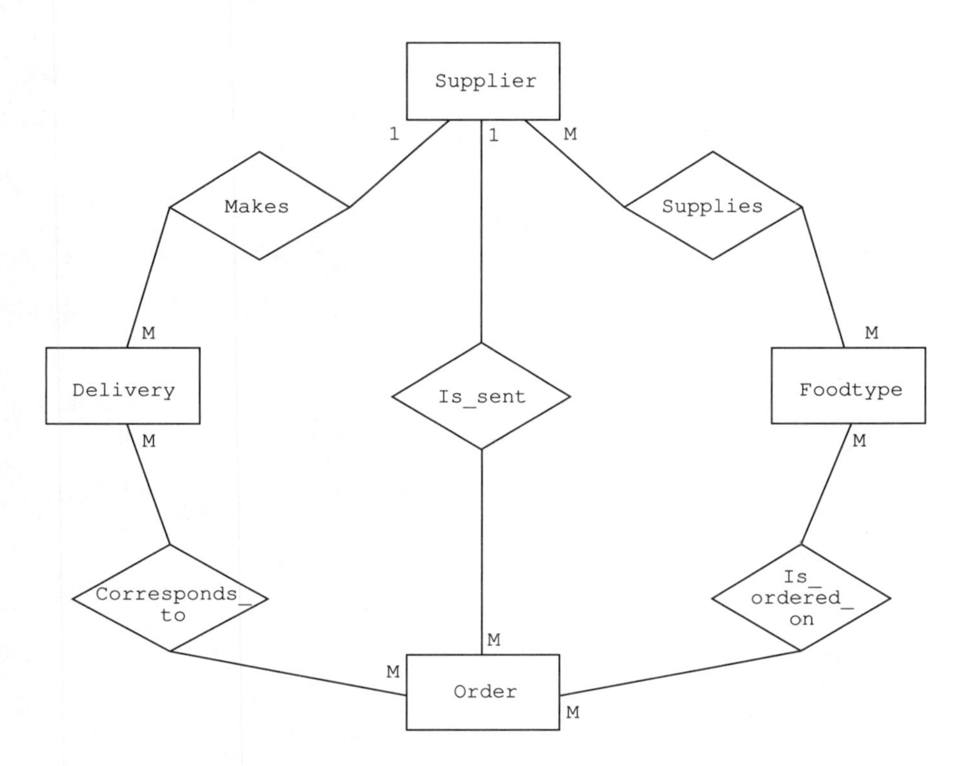

Figure 5.5 Guideline step 4.

it can contain more than one type of food. Other degrees can be derived from the application of 'common sense'; for example, a supplier should be able to be sent more than one order, and therefore be able to send more than one delivery. In the absence of a user to consult, for example when completing coursework or an examination, 'reasonable' assumptions should be made.

Note that the convention is to form relationship names that 'read' from left to right, or from top to bottom of the E-R diagram. Thus in Figure 5.5, the relationship between `Supplier` and `Foodtype` has been named `Supplies`. If, however, the entity `Foodtype` were to be drawn above the entity `Supplier` the relationship name would be `Is_supplied_by`. Perhaps the active version of a relationship name is preferable to the passive version, if only because it usually fits better into the relationship box! But the most important guideline in naming relationships is to ensure the clarity of the diagram; it should be capable of being read and understood by somebody familiar with the technique.

Guideline step 5: See Figure 5.6.

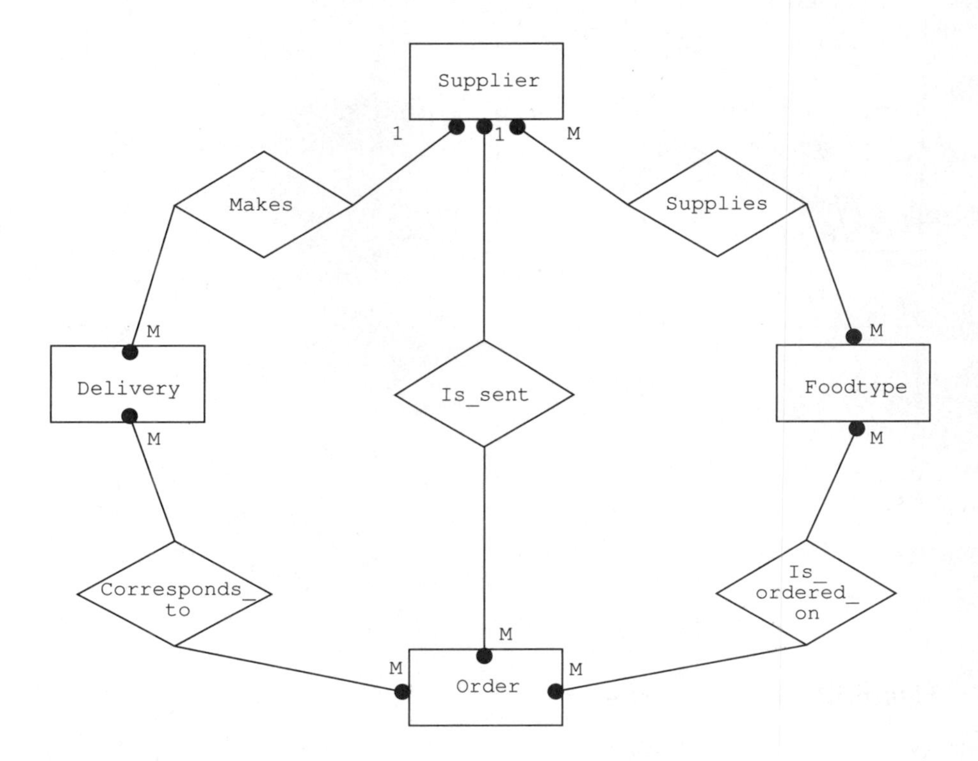

Figure 5.6 Guideline step 5.

Assumptions (continued)

7. Details of suppliers other than those who supply food are also held.
8. At times a foodtype might not have a supplier.
9. A supplier might not have been sent any orders.
10. A supplier might not have made any deliveries.

Once again, some of the membership classes are made explicit in the case scenario; for example, that each order must contain at least one type of food and that it must go to a supplier. For others, reasonable assumptions must be made. Thus it has been assumed that details of a variety of suppliers are held, including some that have not been sent an order. It would have been equally acceptable to assume that only suppliers that have been used are kept on file; in which case the membership class of `Supplier` with `Order` would have been obligatory. Occasionally, therefore, two different assumptions can be equally 'reasonable'; it is vital that you state the assumption you have made and ensure that your diagram is consistent with that assumption.

The entire list of assumptions is included for completeness:

Assumptions

1. Each order is assigned a unique `order_no`.
2. Supplier names are unique.
3. Each delivery is assigned a unique `delivery_no`.
4. A supplier can supply more than one type of food.
5. A foodtype can be supplied by more than one supplier.
6. A delivery could correspond to more than one order, and an order could be split and sent in more than one delivery (for example, if there was insufficient stock to satisfy the whole order immediately).
7. Details of suppliers other than those who supply food are also held.
8. At times a foodtype might not have a supplier.
9. A supplier might not have been sent any orders.
10. A supplier might not have made any deliveries.

The other enterprise rules can be deduced from the scenario.

Exercise 5.1

Identify likely entities, and an identifier for each one, from the following scenario. Make and state any assumptions that you think are necessary.

> A local education authority (LEA) requires a system to hold details of the applicants who apply for courses run by local colleges. Each course is run at only one college, though more than one course is run at each college. Each course and college is allocated a unique code by the LEA. An applicant may apply for several courses. Details are to be kept of the standard qualifications that the majority of applicants have, such as GCSEs.

Exercise 5.2

Identify the relationships between the entities noted in Exercise 5.1. Draw an E-R diagram, adding both the degree and the membership class of each relationship.

Decomposition of M:M relationships

Any M:M relationship can be turned into two 1:M relationships. This is done by converting the relationship into an entity that is at the 'M' end of a 1:M relationship with each of the original entities. See Figure 5.7 for an example.

The new entity needs to be named; in the example in Figure 5.7 it has been given the name `Orderline`. This reflects the fact that the sample order form for the Animal Park contains a number of lines, each of which holds

Figure 5.7 Decomposition of a M:M relationship (1). (a) M:M relationship.

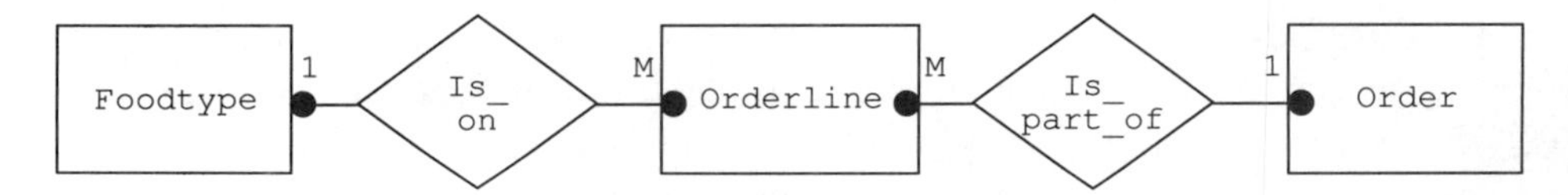

Figure 5.7 (b) M:M relationship decomposed.

information about one type of food. Often it is difficult to think of an appropriate name for the new entity, in which case it is acceptable to form it from the names of the two original entities (see Figure 5.8). The naming of the two new relationships formed by the decomposition can pose a problem; it is acceptable to leave the relationship boxes blank providing this does not impair the clarity of the diagram.

Figure 5.8 Decomposition of a M:M relationship (2). (a) M:M relationship.

Note that the relationship `Supplies` in Figure 5.8 (b) is different from the relationship `Supplies` in Figure 5.8 (a).

Figure 5.8 (b) M:M relationship decomposed.

The new entity needs an identifier. Usually, this can be formed from the identifiers of the two original entities. For example, `order_no` and `food-type_name` together provide an identifier for `Orderline`. Occasionally, an additional attribute is required to form an identifier (see Exercise 5.3).

The membership class of the new entity is obligatory in both of the 1:M relationships in which it takes part. The two original entities retain the same membership class in the new 1:M relationships as they had in the M:M relationship (see Figures 5.7 and 5.8).

There are a number of reasons why M:M relationships should be considered carefully. Decomposition may highlight the existence of a fan trap (see 'Connection Traps' below). There are other unsatisfactory aspects that may be hidden by a M:M relationship, which are dealt with in Chapter 6.

See the discussion of repeating groups in Chapter 6.

Another reason for decomposition is that many database management systems cannot directly support M:M relationships.

Exercise 5.3

Decompose the M:M relationships in the solution to Exercise 5.2.

Guidelines

The guidelines for drawing an E-R diagram that were introduced earlier can now be amended as follows:

1. Select likely entities.
2. Select an identifier for each of the entities.
3. Identify relationships between the selected entities.
4. Sketch an E-R diagram, adding the degree of each relationship.
5. Add membership classes.
6. Decompose any M:M relationships, allocating an identifier to any new entities formed.

Worked example 5.2

Using the specified guidelines, develop an E-R diagram for the estate agent scenario introduced in Exercise 2.2.

See Chapter 2, Exercise 2.2.

Solution:

Guideline steps 1 and 2:

Likely entities	*Identifiers*
`Client`	`client_name`
`Property`	`property_id`
`Potential Buyer`	`potential_buyer_name`
`Buyer`	`buyer_name`
`Invoice`	`invoice_no`

The data stores on the DFD provide clues for possible entities.

For the moment, potential buyers and buyers have been treated as two separate entities. The decision as to whether they should remain two entities or be combined into one entity can be made only when the attributes to be stored for them are considered in detail.

See Chapter 6.

Guideline step 3:
Relationships
A client has a property to sell.
A potential buyer is interested in particular properties.
A buyer buys a property.
An invoice is prepared for a client.

Guideline step 4: See Figure 5.9.

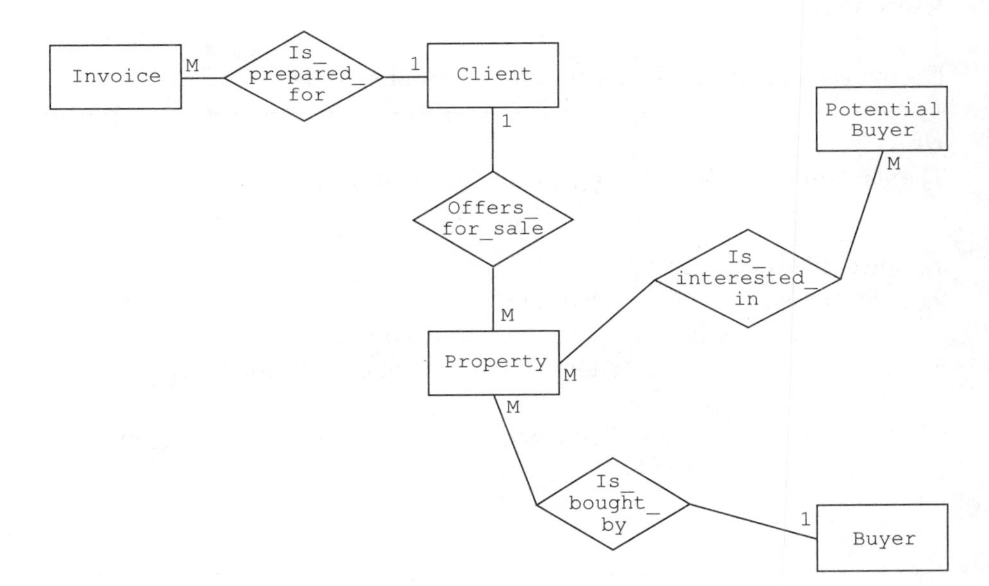

This model will be refined later (see Exercise 5.5).

Figure 5.9 Guideline step 4.

Assumptions
1. A client could sell more than one property (for example, having renovated a large old building, they could be selling several separate flats). Thus a client could be sent more than one invoice.
2. A (wealthy) buyer could buy more than one property.

Guideline step 5: See Figure 5.10.

Assumptions (continued)
3. A property might not have been viewed by anybody.
4. A potential buyer may not have viewed any properties.

Guideline step 6: There is a M:M relationship between `Potential Buyer` and `Property`. On closer examination, the relationship `Is_interested_in` is somewhat ambiguous. Potential buyers provide details of what area, price range and type of property interests them. Thus they

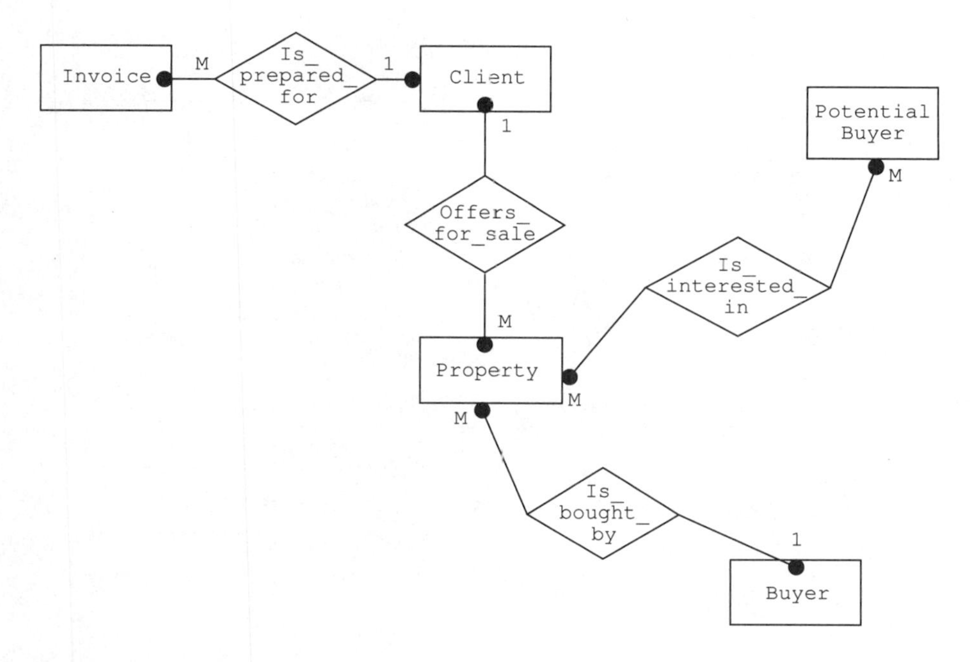

Figure 5.10 Guideline step 5.

are interested in particular categories of property. In addition, the estate agent provides details of particular properties, some of which the potential buyer might proceed to view. It must be determined exactly what information the system is required to store. It will be assumed that details of property viewings are kept.

Assumptions (continued)

5. Information about which properties potential buyers have viewed needs to be stored.
6. Potential buyers might view the same property more than once, but not more than once on any one day.

If they do return more than once to view the property on the same day, this is stored as just one visit.

See Figure 5.11 for an amended diagram.

Likely entities	*Identifiers (continued)*
`Viewing`	`property_id, potential_buyer_name, date_of_viewing`
`Requirement`	`potential_buyer_name, requirement_no`

The entity `Viewing` holds information about which properties have been viewed by which potential clients. The entity `Requirement` holds information about the requirements of each potential buyer. The latter could have more than one requirement, for example they might be interested in several areas or in either a semi-detached or a detached house.

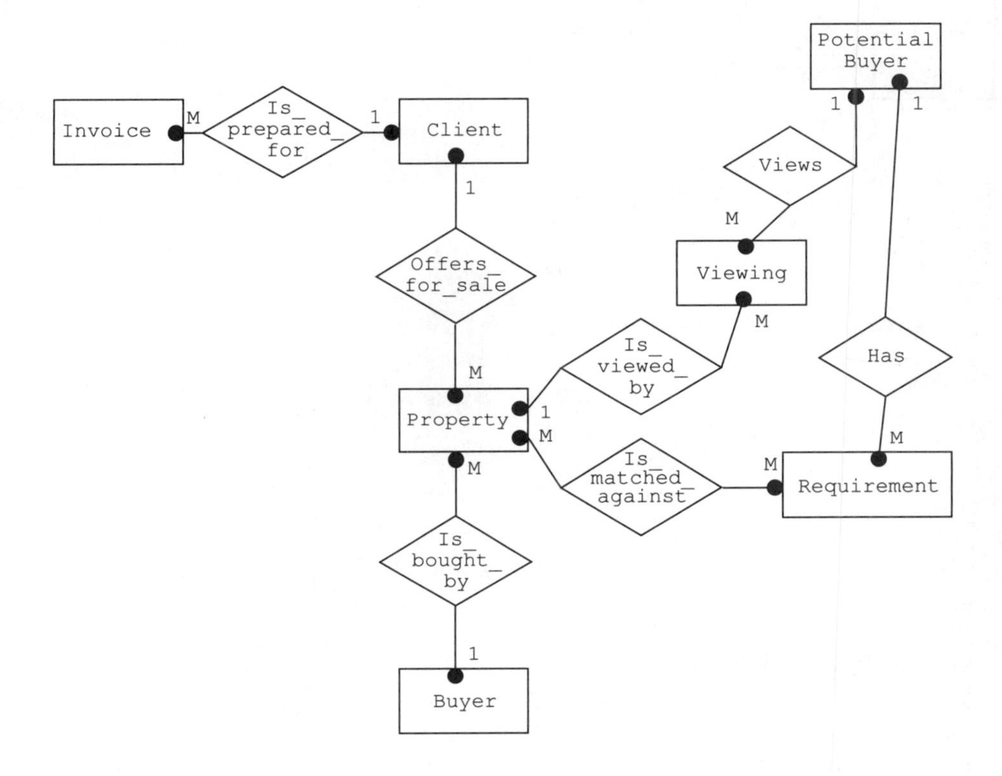

Figure 5.11 Guideline step 6.

Another M:M relationship has emerged from this step, between `Property` and `Requirement`. The relationship `Is_matched_against` represents which properties match potential buyers' requirements, and therefore which details are to be sent to whom. There is no need to redraw the diagram to illustrate the decomposition of this M:M relationship; after the preceding step we can be confident that the meaning of the relationship has been determined. However, the relationship `Is_matched_against` will need an identifier, for example `property_id`, `potential_buyer_name` and `requirement_no`.

The entire list of assumptions is included for completeness:

Assumptions

1. A client could sell more than one property (for example, having renovated a large old building, they could be selling several separate flats). Thus a client could be sent more than one invoice and could make more than one payment.
2. A (wealthy) buyer could buy more than one property.
3. A property might not have been viewed by anybody.
4. A potential buyer may not have viewed any properties.
5. Information about which properties potential buyers have viewed needs to be stored.

6. Potential buyers might view the same property more than once, but not more than once on any one day.

Exercise 5.4

Using the specified guidelines, develop an E-R diagram for the following scenario:

> A mail order company specializes in supplying men's and women's quality fashion clothing. The items of clothing are supplied by manufacturers and textile importers throughout the UK; each item is supplied by just one manufacturer or importer, but most manufacturers and importers supply many different items of clothing. Each item is allocated a unique code by the mail order company. Customers send in orders to the mail order company; each order can be for several items. The mail order company orders stock from the manufacturers and importers using a clothing requisition. A clothing requisition can be for many items of clothing.

Connection traps

See Howe (1989, Chapter 11) for a detailed discussion of connection traps.

The E-R diagram needs to be checked carefully to ensure that it supports all the requirements of the system. The paths must exist through the diagram to satisfy all updates and enquiries. The diagram should be inspected for potential connection traps; these can occur when a path involves three or more entities. They arise through a misunderstanding of the meaning of a relationship and its misrepresentation on the diagram, and they can prevent required information from being held. There are two types of connection trap; the fan trap and the chasm trap.

Fan traps

A **fan trap** involves three entities, and occurs when two 1:M relationships 'fan out' from the entity in the centre. Figure 5.12 represents a system where a lecturer may teach on more than one course and a course may be taught by a number of lecturers; each course includes a number of modules, each of which is offered within just one course. Figure 5.13 shows that the M:M relationship between the entity types `Lecturer` and `Course` hides a fan trap which can be seen once the relationship has been decomposed.

The occurrence diagram in Figure 5.14 illustrates why the term 'fan' is used for this type of connection trap. It also demonstrates that although it is possible to tell which lecturers teach on which courses, and which modules are offered by which courses, it is not possible to answer the queries, 'Which lecturer(s) teaches on a given module?' and 'Which modules does a particular lecturer teach?' Lecturer L1 teaches on course C1, but could

There might be other lecturers and other modules not represented in the occurrence diagram in Figure 5.14.

teach either M1 or M2, or both, or neither. Similarly, module M2 is offered within course C1, but could be taught by any, or none, of lecturers L1, L2 or L3.

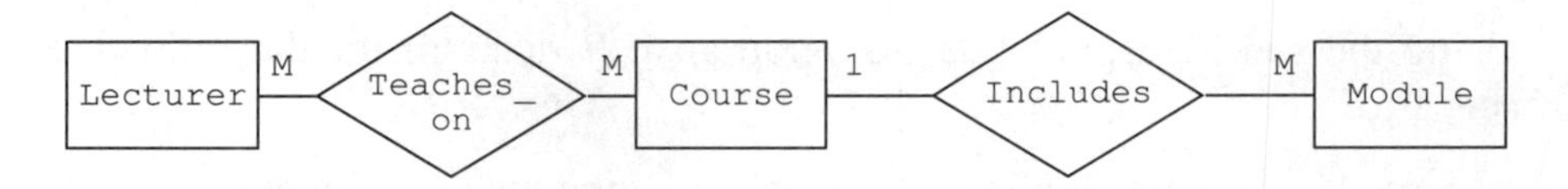

Figure 5.12 The relationships between Lecturer, Course and Module.

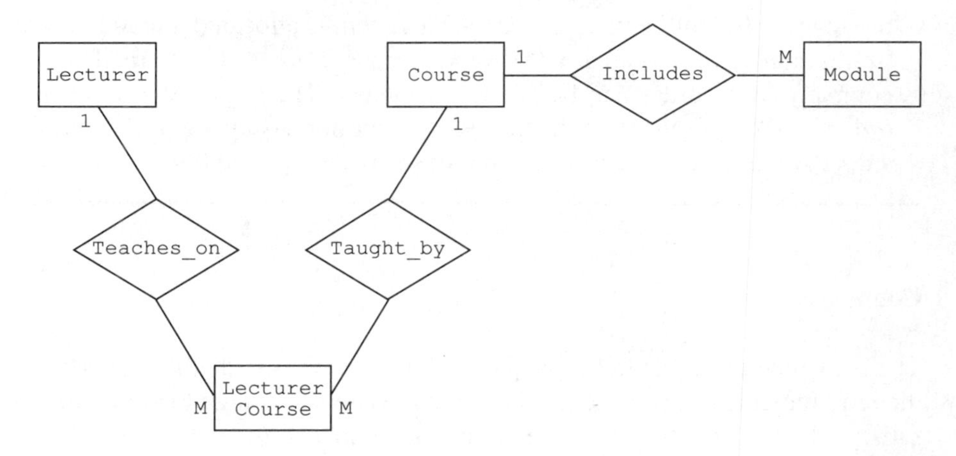

Figure 5.13 A fan trap.

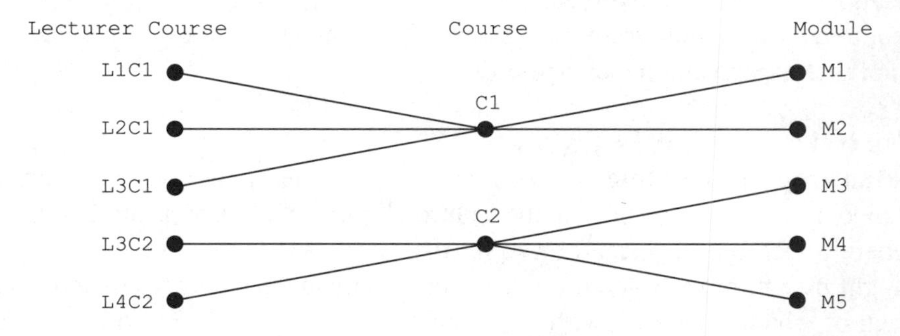

Figure 5.14 Occurrence diagram for Figure 5.13.

To solve this problem, the diagram can be redrawn as shown in Figure 5.15. In effect, the relationships have been re-interpreted and represented more accurately. The occurrence diagram in Figure 5.16 confirms that no information has been lost; it is still possible to tell which lecturers teach on which courses and which modules are offered by which courses. But in addition, the queries that could not be satisfied from the diagram in Figure 5.13 can now be answered.

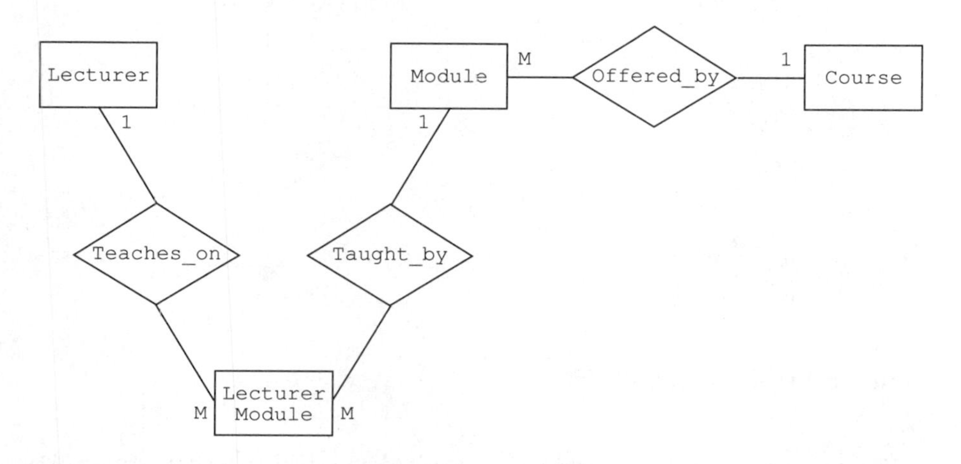

Figure 5.15 Resolution of the fan trap in Figure 5.13.

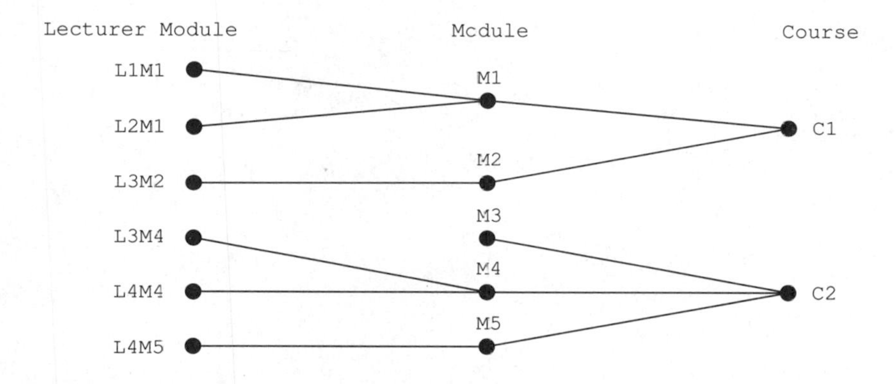

Figure 5.16 Occurrence diagram for Figure 5.15.

Chasm traps

A **chasm trap** occurs when the path through a diagram does not exist for some entity occurrences, and this causes a loss of data. Consider the diagram in Figure 5.17, which represents part of the Somerleyton Animal Park.

Animals of a particular species are always housed in the same area. It is possible to tell for a particular animal, which area it lives in, since an animal lives in just one enclosure and an enclosure is sited within just one area. However, the entity `Species` is non-obligatory in its relationship with the entity `Animal`; so if a particular species is not currently represented in the Park, it is not possible to tell to which area that species would normally be assigned. If this information is needed then this forms a chasm trap.

It is being assumed that a species is kept in only one area.

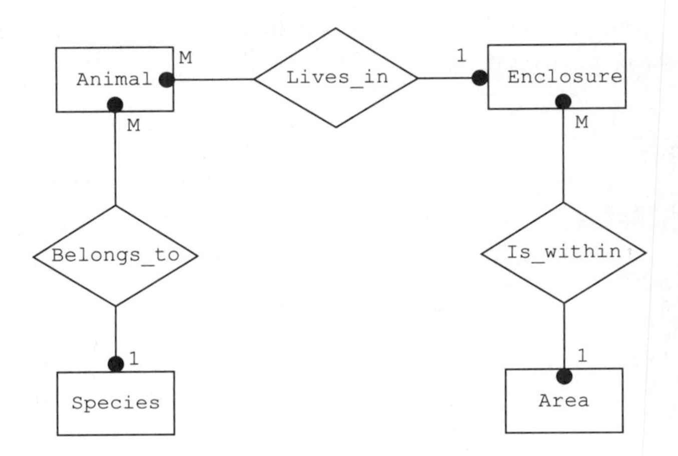

Figure 5.17 A chasm trap.

The solution to a chasm trap is to 'bridge' the chasm by supplying the missing relationship. In the case of the Animal Park, a relationship is needed between `Species` and `Area` (see Figure 5.18).

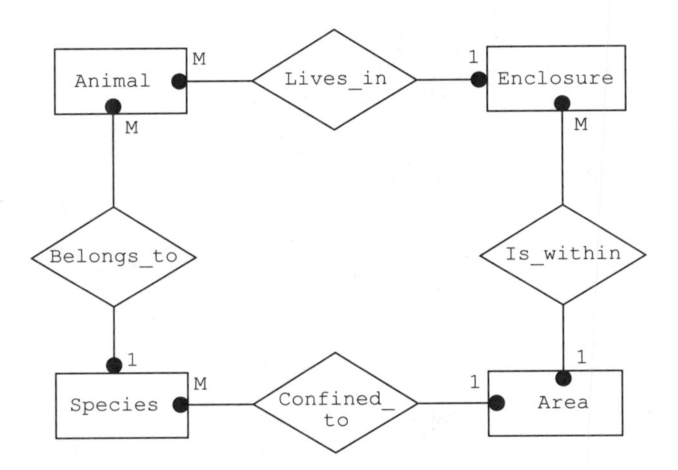

Figure 5.18 Resolution of the chasm trap in Figure 5.17.

Significance of connection traps

Most E-R diagrams will contain potential connection traps. Whether or not those traps are significant depends upon the requirements of the system. If, for example, the Somerleyton Animal Park did not want to store the information about which species are housed in which areas, then Figure 5.17 would not present a problem.

Thus an E-R diagram must first be inspected for potential traps; then the significance or otherwise of any potential traps must be assessed bearing in mind the requirements of the system.

Guidelines

The guidelines for drawing an E-R diagram can now be further amended as follows:

1. Select likely entities.
2. Select an identifier for each of the entities.
3. Identify relationships between the selected entities.
4. Sketch an E-R diagram, adding the degree of each relationship.
5. Add membership classes.
6. Decompose any M:M relationships, allocating an identifier to any new entities formed.
7. Check the diagram for potential connection traps. Assess the significance of any traps found, and eliminate significant traps.

Exercise 5.5

Examine Figure 5.11. Identify potential connection traps and assess their significance. Propose a solution to any significant traps.

Exercise 5.6

Each student at a college is enrolled on one course. The students choose which modules they wish to study. Students from several courses may choose to study the same module. Requirements include the ability to determine which students are enrolled on a particular course; which modules a student has chosen to study, on which course a module may be studied and which modules are offered by a particular course.

Examine Figure 5.19. Identify potential connection traps and assess their significance. Propose a solution to any significant traps.

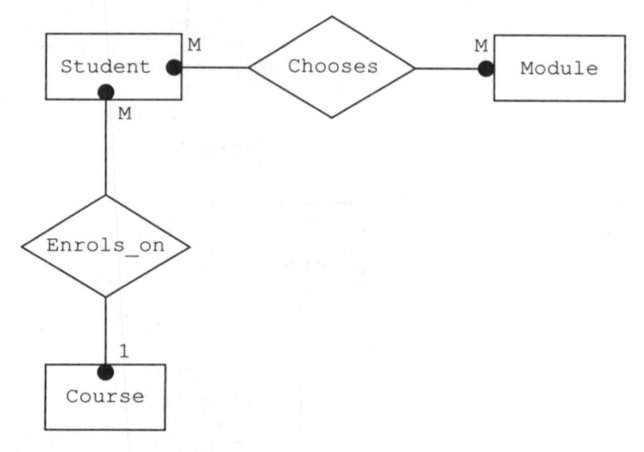

Figure 5.19 Exercise 5.6.

Exercise 5.7

Examine the solution to Exercise 5.6 carefully. Is the modified diagram still free of connection traps?

Alternative notations

Data modelling is an important technique used in all structured methodologies. The notation used in this chapter is similar to that used by Chen (1976) and Howe (1989).

See the SSADM Reference Manual (1990) for a full explanation of the logical data structure notation.

Alternative notation is used by some methodologies. Figure 5.20 contains the E-R diagram from Figure 5.18 redrawn in the logical data structure notation used by SSADM v4. A single line and a 'crow's foot' are used to denote '1' and 'M', respectively; a complete line is used at the obligatory end, and a broken line at the non-obligatory end of a relationship. Relationships may be named at both ends.

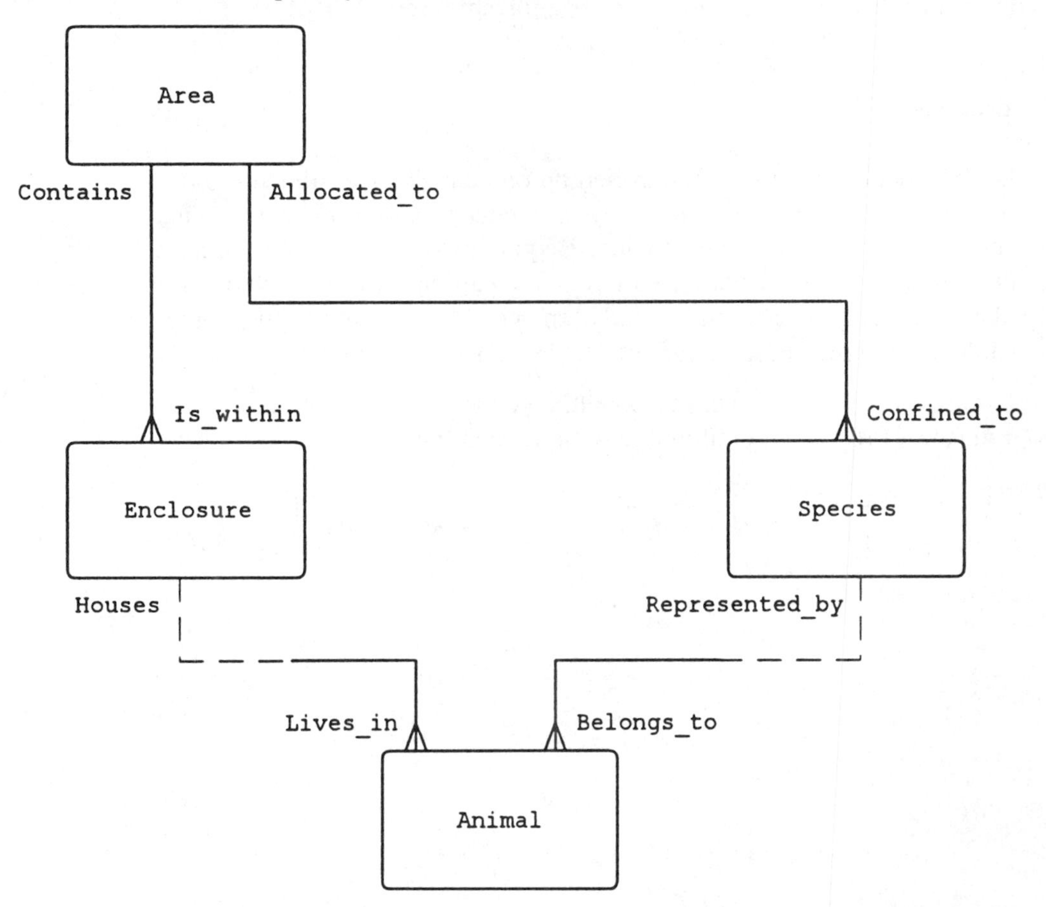

Figure 5.20 Figure 5.18 redrawn as an SSADM v4 logical data structure.

This chapter has introduced a top-down approach to data modelling, using E-R diagrams. Some fundamental concepts have been discussed which are common to all methodologies that apply data modelling, although the diagrammatic notation used can vary.

Chapter 6

Data Modelling 2

OBJECTIVES

In this chapter you will learn:

- □ the components of the relational model;
- □ using guidelines, how to derive a set of well-normalized tables;
- □ using guidelines, how to develop an entity-relationship model.

Introduction

Chapter 5 introduced the entity-relationship diagram as a means to model an enterprise's data by identifying entities and the relationships between them. The top-down approach of the E-R diagram helps the systems analyst to gain an overall perspective of the data within the system under investigation. However, this technique is not sufficient on its own to provide the basis for the design of a computer database or file definitions. It needs to be complemented by a more precise analysis of the attributes of each entity.

See Chapter 5 for a definition of an attribute.

This chapter introduces the concepts of **relational data modelling**, and explains how to derive a set of well-normalized tables. These, together with the E-R diagram, form an **E-R model** that provides a logical view of the data required to support the information system under development.

The relational model

The **relational model** is a conceptually simple and flexible way in which to model the data within an information system. In this model, data is held in a set of two-dimensional relations. The term 'relation' is derived from the mathematical discipline of set theory and the model has a sound mathematical basis, which it is not the purpose of this text to explore.

See Maier (1983) and Atzeni and De Antonellis (1993). Properties that tables must have to be relations are described below. See Date (1990) for further details.

The term 'table' is often used as a synonym for 'relation'. However, it should be noted that 'relation' is a mathematical concept, whereas 'table' is a less formal term with a broader meaning. Following the common practice of treating the two terms as though they were interchangeable, 'table' will be used as the preferred term in this text.

Tables

Figure 6.1 shows a table occurrence, and illustrates the terminology used for the components of a table.

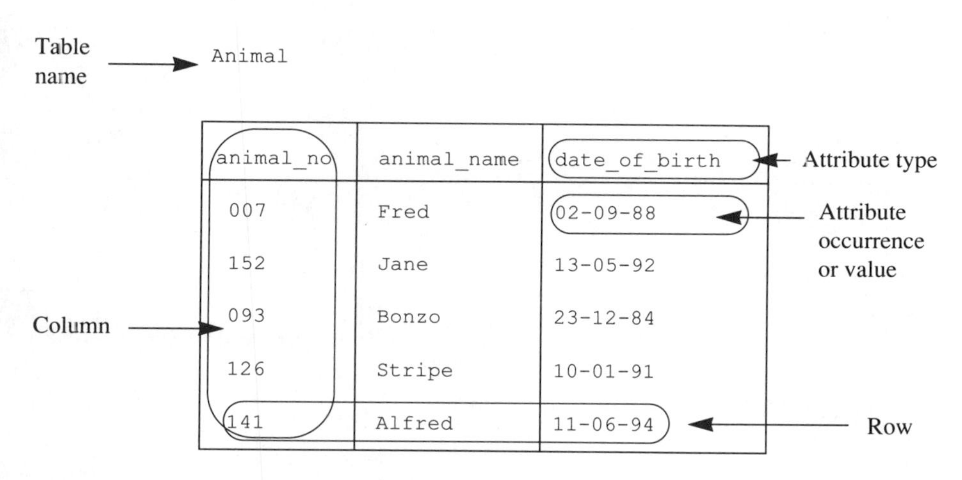

animal_no	animal_name	date_of_birth
007	Fred	02-09-88
152	Jane	13-05-92
093	Bonzo	23-12-84
126	Stripe	10-01-91
141	Alfred	11-06-94

Figure 6.1 A table occurrence.

The table name is `Animal`. The table contains three columns, the column headings being the attribute types `animal_no`, `animal_name` and `date_of_birth`. The column contents consist of attribute occurrences.

Note that `Animal` is an entity selected for the Somerleyton Animal Park case scenario.

See Chapter 5, Figures 5.1 and 5.2.

A table name is always the name of either an entity type or a relationship type. The column headings correspond to the attribute types associated with that entity or relationship. A single occurrence of the entity or relationship forms a row in the table.

The diagram in Figure 6.1 is a table occurrence because it contains values for the attribute types. Figure 6.2 illustrates a table type. The table name is `Animal`; the names of the attribute types it contains are listed within brackets, separated by commas. The identifier of the table has been underlined.

See Chapter 5 for a definition of the term 'identifier'.

```
Animal (animal_no, animal_name, date_of_birth)
```

Figure 6.2 A table type.

Properties of tables

Figure 6.1 also illustrates some of the properties of tables:

1. The order of the rows is not significant. The rows in Figure 6.1 are not in any particular order, and the order may be changed without loss of

information. Compare this with the 'table' in Figure 6.3 where the order of the rows cannot be changed without altering the information content. The table needs to be amended as shown in Figure 6.4, where the order of the rows is no longer significant.

Food Supplier

supplier_name	foodtype	cost_per_kilo
Waltons	apricots	0-70
	bananas	0-50
	oranges	0-60
Fresh Fruit Co.	apples	0-50
	bananas	0-50

Figure 6.3 Order of rows is significant.

Food Supplier

supplier_name	foodtype	cost_per_kilo
Waltons	apricots	0-70
Waltons	bananas	0-50
Waltons	oranges	0-60
Fresh Fruit Co.	apples	0-50
Fresh Fruit Co.	bananas	0-50

Figure 6.4 Order of rows is not significant.

2. The order of the columns is not significant. The columns of the tables in both Figures 6.1 and 6.4 may be interchanged without loss of information. The column or columns containing the attributes forming the identifier of the table are often put on the left, but this is merely a convention.
3. Each row/column intersection contains only one value. All of the tables illustrated so far have had this property; Figure 6.5 contains one that does not. The significance of this aspect of tables is discussed below. Meanwhile, Figure 6.6 shows how the `Animal Food` table can be re-designed so that it has this property.
4. Each row in a table must be capable of being uniquely identified. The rows in the tables in Figures 6.1, 6.4 and 6.6 can all be distinguished from the other rows by an attribute or a combination of attributes. (Note that this is not true of the 'table' in Figure 6.3.)

Animal Food

animal_no	foodtype
007	bananas, oranges
152	oranges

Figure 6.5 More than one value at a row/column intersection.

Animal Food

animal_no	foodtype
007	bananas
007	oranges
152	oranges

Figure 6.6 One value at a row/column intersection.

Exercise 6.1

Examine the table in Figure 6.7. Does it possess the four properties of tables described in the previous section? If not, suggest any necessary amendments.

Applicant

applicant_name	course_no
Wexford	C152
Morse	C155
	C201
Bond	C397
Marple	C721
Bond	C397

Figure 6.7 Exercise 6.1.

Redundant and duplicated data

Duplicated data

In the process of ensuring that a table has the four properties described above, some duplication of data has been introduced. In Figure 6.6, for example, the attribute value 007 has been repeated to ensure that each row/column intersection contains only one value. The duplication is necessary; if one of the values were to be removed, the order of the rows would become significant. Similarly, in Figure 6.8 the values Dawson, 15-11-93 and 152 are duplicated; but if any one of these values were to be removed, some loss of information would occur.

Animal Treatment

vet_name	date	animal_no	animal_name
Dawson	10-09-93	152	Jane
Dawson	15-11-93	007	Fred
Mitchell	15-11-93	093	Bonzo
Dawson	30-05-94	152	Jane

This value can be removed with no loss of information

Figure 6.8 Duplicated and redundant data.

It is assumed that animal numbers are unique, and that the same animal_name is associated with a given value of animal_no.

Redundant data

Figure 6.8 contains another duplicated value, Jane. However this time, if one of the occurrences of Jane were to be removed, no loss of information would occur. It is possible to tell from just one occurrence of the value that the name of animal 152 is Jane. The value Jane has been duplicated unnecessarily; in other words, it is **redundant**.

See the following section for how to eliminate redundant data.

There is therefore an important distinction between duplicated data and redundant (unnecessarily duplicated) data. One aim of data analysis is to produce a logical data model which does not contain any redundant data. The techniques described in the rest of this chapter help to achieve that aim.

Exercise 6.2

Examine Figure 6.9 and identify which data is duplicated, and which is redundant.

Applicant

applicant_no	applicant_name	course_no	course_name
A100	Wexford	C152	Rose growing
A101	Morse	C155	Classic car maintenance
A101	Morse	C201	Singing for pleasure
A102	Bond	C397	Wine making
A103	Marple	C721	Crochet for beginners
A104	Bond	C397	Wine making

Figure 6.9 Exercise 6.2.

Problems of redundancy

Holding redundant data in a database can lead to a number of problems. Consider Figures 6.9 and 6.10, which contain information about applicants for evening classes and the staff who take them.

Course Staff

course_no	course_name	staff_name	staff_address
C152	Rose growing	Smith	Rose Cottage, High Street
C155	Classic car maintenance	Wilson	13 Ash Way
C201	Singing for pleasure	Smith	Rose Cottage, High Street
C721	Crochet for beginners	Brockford	152 Main Road
C397	Wine making	Walters	64 Glove Road

Figure 6.10 Table occurrence for Course Staff.

If it were decided to alter the name of course number C397 to 'Wine and beer making', all instances of the course name 'Wine making' in the database would need to be updated. If the Applicant table is updated but the Course Staff table is not, then the database will contain inconsistent data. The potential for errors in a large database with many thousands of records is clear. Holding redundant data is a waste of storage space, and the work involved in updating it is unproductive maintenance, using up valuable processing time.

Problems can also arise with the insertion and deletion of data. See Atzeni and De Antonellis (1993) and Mannila and Raiha (1992).

Tables that contain no redundant data will overcome problems with updating, inserting and deleting data, and will minimize the storage space and processing time required to maintain the database.

Enterprise rules

Enterprise rules have already been discussed in Chapter 5 in relation to the decisions that a systems analyst has to make in creating an accurate E-R diagram.

See Chapter 5 ('Enterprise rules').

It is equally important to establish the rules that govern the data when defining tables that are free from redundancy. In the discussion above relating to the redundant data in Figure 6.8, it was assumed that one of the enterprise rules for Somerleyton Animal Park is that each animal has a unique animal_no and it is not possible for two animals to be assigned the number 152 as well as the same name.

Two animals could be assigned the same name providing they are each assigned a unique number.

Similarly, the solution to Exercise 6.1 assumed that the two occurrences of Bond in Figure 6.7 referred to two different applicants, there being perhaps an enterprise rule to the effect that each applicant can only apply once for the same course.

The duplication of the values A101 and Morse does suggest that an applicant is assigned one number, no matter how many applications he or she makes. However, it could still be the case that where the same applicant applies more than once for the same course, different numbers are assigned to distinguish the applications.

Examining a table fragment such as that in Figure 6.7 in order to derive enterprise rules may lead to errors. Perhaps the same individual called Bond *did* apply twice for the same course to be sure of securing a place, or perhaps one application was for a September start and one was for a January start. Examination of the `Animal Treatment` table in Figure 6.8 could lead to the deduction of the enterprise rule that a veterinary surgeon treats only one animal on any one day; such a rule is surely unrealistic, and is implied only because such a small subset of the data is present.

See Chapter 8.

Enterprise rules cannot be determined solely by examining subsets of data. The systems analyst needs to check his/her understanding with the user to ensure that correct assumptions are being made.

Normalization

Normalization is a technique that can be applied to data to ensure that a set of tables is derived that contains no redundant data. Tables can be unnormalized, or can be refined through several stages, or 'normal forms'. Five normal forms have been identified; normalizing data to third normal form (3NF) is regarded as sufficient for most practical purposes, and this text will not deal with fourth and fifth normal forms.

See Kent (1983) for a discussion of all five normal forms.

Another expression of normalization that is sometimes used is Boyce/Codd normal form (BCNF), which is a stronger definition of 3NF, and this will also be explained below.

First normal form: removal of repeating groups

The 'table' in Figure 6.11a does not conform to the properties of tables explained above because the order of the rows is significant. The attributes `potential_buyer_name` and `date_of_viewing` are said to form a **repeating group** of data. Figure 6.11b shows how such a repeating group is indicated in a table type.

Property

property_id	client_name	potential_ buyer_name	date_of_viewing
P101	Smith	Jones	17-04-94
		Perkins	19-04-94
		Patel	09-03-94
		Jones	03-05-94
P106	Parker	Jones	29-04-94
		Mitchell	23-04-94

Figure 6.11 Table containing a repeating group. (a) Table occurrence.

Property (<u>property_id</u>, client_name, (potential_buyer_name, date_of_viewing))

Figure 6.11 (b) Table type.

Repeating groups are also to be found in Figures 6.3, 6.5 and 6.7.

One way of eliminating the repeating group and ensuring that the table has the property that the order of the rows is not significant has already been demonstrated, and is illustrated for the `Property` table in Figure 6.12. However, this solution has introduced some redundancy as the values `Smith` and `Parker` are now unnecessarily duplicated.

Property

property_id	client_name	potential_buyer_name	date_of_viewing
P101	Smith	Jones	17-04-94
P101	Smith	Perkins	19-04-94
P101	Smith	Patel	09-03-94
P101	Smith	Jones	03-05-94
P106	Parker	Jones	29-04-94
P106	Parker	Mitchell	23-04-94

Figure 6.12 Elimination of repeating group, but at the expense of introducing redundancy.

A technique for eliminating the repeating group without introducing redundancy is to split off the repeating group into a table of its own. The new table will need to incorporate the identifier of the original table so that a link may be made between them. Figure 6.13a shows the resultant table occurrences and Figure 6.13b the table types when this technique is applied to the `Property` table.

Property

property_id	client_name
P101	Smith
P106	Parker

Viewing

property_id	potential_buyer_name	date_of_viewing
P101	Jones	17-04-94
P101	Perkins	19-04-94
P101	Patel	09-03-94
P101	Jones	03-05-94
P106	Jones	29-04-94
P106	Mitchell	23-04-94

Figure 6.13 Figure 6.11 split into two tables. (a) Table occurrences.

The identifiers for Property, and Viewing were selected in Chapter 5, Worked example 5.2.

```
Property      (property_id, client_name)
Viewing       (property_id, potential_buyer_name, date_of_viewing))
```

Figure 6.13 (b) Table types.

Figure 6.13a contains some duplicated data; property_id appears in both the Property table and the Viewing table. However, this duplication is necessary; to omit property_id from the Viewing table would result in the loss of information about each viewing. Note that as Figure 6.11b clearly illustrates, the Viewing table formed an *embedded* table within the original Property table.

A table that contains no repeating groups is said to be in first normal form (1NF).

Exercise 6.3

Examine the solution to Exercise 6.1 in Figure E6.1. In solving the original problems with this table, some redundancy has been introduced. Split the table in Figure E6.1 into two tables to eliminate the redundant data. Provide both the table occurrences and the table types.

Exercise 6.4

See Chapter 5, Exercise 5.1.

Referring to the scenario given in Exercise 5.1, decide whether the following table is in 1NF and, if not, take the necessary steps to put it into 1NF:

```
Applicant (applicant_no, applicant_name, applicant_address,
           date_of_birth, qualification_code, course_code)
```

Determinancy

Both second and third normal forms involve the concept of **determinancy.** Examine the Applicant table in Figure 6.9. Each applicant is assigned a unique applicant_no. Thus applicant_no A101 is always associated with one applicant_name (Morse). Similarly, each course is assigned a unique course_no and so course_no C397 is always associated with one course_name (Wine making). Applicant_no is said to be the *determinant* of applicant_name and course_no the *determinant* of course_name. Applicant_name and course_name are *determined by* applicant_no and course_no respectively.

Values of B can still be updated, for example

In more general terms, an attribute A is said to be the determinant of attribute B if a given value of A is associated with just one value of B at any given time. B is said to be determined by A, or functionally dependent upon A.

'Wine making' could be changed to 'Cheese making'. However, C397 can have only one course name associated with it at any one time.

In the Applicant table, is applicant_name a determinant of applicant_no? The applicant_name Morse is associated with the same value of applicant_no in each of its two occurrences in the table. However, the applicant_name Bond is associated with two different values of applicant_no, A102 and A104. Applicant_name is not therefore a determinant of applicant_no. To establish the determinancies between data it is necessary to have a thorough knowledge of the enterprise rules.

See the section above on 'Enterprise rules'.

In the case of the evening classes, each applicant is assigned a unique number; however, applicant names may be duplicated.

Determinancy diagrams

Figure 6.14 illustrates how determinancies between attributes may be represented diagrammatically by a **determinancy diagram.** It denotes that applicant_no is a determinant of applicant_name.

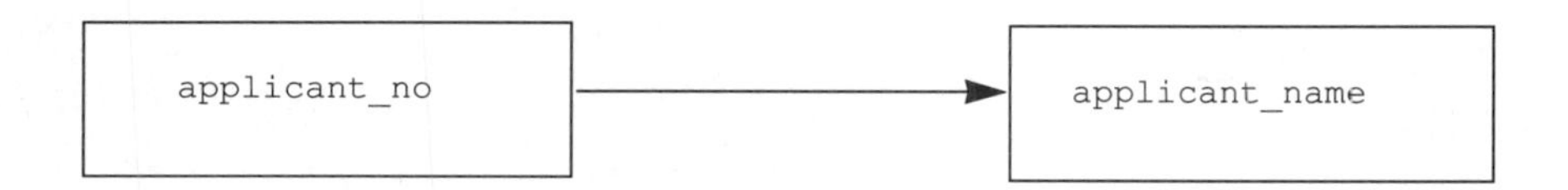

Figure 6.14 Determinancy diagram.

The scenario below is an elaboration on that first introduced in Chapter 5.

> A lecturer may teach on more than one course and a course may be taught by a number of lecturers. Each course includes a number of modules, each of which is offered within just one course. A module may be taught by more than one lecturer, each of whom will teach that module for a set number of weeks. A module is always taught in the same room, though several modules are allocated to a room. Each course and module has a unique number and a title. Students are enrolled on one course, but may choose which modules to take. They are awarded a grade for each module completed. Students and lecturers are allocated a unique number and their name and address are recorded.

Figure 6.15 illustrates that student_no determines student_name, student_address and course_no. Course_no determines course_title. Student_no **transitively** determines course_title.

The grade obtained by a student is determined both by the student_no and the module_no. Figure 6.16 shows how this composite determinancy is depicted using a determinancy diagram.

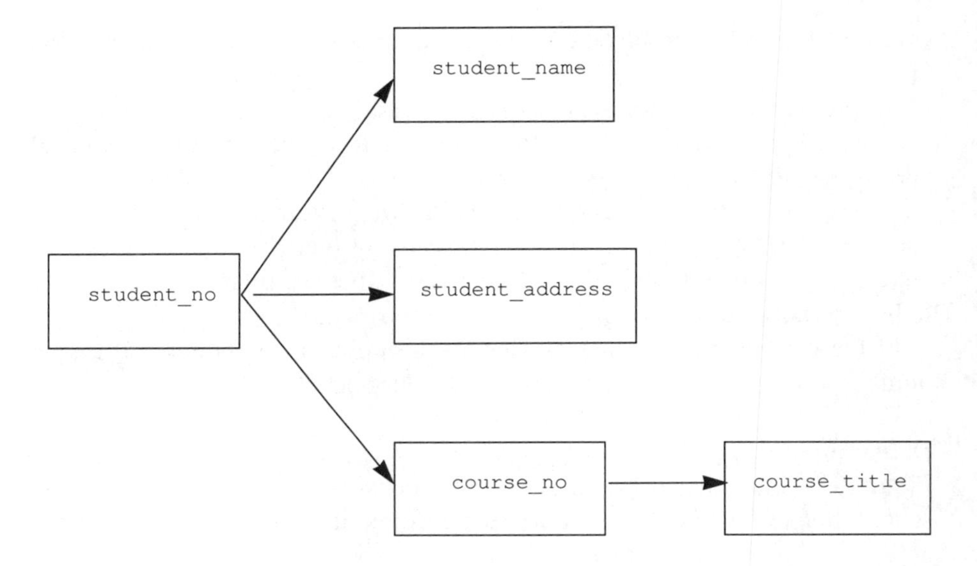

Figure 6.15 Determinancy diagram for `Student`.

Figure 6.16 Composite determinant.

Exercise 6.5

Examine Figure 6.17 and explain what it represents.

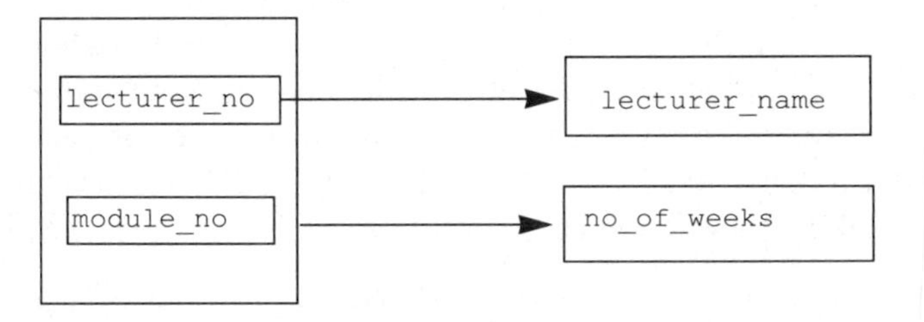

Figure 6.17 Exercise 6.5.

The complete determinancy diagram for the scenario is shown in Figure 6.18.

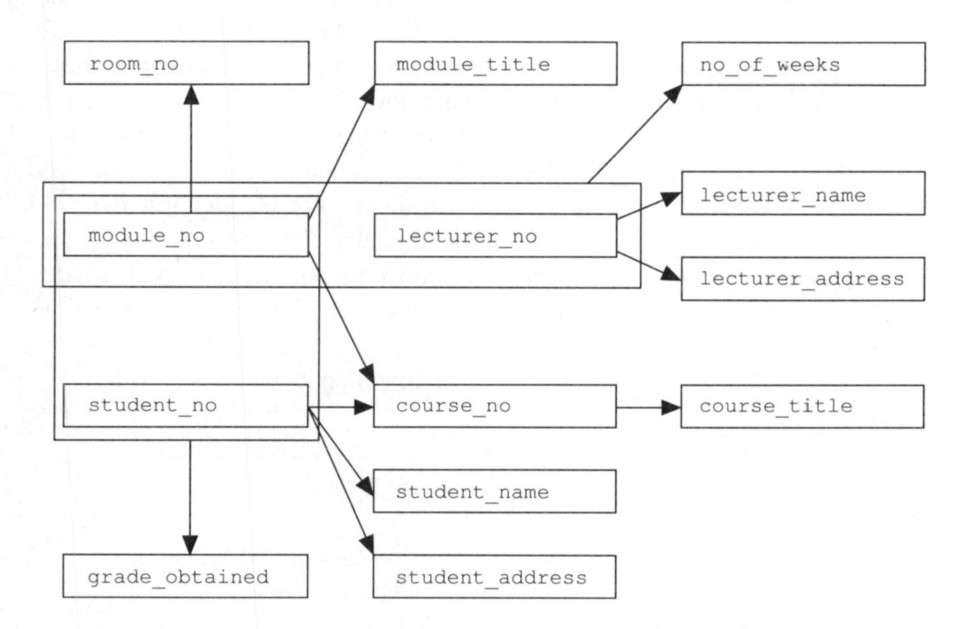

Figure 6.18 Determinancy diagram for `Course`, `Student`, `Lecturer` and `Module`.

Determinancy diagrams provide a way in which to express enterprise rules clearly and without ambiguity, avoiding the confusion that can creep in when descriptive text is used. As we shall see later, they can be used to construct tables that are free from redundant data.

See Howe (1989) for further discussion of determinancy diagrams.

Exercise 6.6

The mail order company described in Chapter 5, Exercise 5.4 stores the following information about each customer order that it receives: a unique customer order number, the date the order is received, the customer's unique number, the customer's name, address and telephone number, the unique code for each item ordered and its description and price, the quantity ordered and the delivery address if different from the customer's address.

Draw a determinancy diagram for the customer order. Make plausible assumptions where necessary.

Second normal form

Having introduced the concept of determinancy, we can now resume the discussion of normalization and examine second normal form (2NF). The tables for `Property` and `Viewing` illustrated in Figure 6.13 are in 1NF, repeating groups having been eliminated. Further analysis reveals that more data needs to be stored about each viewing and about each property that is for sale. For viewings, the estate agent wants to record the telephone number of the potential buyer and the name of the member of staff who was responsible for organizing the viewing arrangements. A viewing may be arranged by any of the staff within the agents. For each property, the address, type of property, area and price must be recorded, as well as the address and telephone number of the client and details about viewing arrangements.

As stated in Chapter 5, Worked example 5.2, it is assumed that a potential buyer might view the same property more than once, but not more than once on any one day. More than one visit on the same day would count as one viewing.

The `Viewing` and `Property` tables will be amended as follows:

```
Viewing  (property_id, potential_buyer_name, date_of_viewing,
         potential_buyer_tel_no, arranged_by)

Property (property_id, property_address, property_type, area,
         price, client_name, client_address, client_tel_no,
         viewing_arrangements)
```

To be in 2NF, a table must be in 1NF, and each attribute that is not part of the identifier (i.e. each non-key attribute) must be determined by the whole of the identifier. A determinancy diagram for the Viewing table is provided in Figure 6.19.

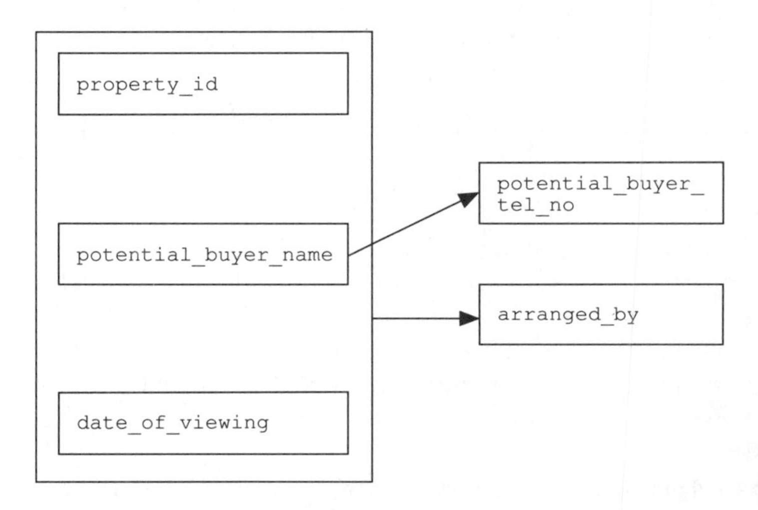

Figure 6.19 Determinancy diagram for `Viewing`.

This illustrates that while the whole of the identifier is required to determine who arranged the viewing, only the `potential_buyer_name` is required to determine the `potential_buyer_tel_no`. The table is therefore not in 2NF. To put it into 2NF, a separate table needs to be created to hold the details of the potential buyer. It is clear from the definition of 2NF given above that only tables with composite identifiers need to be considered at this stage. The `Property` table has a single identifier. The complete 2NF tables are therefore as follows:

```
Property (property_id, property_address, property_type, area,
         price, client_name, client_address, client_tel_no,
         viewing_arrangements)

Viewing  (property_id, potential_buyer_name, date_of_viewing,
         arranged_by)

Potential Buyer (potential_buyer_name, potential_buyer_tel_no)
```

Exercise 6.7

The table type for the table occurrence in Figure 6.9 (now renamed `Application`) is as follows:

```
Application   (applicant_no, course_no, applicant_name,
              course_name)
```

Draw a determinancy diagram for this table. Decide whether it is in 2NF, and if not, take the necessary steps to put it into 2NF.

Third normal form

To be in 3NF, a table must be in 2NF and the non-key attributes must be mutually independent. That is, there must be no functional dependencies between them. A determinancy diagram will demonstrate whether this is the case.

As the tables `Viewing` and `Potential Buyer` contain only one non-key attribute, only the `Property` table needs to be considered at this stage. A determinancy diagram for `Property` is shown in Figure 6.20, where it has been assumed that the attribute `viewing_arrangements` is determined by `property_id`. (It is possible that `viewing_arrangements` is also determined by `client_name`. A client may offer more than one property for sale and if the second assumption is made then the viewing arrangements for all of the properties belonging to one client will be the same. The assumption reflected in Figure 6.20 allows for more flexibility in that the viewing arrangements for each property owned by a client may be different.)

In reality, the systems analyst would need to check this assumption with the user.

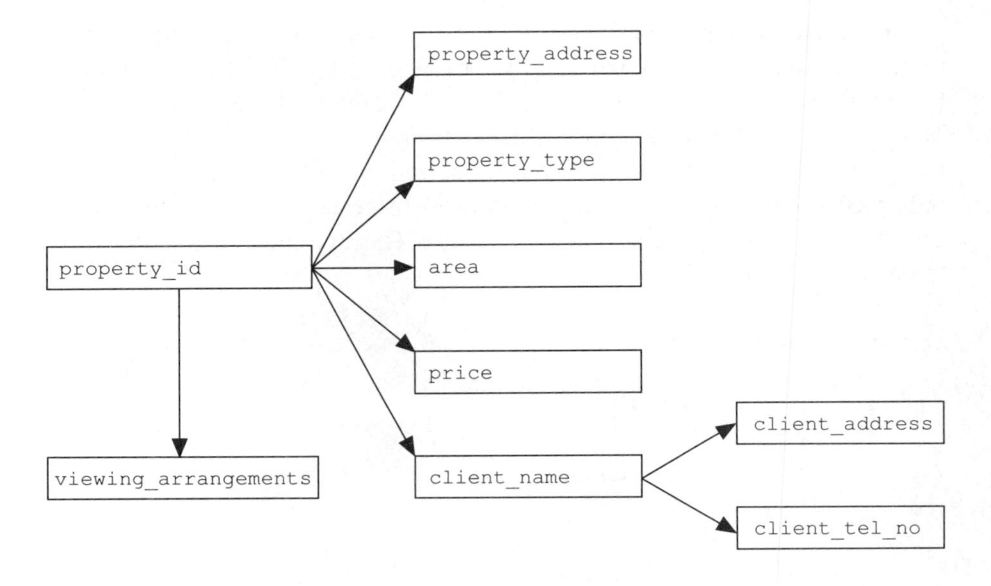

Figure 6.20 Determinancy diagram for `Property`.

The determinancy diagram in Figure 6.20 shows that one of the non-key attributes, `client_name`, is a determinant of two other non-key attributes, `client_address` and `client_tel_no`. Hence the `Property` table is not in 3NF. Note that while `property_id` *transitively* determines `client_address` and `client_tel_no`, it does not *directly* determine them.

To put the `Property` table into 3NF, a separate table must be created to hold the client details. The complete set of 3NF tables is as follows:

```
Property (property_id, property_address, property_type, area,
         price, client_name, viewing_arrangements)

Client  (client_name, client_address, client_tel_no)

Viewing  (property_id, potential_buyer_name, date_of_viewing,
         arranged_by)

Potential Buyer (potential_buyer_name, potential_buyer_tel_no)
```

Exercise 6.8

A table that is in 2NF is also in 1NF, and a table that is in 3NF is also in 1NF and 2NF.

The tables referred to in Exercise 6.4 have been further augmented, and the revised versions appear below. Ensure that they are still in 1NF, then using determinancy diagrams check whether they are in 3NF; if not, take the necessary steps to put them in 3NF:

```
Applicant (applicant_no, applicant-name, applicant_address,
          date_of_birth, referee_name, referee_status,
          referee_address)

Applicant Qualification (applicant_no, qualification_code,
                         qualification_title, grade_obtained)

Application     (applicant_no, course_code, date received)
```

Candidate identifiers

Suppose that in the table below, both course_code and course_title are unique:

```
Course   (course_code, course_title, course_length, written_by)
```

The determinancy diagram is shown in Figure 6.21.

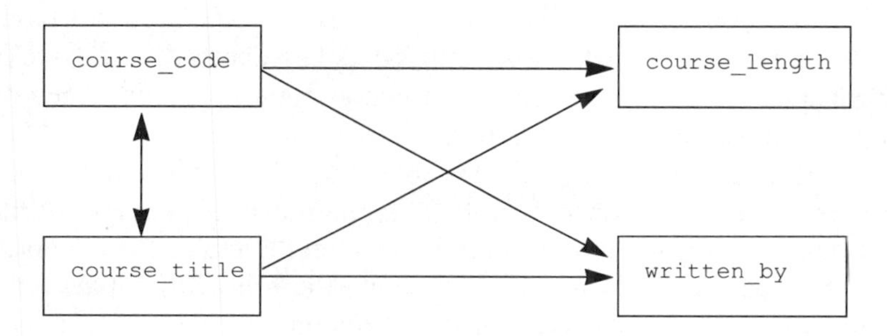

Figure 6.21 Determinancy diagram for Course.

There are two **candidate identifiers** for the table, course_code and course_title. Our definitions of 2NF and 3NF should have allowed for more than one candidate identifier. The simplest way of allowing for candidate identifiers is to use the definition of Boyce/Codd normal form, which follows.

Boyce/Codd Normal Form

Boyce/Codd Normal Form (BCNF) is a stronger definition than 3NF because it covers the case where there is more than one candidate identifier for a table, even when the candidate identifiers overlap.

See Date (1990) for an example of overlapping candidate identifiers.

BCNF may be defined as follows: 'Every determinant must be a candidate identifier'. If this rule is applied to the Course table, then both of the determinants course_code and course_title are candidate identifiers, and so the table is in BCNF.

See Date (1990) for discussion of the inadequacies of 3NF.

Returning to the example of the Property table used in the discussion of 3NF above, and for which there is a determinancy diagram in Figure 6.20, it can be seen that the determinant client_name is not a candidate identifier for the Property table. Hence the table is not in BCNF. In this case, it is necessary to remove the attributes that are functionally dependent upon client_name. The complete set of 3NF tables provided above is also in BCNF.

Howe (1989) uses the term 'well-normalized' for tables that are in BCNF.

First, second and third normal form are discussed in this text because they are frequently referred to in other sources in the context of data analysis. However, to arrive at well-normalized tables it is necessary only to remove repeating groups (put the tables into 1NF), and then apply the Boyce/Codd rule. The use of determinancy diagrams will aid the identification of determinants.

Approach to the development of an Entity-Relationship model

The E-R model

An E-R model comprises an E-R diagram and a set of well-normalized tables. So far, for clarity, these two components have been discussed separately in Chapters 5 and 6, respectively. It is now time to see how they may be developed in parallel to produce an E-R model.

The E-R diagram and its associated tables must be consistent with each other. In particular, a relationship on the diagram must be represented in the tables; it must be possible to make a link between the tables for the entities involved. This can be illustrated by examining the following tables for the estate agent example, and the associated E-R diagram in Figure 6.22:

```
Property (property_id, property_address, property_type, price,
         client_name, viewing_arrangements)

Client  (client_name, client_address, client_tel_no)

Viewing  (property_id, potential_buyer_name, date_of_viewing,
         arranged_by)

Potential Buyer (potential_buyer_name, potential_buyer_tel_no)
```

The diagram shows a relationship between Client and Property. In the tables, the identifier of Client, client_name, is present in (is **posted into**) the Property table. Similarly, the diagram shows that Viewing has a relationship with Property and with Potential Buyer. The identifiers of the latter two are posted into the former. The posted identifier is referred to as a **foreign key**; a foreign key is an attribute or collection of attributes within a table that exists as the identifier of another table.

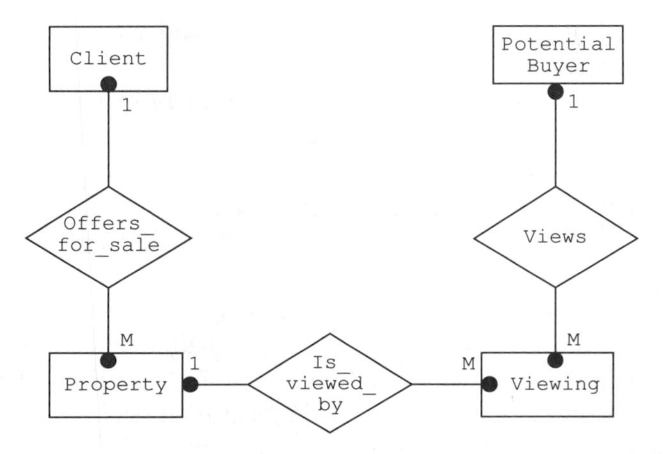

Figure 6.22 E-R diagram.

Note that it is always the identifier of the table for the entity at the '1' end of the relationship that is posted into the table for the entity at the 'M' end. If the identifier of the table for the entity at the 'M' end were to be posted into the table for the entity at the '1' end it would form a repeating group. For example, each property may be viewed more than once, so there would be a repeating group of the identifier of `Viewing` within the `Property` table.

In all of these examples, the membership class of the entity at the 'M' end of the relationship is obligatory. When the membership class of the entity at the 'M' end is non-obligatory, the posting of the identifier of the entity at '1' end would result in null values for some occurrences; the same problem arises in a 1:1 relationship when the membership class on both sides is non-obligatory.

When there is a M:M relationship, it cannot be represented in the tables by posting the identifier of either entity into the table for the other. In each of these instances, a separate table is needed to represent the relationship. The identifier of an entity at the non-obligatory end of a 1:1 relationship may be posted into the table for the entity at the obligatory end. If a 1:1 relationship is obligatory on both sides, only one table is needed. These issues will receive further illustration and comment in Worked example 6.1.

Although null values are permissible for foreign keys, we prefer to avoid them at this stage of data modelling. This prevents us digressing into a consideration of what are essentially design issues such as what proportion of an entity's occurrences will have null values for a particular foreign key. These issues can be considered during the construction of the second and third level data models. See Howe (1989).

Guidelines for the development of an E-R model

These guidelines incorporate the guidelines for drawing an E-R diagram provided in Chapter 5. Additional steps allow for the development of well-normalized tables that are consistent with the diagram.

1. Select likely entities.
2. Select an identifier for each of the entities. Write down a table containing just this identifier for each entity.
3. Identify relationships between the selected entities.

4. Sketch an E-R diagram, adding the degree and membership class of each relationship.
5. Decompose any M:M relationships, allocating an identifier to any new entities thus formed. Add a table containing the identifier for any new entities.
6. Ensure that each relationship on the diagram is represented in the tables, either by posting identifiers or by creating a table for the relationship, as appropriate.
7. Make a list of the attributes that need to be stored.
8. Allocate the attributes to the tables. Ensure that the tables are in BCNF, by eliminating repeating groups and applying the Boyce/Codd rule, drawing determinancy diagrams to aid this step if necessary.
9. Add an entity and the necessary relationships to the diagram for any extra tables produced by step (8). Include degrees and membership classes.
10. Check the diagram for potential connection traps. Assess the significance of any traps found, and eliminate significant traps.
11. Check that the tables are still in BCNF after any amendment made during step (10). Repeat steps (8)–(11) as necessary.

Attributes will be identified during fact finding, when information will be gathered about the data that needs to be stored and processed by the information system, and about the reports that it needs to produce. If a start has been made on the data dictionary (see Chapter 4), inspection of it may also help to provide a list of attributes.

It can be seen that the development of an E-R model is an iterative process, some of the steps having to be repeated to check that the diagram and tables are consistent and accurate. Since the purpose of the exercise is to produce a logical data model that supports the requirements of the information system, these requirements must be constantly borne in mind. For example, in step (10) the significance of a connection trap can only be assessed in the light of system requirements; if certain information is not required then the inability to retrieve it is not a problem. The entire model should be constructed with the requirements in mind and should be checked against these to ensure its accuracy and relevance.

Worked example 6.1

Using the guidelines specified, develop an E-R model for the Somerleyton Animal Park case scenario.

An E-R diagram for the ordering supplies section (subsystem) was developed in Chapter 5, Worked example 5.1, and the same assumptions will be made. This Worked example will develop an E-R model for the whole of the proposed system.

Solution:

Guideline steps 1 and 2:

Likely entities	*Identifiers*
Order	order_no
Supplier	supplier_name
Foodtype	foodtype_name
Delivery	delivery_no

```
Invoice        invoice_no
Animal         animal_no
Species        species_no
Area           area_no
Enclosure      enclosure_no
Keeper         employee_no
Visit          visit_no
```

Tables

```
Order          (order_no,
Supplier       (supplier_name,
Foodtype       (foodtype_name,
Delivery       (delivery_no,
Invoice        (invoice_no,
Animal         (animal_no,
Species        (species_no,
Area           (area_no,
Enclosure      (enclosure_no,
Keeper         (employee_no,
Visit          (visit_no,
```

Assumptions

1. Each order is assigned a unique `order_no`.
2. Supplier names are unique.
3. Each delivery and each invoice is assigned a unique `delivery_no` and `invoice_no`, respectively.
4. Each animal and each species is assigned a unique `animal_no` and `species_no`, respectively.
5. Each area and each enclosure is assigned a unique `area_no` and `enclosure_no`, respectively.
6. Each keeper is identified by a unique `employee_no`.
7. Each visit is assigned a unique `visit_no`.

Guideline step 3:
Relationships

An order is sent to a supplier.
A foodtype is ordered on an order.
A supplier supplies foodtypes.
A supplier makes a delivery.
A delivery corresponds to an order.
A supplier sends an invoice.
An invoice matches an order.
An animal represents a particular species.
An animal lives in an enclosure.
Each area contains several enclosures.

A keeper is assigned to an area.
An area is in the charge of a headkeeper.
A species is confined to one area.
An animal eats certain types of food.
Information about species is of interest to Visits.

Guideline step 4: See Figure 6.23.

Assumptions (continued)

See Chapter 5, Worked example 5.1 for discussion relating to assumptions (8)–(14).

8. A supplier can supply more than one type of food.
9. A foodtype can be supplied by more than one supplier.
10. A delivery could correspond to more than one order and an order could be split and sent in more than one delivery (for example, if there was insufficient stock to satisfy the whole order immediately).
11. Details of suppliers other than those who supply food are also held.
12. At times a foodtype might not have a supplier.
13. A supplier might not have been sent any orders.
14. A supplier might not have made any deliveries.
15. An invoice is sent for each order.

There are two relationships between `Keeper` and `Area`. The case scenario states that each area is under the care of a headkeeper, though more than one keeper helps to look after the larger areas, and a keeper only works on one area at a time. It is assumed that the system will record which keepers are assigned to each area at any one time, and which keeper is in charge of which area. It is further assumed that `Keeper` and `Headkeeper` are represented by the same entity. This assumption can be checked when the tables are completed; two entities may be justified if different data needs to be kept about each one.

Assumptions (continued)

16. The system will only record which area a keeper is currently assigned to; s/he must be assigned to an area.
17. The same data will be kept about keepers and headkeepers and thus they may be represented by the same entity.
18. An area must have a keeper assigned to be in charge of it.
19. An enclosure may have no animals assigned to it; for example, during periods of rebuilding/maintenance.
20. An animal need not be assigned to an enclosure; for example, it may be on loan to another animal park.
21. Information is kept about species that may not be currently represented by any animals in the animal park.
22. A species is kept in just one area.

It seems likely that if Visits request the provision of information sheets, they will express an interest in particular species. Otherwise, they could be given a more general handout about the contents of the Animal Park.

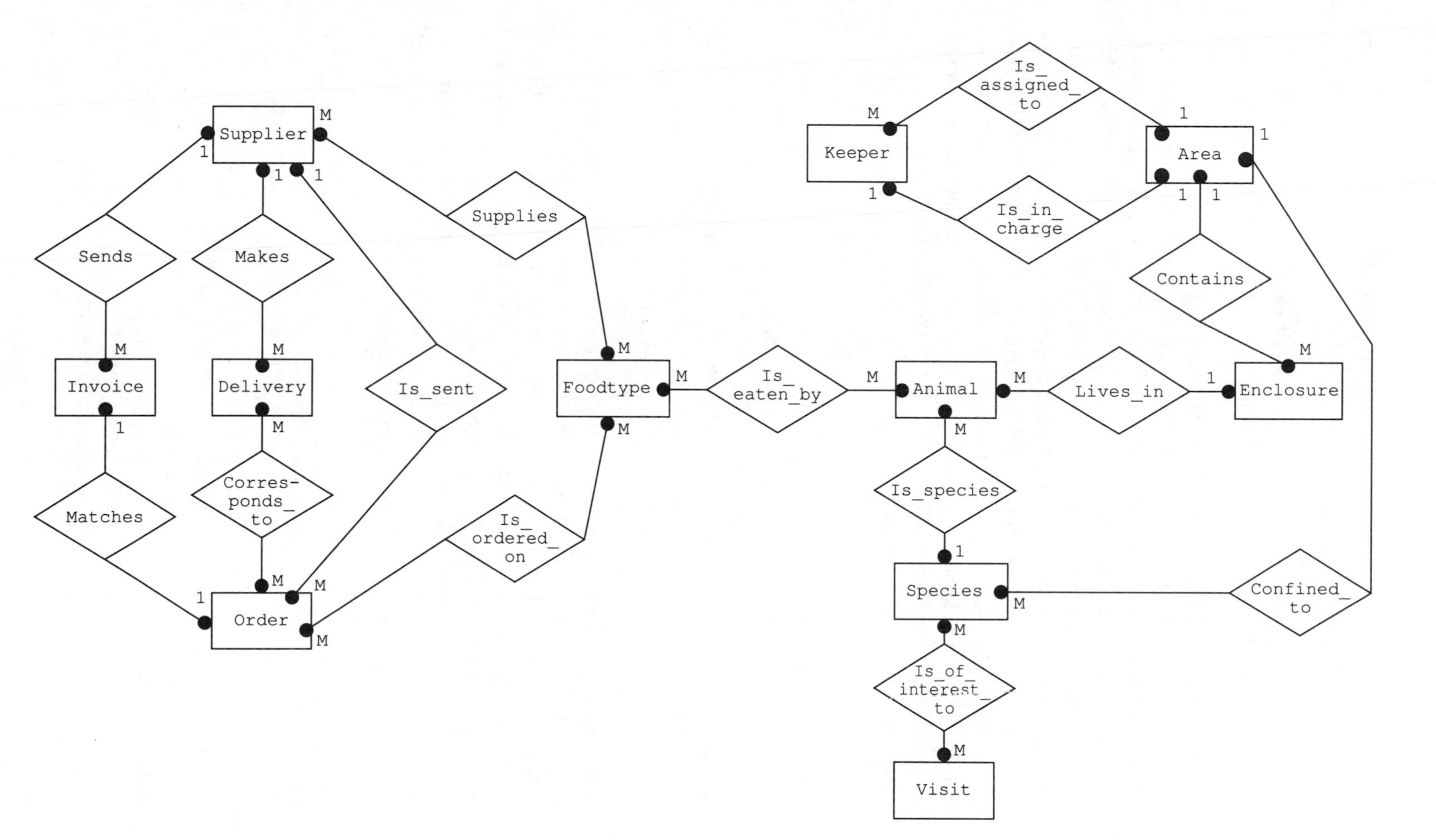

Figure 6.23 Guideline step 4.

23. Visits may (but do not have to) request information relating to particular species and information is compiled especially for such requests.

Guideline step 5: Figure 6.24 reproduces part of Figure 6.23 to show the decomposition of the M:M relationships.

This convention will be followed in future diagrams.

In decomposing the M:M relationship between Delivery and Order, a refinement in the identification of relationships in step (3) is uncovered. A delivery may contain more than one delivery item; a delivery item corresponds to an orderline, not directly to an order. The M:M relationship between the entity types Delivery Item and Orderline is justified by assumption (10). There is still a M:M relationship on the diagram. This is acceptable, providing we make sure that there is a table to represent the relationship. It is useful to place an asterisk in the box of those relationships that do not need a table to represent them. As explained above, a table is needed for a relationship if it is M:M or if the membership class of the entity at the 'M' end is non-obligatory.

Additional tables:

```
Foodtype Supplier      (supplier_name, foodtype_name,
Orderline              (order_no, foodtype_name,
Delivery Item          (delivery_no, foodtype_name,
Corresponds_to         (delivery_no, order_no, foodtype_name,
Animal Food            (animal_no, foodtype_name,
Visit Species          (species_no, visit_no,
```

Guideline step 6: There is one M:1 relationship where the membership class of the entity type at the 'M' end is non-obligatory; an extra table is required to represent the relationship Lives_in. The area_no within Keeper indicates to which area a keeper is currently assigned. The employee_no within Area indicates which headkeeper is in charge of that area.

See the section above on 'The E-R model'.

The relationship between Invoice and Order is 1:1. The membership class of Order is non-obligatory, as an invoice may not yet have been received, so the identifier of Order must be posted into Invoice.

Tables

```
Order                  (order_no, supplier_name,
Supplier               (supplier_name,
Foodtype               (foodtype_name,
Delivery               (delivery_no, supplier_name,
Invoice                (invoice_no, supplier_name, order_no,
Animal                 (animal_no, species_no,
Species                (species_no, area_no,
Area                   (area_no, employee_no,
Enclosure              (enclosure_no, area_no,
```

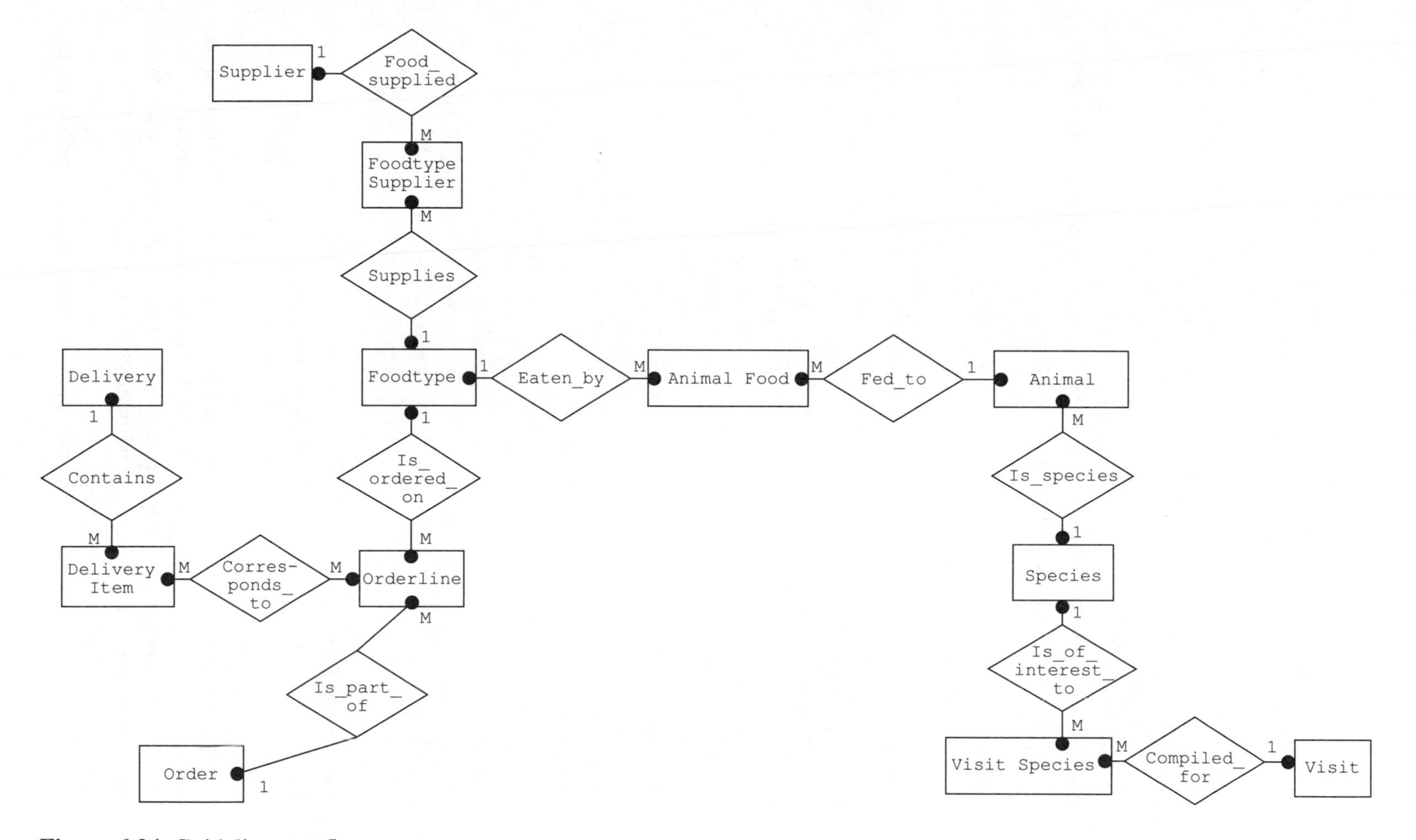

Figure 6.24 Guideline step 5.

Keeper (employee_no, area_no,
Visit (visit_no,
Foodtype Supplier (supplier_name, foodtype_name,
Orderline (order_no, foodtype_name,
Delivery Item (delivery_no, foodtype_name,
Corresponds_to (delivery_no, order_no, foodtype_name,
Animal Food (animal_no, foodtype_name,
Visit Species (species_no, visit_no,
Lives_in (animal_no, enclosure_no,

Guideline step 7: Note that attributes that have already been included as identifiers of the tables listed so far are omitted from the list below:

Attributes

species_name
animal_name
date_of_birth
special_notes
feeding_notes
date_gave_birth
mate_animal_no
birth_notes
offspring_animal_no
treatment_date
vet_name
treatment_details
start_date_of_loan
end_date_of_loan
park_loaned_to/from
loan_purpose
comments
date_reintroduced_to_wild
reintroduction_comments
date_died
cause_of_death
species_details
daily_quantity
stock_level
quantity_ordered
supplier_address
supplier_tel_no
supplier_fax_no
main_contact
standard_discount
price
order_date
delivery_date
order_status
quantity_delivered
total_discount_offered
visit_date
party_name
contact_tel_no
employee_name
employee_address
employee_tel_no
employee_grade
invoice_no
invoice_date
invoice_value
invoice_status

Guideline step 8: It is a straightforward matter to assign some of the attributes such as animal_name, date_of_birth and species_name. However, two sets of repeating groups may be readily identified; an animal may breed more than once and may be treated more than once by the veterinary surgeon. So two further tables, Treatment and Breeding History are included. Each breeding episode could result in more than one offspring; as the details of offspring will be in every respect the same as those for any other animal (that is, animal_no, animal_name, date_of_birth, species_no), Breeding History has a second relationship with Animal, as illustrated in Figure 6.25.

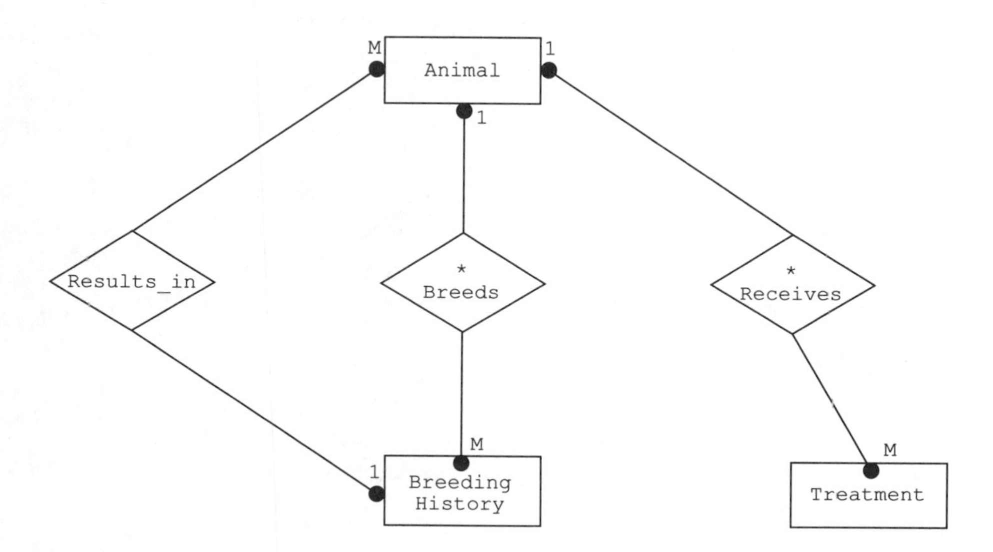

Figure 6.25 Guideline step 8.

It has been assumed that not every breeding episode will necessarily be successful, and not every animal in the park will have been born there, so the relationship is non-obligatory on both sides. A table is therefore needed to represent the relationship Results_in. The animal_no in the identifier of the Breeding History table is that of the mother animal; the father being recorded as mate_animal_no.

This could be shown as an additional relationship between Animal and Breeding History on the E-R diagram.

Another repeating group of data about Animal can be identified. An animal could be loaned to another animal park more than once, and conversely an animal could be borrowed from another park more than once. So a table to hold data about loans is created, called Animal Loan. Each loan episode is assigned a unique loan number irrespective of whether the loan is of a Somerleyton animal to another park or the loan of an animal from another park to Somerleyton.

For example, for breeding purposes.

The final set of attributes relating to the entity type Animal concerns its death or reintroduction to the wild. Both of these are determined by animal_no and so it appears that they should go into the Animal table. However, the values of these attributes will be null for all of the park's current animals. As indicated earlier we prefer to avoid null values at this stage and so a separate table is created to store these attributes. As it is assumed that an animal is *either* returned to the wild *or* its death in the park is recorded, one table called Animal Demise is created.

The tables and the further assumptions that support them appear below.

Tables

Order (order_no, supplier_name, order_date, order_status, total_discount_offered)

Supplier (supplier_name, supplier_address, supplier_tel_no, supplier_fax_no, main_contact, standard_discount)

Foodtype (foodtype_name, stock_level)

Invoice (invoice_no, supplier_name, order_no, invoice_date, invoice_value, invoice_status)

Delivery (delivery_no, supplier_name, delivery_date)

Animal (animal_no, species_no, animal_name, date_of_birth, special_notes)

Treatment (animal_no, treatment_no, treatment_date, vet_name, treatment_details)

Breeding History (animal_no, breeding_no, mate_animal_no, date_gave_birth, birth_notes)

Results_in (animal_no, breeding_no, offspring_animal_no)

Animal Loan (loan_no, animal_no, start_date_of_loan, end_date_of_loan, park_loaned_to/from, loan_purpose, comments)

Animal Demise (animal_no, date_died/date_reintroduced_to_wild, reintroduction_comments/cause_of_death)

Species (species_no, area_no, species_name, species_details)

Area (area_no, employee_no)

Enclosure (enclosure_no, area_no)

Keeper (employee_no, employee_name, employee_address, employee_tel_no, area_no)

Visit (visit_no, party_name, contact_tel_no, visit_date)

Foodtype Supplier (supplier_name, foodtype_name)

Orderline (order_no, foodtype_name, quantity_ordered, price)

Delivery Item (delivery_no, foodtype_name, quantity_delivered)

Corresponds_to (delivery_no, order_no, foodtype_name)

Animal Food (animal_no, foodtype_name, daily_quantity, feeding_notes)

Visit Species (species_no, visit_no)

Lives_in (animal_no, enclosure_no)

Assumptions (continued)

24. Not every breeding episode will result in offspring.
25. Not every animal was born in the animal park.
26. Minimal details of the keepers are kept in this system; it is assumed that the payroll is the subject of a separate system.
27. As the same data is kept about all keepers, only one table is necessary,

as previously assumed. The `employee_no` in `Area` allows us to find out who are the headkeepers.

28. A supplier offers a standard discount. However, a level of discount that could be more (but never less) than this can be negotiated for each order.
29. The attribute `feeding_notes` is dependent upon both `animal_no` and `foodtype`, as notes are made about the method of feeding a particular foodtype to a particular animal.
30. An animal is either reintroduced to the wild or dies while in Somerleyton or another park, but not both.

Guideline step 9: See Figure 6.26.
Entities and relationships have been added for each of the additional tables created in step (8).

Guideline step 10: There are a number of potential fan traps in the E-R diagram in Figure 6.26. For example, there is one between `Keeper`, `Area` and `Enclosure`. Although it is possible to tell to which area a keeper is assigned, it is not possible to tell for which enclosures he or she is responsible. The case scenario states that a large area may have more than one keeper assigned to it, but does not say whether each keeper is assigned to a particular set of enclosures within that area. If this is the case, and it is a requirement to be able to determine to which enclosures a keeper is assigned, then the fan trap is a significant one and the relationship `Is_assigned_to` should be between `Keeper` and `Enclosure`. It will be assumed that keepers are assigned only to an area, and may look after any of the enclosures within that area; the diagram can therefore be left as it is.

Assumptions (continued)

31. A keeper is assigned to an area, not to specific enclosures within that area.

Another potential fan trap exists between `Animal`, `Species` and `Visit Species`. It has been assumed that visiting parties may want information about animals of a particular species, and not on individual animals. This assumption is consistent with the information supplied in the case scenario, so this fan trap is assessed as not significant. Each of the other fan traps can similarly be assessed as not significant in the light of the requirements of the information system.

See Chapter 1 for the requirements.

There is one requirement that the model does not support. This is the need to determine the stocks of each foodtype held within each area. There is a chasm trap between `Area` and `Foodtype`; there exists a M:M relationship between these two entities. The trap can be resolved by the provision of an entity `Animal Food Stock` with a composite identifier formed by the attributes `area_no` and `foodtype_name`. The attribute `stock_level` has been

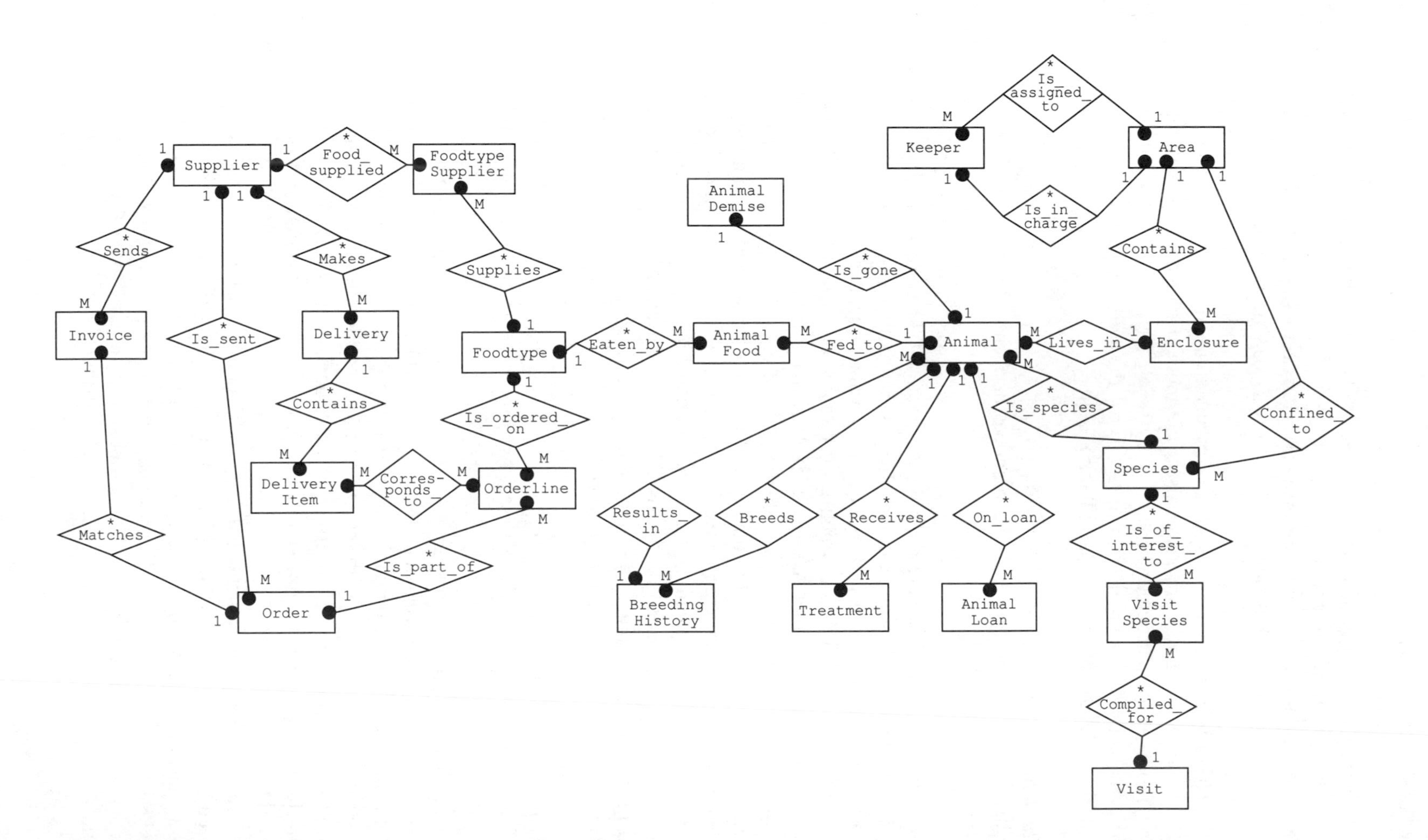

Figure 6.26 Guideline step 9.

erroneously assigned to the entity Foodtype; it instead should be assigned to the new entity Animal Food Stock. Figure 6.27 contains the revised E-R diagram, and the complete set of tables is reproduced below.

Tables

Order (order_no, supplier_name, order_date, order_status, total_discount_offered)

Supplier (supplier_name, supplier_address, supplier_tel_no, supplier_fax_no, main_contact, standard_discount)

Foodtype (foodtype_name)

Invoice (invoice_no, supplier_name, order_no, invoice_date, invoice_value, invoice_status)

Delivery (delivery_no, supplier_name, delivery_date)

Animal (animal_no, species_no, animal_name, date_of_birth, special_notes)

Treatment (animal_no, treatment_no, treatment_date, vet_name, treatment_details)

Breeding History (animal_no, breeding_no, mate_animal_no, date_gave_birth, birth_notes)

Results_in (animal_no, breeding_no, offspring_animal_no)

Animal Loan (loan_no, animal_no, start_date_of_loan, end_date_of_loan, park_loaned_to/from, loan_purpose, comments)

Animal Demise (animal_no, date_died/date_reintroduced_to_wild, reintroduction_comments/cause_of_death)

Species (species_no, area_no, species_name, species_details)

Area (area_no, employee_no)

Enclosure (enclosure_no, area_no)

Keeper (employee_no, employee_name, employee_address, employee_tel_no, area_no)

Visit (visit_no, party_name, contact_tel_no, visit_date)

Foodtype Supplier (supplier_name, foodtype_name)

Orderline (order_no, foodtype_name, quantity_ordered, price)

Delivery Item (delivery_no, foodtype_name, quantity_delivered)

Corresponds_to (delivery_no, order_no, foodtype_name)

Animal Food (animal_no, foodtype_name, daily_quantity, feeding_notes)

Visit Species (species_no, visit_no)

Lives_in (animal_no, enclosure_no)

Animal Food Stock (area_no, foodtype_name, stock_level)

Guideline step 11: The tables are still all in BCNF. However the reassignment of the attribute stock_level has left the Foodtype table with just one attribute. Further analysis is required to determine whether this table is needed. Perhaps it forms a list of approved foodtypes, or perhaps there is

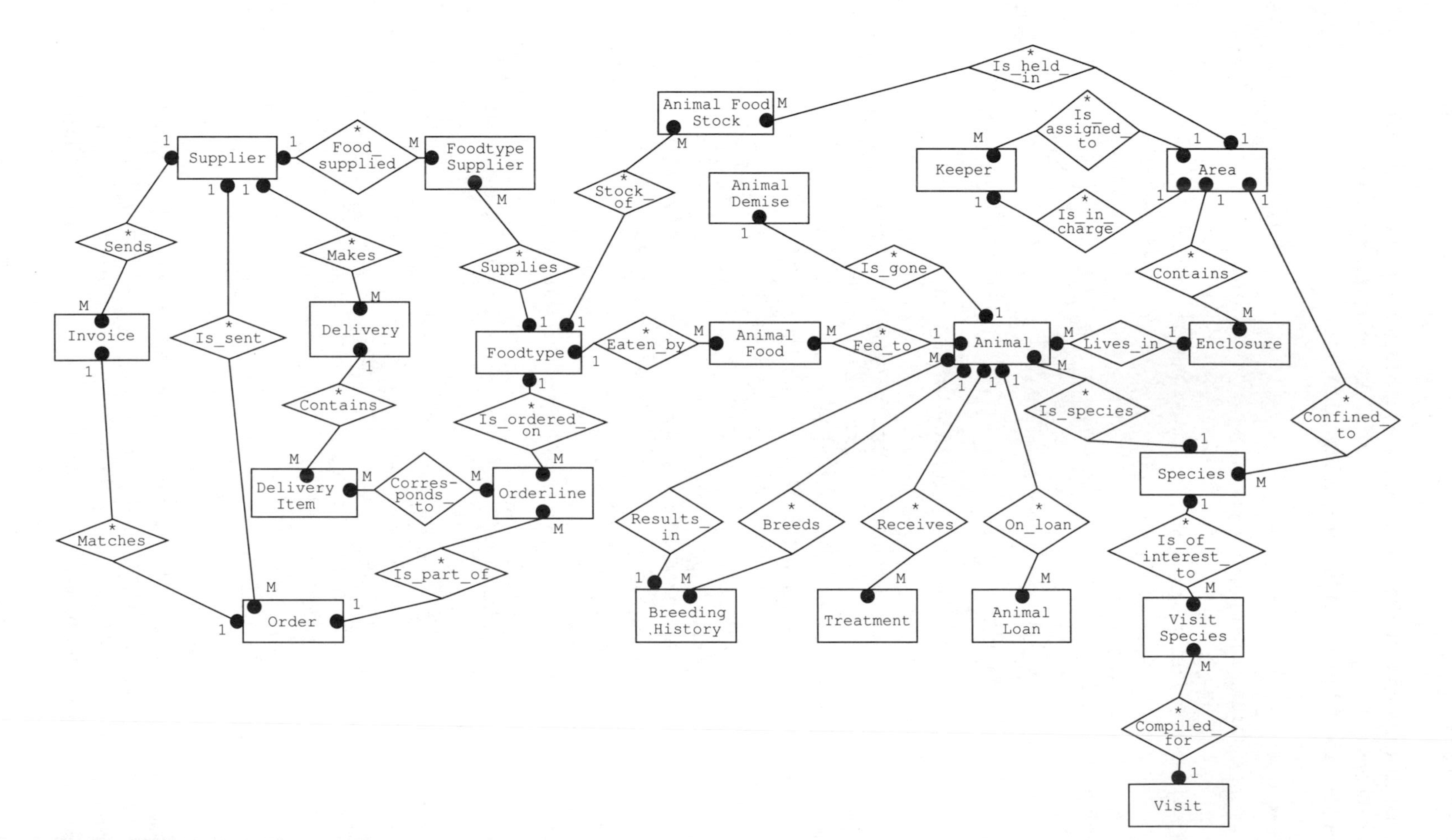

Figure 6.27 Guideline step 10.

further data about foodtypes that needs to be kept. If neither of these is applicable the table is superfluous and could be removed. For the moment, it will be left in the model.

The complete E-R model comprises the E-R diagram in Figure 6.27, the tables listed in guideline step (10) above and the accompanying assumptions which are reproduced in full below for completeness.

Assumptions

1. Each order is assigned a unique `order_no`.
2. Supplier names are unique.
3. Each delivery and each invoice is assigned a unique `delivery_no` and `invoice_no`, respectively.
4. Each animal and each species is assigned a unique `animal_no` and `species_no`, respectively.
5. Each area and each enclosure is assigned a unique `area_no` and `enclosure_no`, respectively.
6. Each keeper is identified by a unique `employee_no`.
7. Each visit is assigned a unique `visit_no`.
8. A supplier can supply more than one type of food.
9. A foodtype can be supplied by more than one supplier.
10. A delivery could correspond to more than one order and an order could be split and sent in more than one delivery (for example, if there were insufficient stock to satisfy the whole order immediately).
11. Details of suppliers other than those who supply food are also held.
12. At times a foodtype might not have a supplier.
13. A supplier might not have been sent any orders.
14. A supplier might not have made any deliveries.
15. An invoice is sent for each order.
16. The system will only record which area a keeper is currently assigned to; s/he must be assigned to an area.
17. The same data will be kept about keepers and headkeepers and thus they may be represented by the same entity.
18. An area must have a keeper assigned to be in charge of it.
19. An enclosure may have no animals assigned to it; for example during periods of rebuilding/maintenance.
20. An animal need not be assigned to an enclosure; for example it may be on loan to another animal park.
21. Information is kept about species that may not be currently represented by any animals in the animal park.
22. A species is kept in just one area.
23. Visits may (but do not have to) request information relating to particular species and information is compiled especially for such requests.
24. Not every breeding episode will result in offspring.
25. Not every animal was born in the animal park.

26. Minimal details of the keepers are kept in this system; it is assumed that the payroll is the subject of a separate system.
27. As the same data is kept about all keepers, only one table is necessary, as previously assumed. The `employee_no` in `Area` allows us to find out who are the headkeepers.
28. A supplier offers a standard discount. However, a level of discount that could be more (but never less) than this can be negotiated for each order.
29. The attribute `feeding_notes` is dependent upon both `animal_no` and `foodtype`, as notes are made about the method of feeding a particular foodtype to a particular animal.
30. An animal is either reintroduced to the wild or dies while in Somerleyton or another park, but not both.
31. A keeper is assigned to an area, not to specific enclosures within that area.

Exercise 6.9

See the Albany Hotel case scenario in Chapter 1.

Using the specified guidelines, develop an E-R model for the Albany case scenario.

Current vs. required data model

See the second section in Chapter 5.

Process modelling as described in Chapters 2 and 3 proceeded from the development of a current physical model to that of a current logical model and then a required logical model. The model produced as a result of data analysis is by definition a logical model. The data used by an organization and the enterprise rules which govern it remain the same whether a manual or computerized system is in operation.

The requirements for the new information system may necessitate the storage of extra data that is not currently held; for example, the storage of details about *all* of the species in Somerleyton Animal Park, and not just about big cats and reptiles as in the current system. Often, the requirements include the better use of data that is currently kept in manual form or in the inflexible file structures of a previous system. Typically, this may include the analysis of data for management information, or the ability for a user to make *ad hoc* queries of a database system where previously special purpose programs needed to be written. Hence the requirements will affect the structure of the data as depicted in the E-R model; in guideline step (10) above the question was raised as to whether visiting parties would want information about individual animals or about species, and the answer to this affects the relationships between the entity types `Animal`, `Species` and `Visit Species`.

It has been implied in this chapter that the E-R model may be developed in one go. In reality, it is more likely that preliminary models will be sketched during the early stages of analysis, and will gradually be extended and refined as analysis proceeds. The early sketches will tend to reflect the data held in the current system; the later ones will incorporate the requirements of the new information system.

Worked example 6.1 developed an E-R model for the required system for Somerleyton Animal Park; hence, it included entities and attributes relating to visits and to the information to be prepared for visits.

Other approaches to data modelling

SSADM

Within SSADM v4, the data model is called the **logical data model** (LDM) and comprises a diagram called the **logical data structure** (LDS), a set of **entity descriptions** and a set of **relationship descriptions.**

See Chapter 5, Figure 5.20 for an example of a LDS.

The LDM is developed through several steps. An overview LDS is developed either during the feasibility study, if there is one, or during the 'Establish Analysis Framework' step. This overview LDS should contain only about 8–12 of the most important entities as identified by the users, together with brief entity descriptions. In a later step, this overview model is extended to form a **current environment LDM.** This model includes more entities; the attributes associated with them are added to the entity descriptions, but only the data in the current system is considered. Any additional requirements are recorded in the requirements catalogue.

The current environment LDM is transformed during a later step into a **required system LDM.** This is done by examining the requirements catalogue and amending the LDM to support the requirements of the new information system. The entity descriptions are also completed at this point. Finally, to complement the top-down data modelling carried out up to this point, a further step takes a bottom-up approach by applying relational data modelling to the most important or the most complex input/output data structures. These are normalized to 3NF and after any necessary amendments to the LDM the **enhanced required data model** is produced.

In this way SSADM v4 develops a current logical then a required logical data model in a more systematic way than the approach proposed above.

Gane and Sarson

Gane and Sarson do not advocate the development of an E-R model in the manner described in this text. Instead, their data model consists of a set of 3NF relations (tables) derived from the data stores in the process model. For each data store in turn, the data flows in and out of the store are examined carefully and the constituent data structures and data elements are identified. Having thus determined the contents of each store, the technique of

The E-R diagram, which is useful in providing an overview and in preventing the systems analyst from becoming lost in the detail of bottom-up data analysis, is not used in this approach.

normalization as described in this chapter is applied to derive a set of 3NF tables.

Alternative database implementations

It is beyond the scope of this text to discuss the steps involved in mapping the data model onto target software. See Chapter 5 for a brief description of DBMSs and for further references.

See Cardenas (1985), Date (1990) and Hughes (1991).

This and previous chapters have described an approach to relational data modelling. The resultant data model may be easily mapped onto a relational database management system. There are many commercial database management system (DBMS) packages available that directly support the relational model.

There are alternative DBMSs, including hierarchical, network and object oriented. A discussion of the distinction between them is beyond the scope of this text.

The production of a relational data model is still applicable even when the target software is a hierarchical or network DBMS, although more steps will need to be taken to map the model onto the target software.

Summary of data modelling

Data is a fundamental resource of every organization. An information system stores, processes and retrieves data. It is therefore of the utmost importance for the systems analyst to ensure that the structure of the data to support a new information system is fully and accurately determined. An approach to the development of a logical data model has been described and illustrated. Its production is an iterative process. Early models reflecting the data currently held are successively refined until a model free from redundancy that supports the new system requirements has been produced. The data model will be developed in parallel with the process model described in Chapters 2 and 3. Chapter 8 will discuss how all of the models produced by the systems analyst are cross-checked to ensure their consistency.

Chapter 7

Entity Life Histories

OBJECTIVES

In this chapter you will learn:

- □ how an Entity Life History (ELH) may be used to model the time view of a system;
- □ using guidelines, how to develop an ELH.

Introduction

Chapters 2 and 3 have explained how to model the process view, and Chapters 5 and 6 the data view, of an information system. This chapter is concerned with the third view of an information system that is modelled by structured systems analysis methods; that of time. The **time view** can be modelled using a technique called the **entity life history** (ELH).

The need for a third model

A third view is necessary because neither the set of levelled data flow diagrams (DFDs) nor the entity-relationship (E-R) model takes into account the concept of time. The DFDs specify the processes that need to be executed by the information system. The E-R model provides a static view of the data that is needed to execute those processes. The ELH forms a dynamic view of the data by modelling the effect that events have on the entities in the E-R model. It models all the possible combinations of events that could affect an occurrence of an entity during its 'life' in the information system, and the permissible sequence of those events.

In this context, an **event** is something that triggers a process into updating system data. An **effect** is the change caused by an event, such as the creation, deletion or modification of an entity occurrence. For example, when an animal in Somerleyton Animal Park is treated by the veterinary surgeon, a new occurrence of the entity `Treatment` needs to be created. The event is the treatment of an animal, the effect is the creation of a new occurrence of `Treatment`.

See the E-R model developed in Chapter 6, Worked example 6.1.

See Chapter 8 for a discussion of the cross-checking of the three views of an information system.

The ELH can also be used to cross-check the DFDs against the E-R model to ensure their consistency and completeness. This is achieved by checking that all the events needed to create, modify and delete an occurrence of each entity are supported by the processes in the set of levelled DFDs.

ELH notation

Structure diagrams are used within Jackson System Programming (JSP) and Jackson System Development (JSD) (see Cameron, 1989). They are now also used extensively within SSADM version 4.

The structure diagram

An ELH is drawn using the structure diagram notation first introduced by Jackson (1975). Figure 7.1 illustrates this notation.

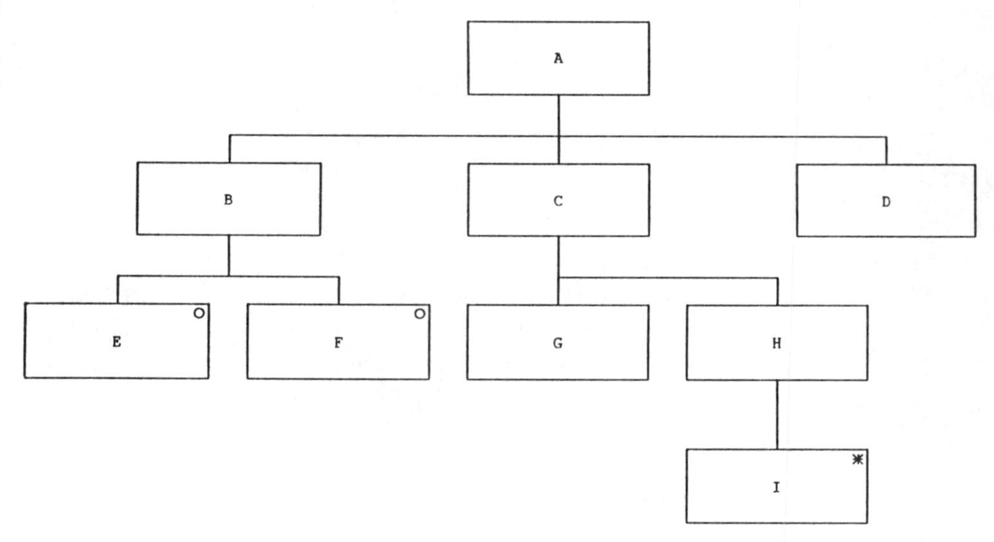

Figure 7.1 ELH notation.

See Chapter 4 for a discussion of the three constructs sequence, selection and iteration.

Box A is called the **root** and boxes E, F, G, I and D are called the **leaves** of the inverted tree. Boxes B, C and H are neither roots nor leaves but **structure boxes**, the purpose of which is explained below. Three types of activity can be represented. The first of these is **sequence** and is illustrated in Figure 7.1 by boxes B, C and D. The notation denotes that A consists of a sequence of B, followed by C, followed by D. A second sequence in the same figure is that C consists of G followed by H. The second type of activity that can be represented is **selection** and is illustrated by E and F. The circle in the upper right hand corner of these boxes denotes that B consists of either E or F, but not both. The third type of activity is **iteration** and is illustrated by I. The asterisk in the upper right-hand corner of this box denotes that H consists of an iteration of zero or more of I.

Different types of activity must not be mixed on the same level of the same branch of the tree. For example, Figure 7.2 illustrates an incorrect use of the notation.

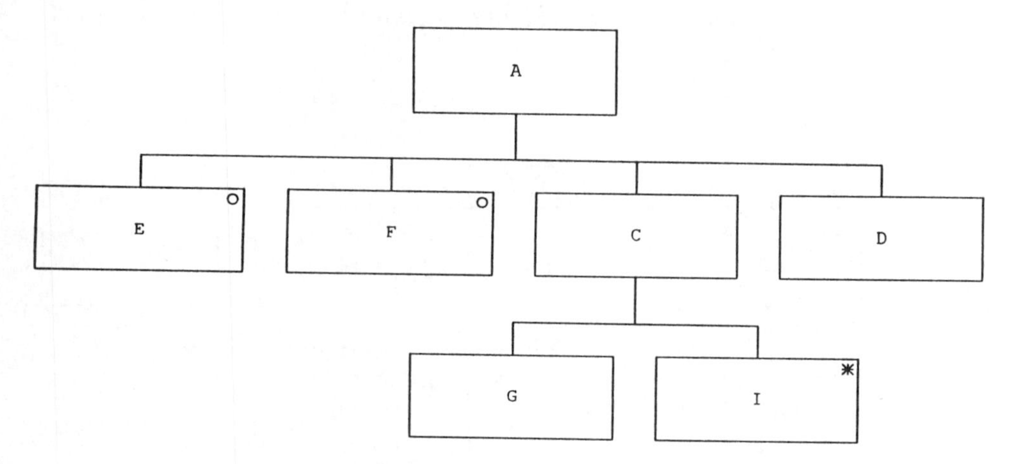

Figure 7.2 Incorrect use of notation.

Boxes E and F represent a selection, but this has been incorrectly placed in a sequence with boxes C and D. Box I is an iteration and has been incorrectly placed in a sequence with box G.

In Figure 7.1, boxes B, C and H are called 'structure boxes'; they are there to preserve the integrity of the notation, that is to ensure that sequence, selection and iteration are not mixed on the same level of one branch of the tree.

It is also possible to have an empty or null box; for example, the structure diagram in Figure 7.3 denotes that A consists of either B or C or nothing at all.

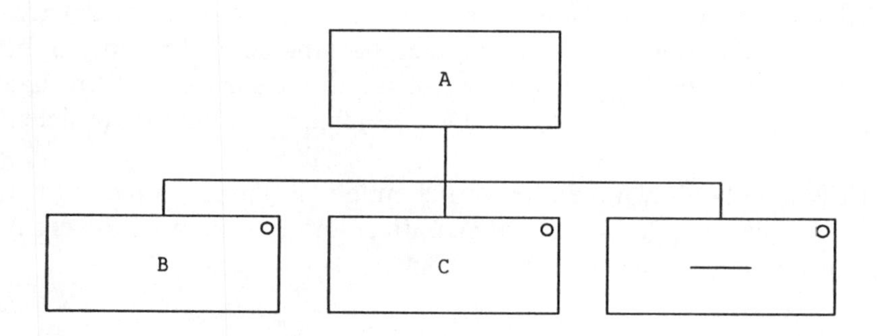

Figure 7.3 Use of the null box.

Exercise 7.1

Examine the structure diagram in Figure 7.4, which contains some incorrect use of the notation, and correct the errors you find.

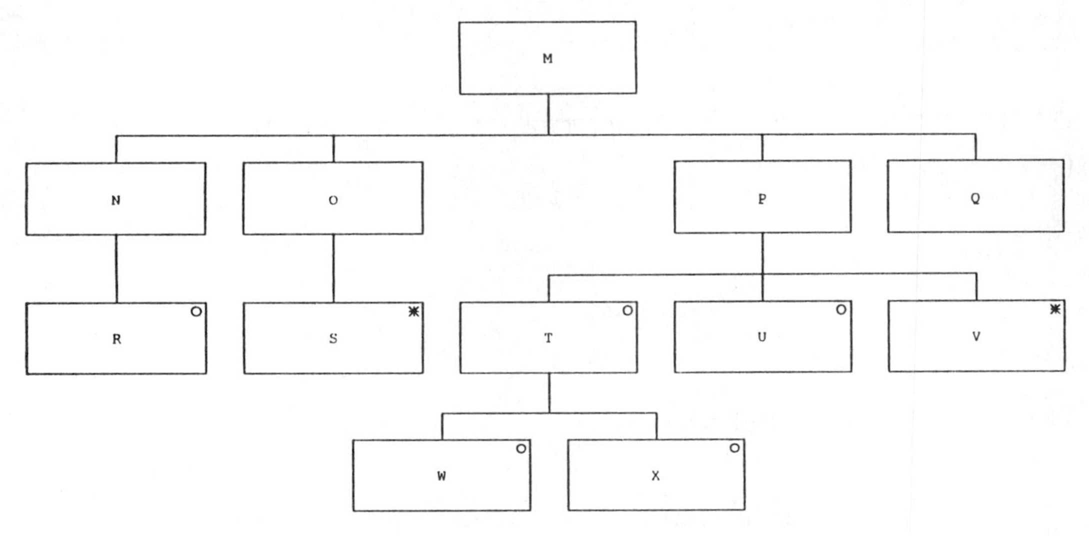

Figure 7.4 Exercise 7.1.

The ELH structure diagram

In an ELH the root contains the name of an entity type and the leaves the names of events that affect an occurrence of that entity. The structure boxes do not have to contain any names, but for clarity it is helpful to place in them names that reflect accurately the events at the end of that branch of the tree.

Figure 7.5 illustrates the typical structure of an ELH. To the left of the diagram are the events that cause an occurrence of the entity to be created, to the right the events that cause an occurrence of the entity to be deleted. In between are the events that will cause an occurrence of the entity to be modified, for example by updating one of its attributes. Hence there is a sequence of create events, followed by amendments, followed by delete events.

An ELH may be drawn for each of the entities in the E-R model. Each ELH models all possible events that can affect an occurrence of the entity and the permissible sequence of those events.

An example ELH

See Chapter 2, Exercise 2.2; Chapter 5, Worked example 5.2; and Chapter 6.

Figure 7.6 contains an ELH for the entity type `Property` from the estate agent case scenario introduced in earlier chapters.

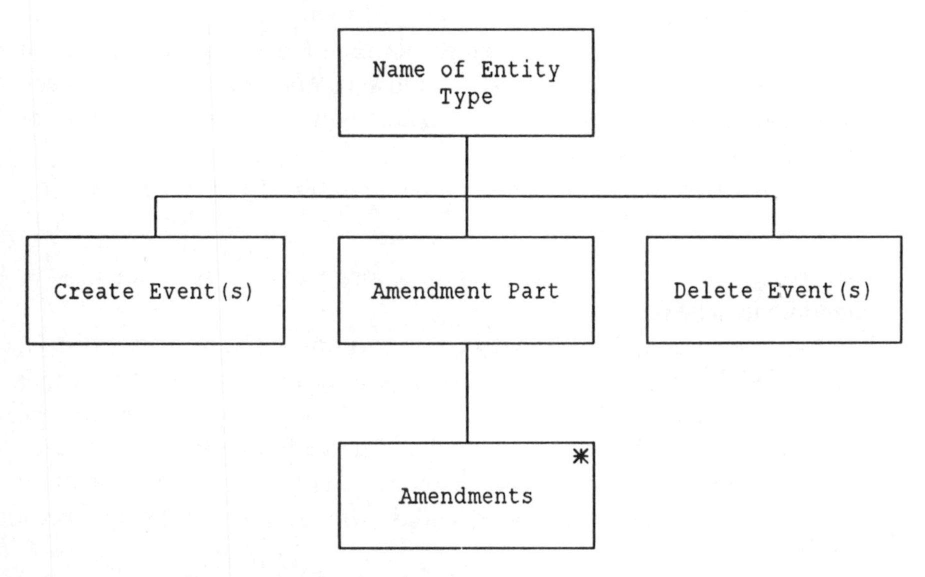

Figure 7.5 Typical ELH structure.

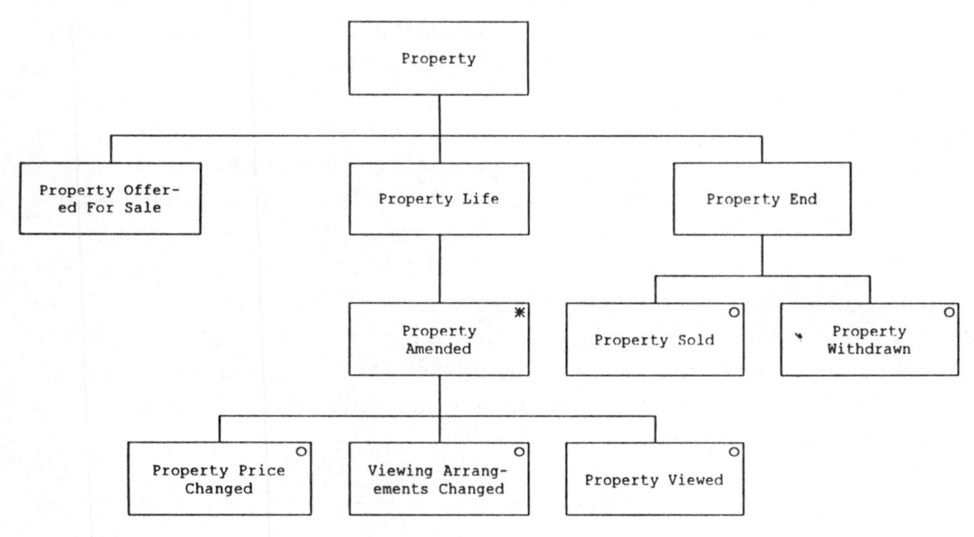

Figure 7.6 ELH for `Property`.

The ELH models the following 'life' for an occurrence of the entity type `Property`: an occurrence of `Property` is added to the information system when a property is offered for sale. While it exists in the system, zero or more amendments may be made to it. An amendment may be either a change in the price of the property, a change in the viewing arrangements or a viewing of the property. The occurrence is retained until either the property is sold or it is withdrawn from the estate agent; either of these events causes its deletion from the system.

Note that although the viewing of a property will not cause a change to the attributes of `Property`, it will cause the creation of an occurrence of `Viewing`, with which `Property` has a relationship. The relationship between these two occurrences needs to be established, which is the justification for including this event on the `Property` ELH.

See the E-R diagram in Chapter 5, Figure 5.11.

The sequences of events supported by the `Property` ELH include the following:

- A property is offered for sale and withdrawn before any viewings or amendments take place.
- A property is offered for sale; viewed several times; the price is dropped; it is viewed several more times; the price is dropped again; after more viewings it is finally sold.
- A property is offered for sale; after some time with no viewings the price is dropped; the viewing arrangements are changed; after several more changes of price the property is withdrawn after never having been viewed.

The ELH imposes a permissible sequence of events: a property cannot be viewed before it has been offered for sale; the price or viewing arrangements can be changed at any time, before or after any viewings; the property cannot be viewed again after it is sold or withdrawn.

At this point the life appears simple and somewhat unrealistic. The possibility of offers being made for a property has not been catered for, nor has the case where a property may be under offer but still able to be viewed until the time that contracts are exchanged. Yet the life as illustrated is consistent with the previous models and the case scenario as originally presented. The answer is that the latter was deliberately simplified because of the point in this text at which it was introduced. The ELH has served to highlight the omissions; in a real life project the systems analyst would now return to the user to double check the procedures for handling offers and the viewing of properties under offer, and subsequently enhance all of the models if necessary.

We shall not undertake the enhancement of the estate agent case scenario models in this text. Chapter 8 contains a discussion of the issues raised when the three views of a system highlight omissions, with reference to the Somerleyton Animal Park case scenario.

Exercise 7.2

Examine the ELH in Figure 7.7. Which of the following sequences of events are permissible?

(i) A student enrols, then withdraws.
(ii) A student enrols, is assigned a personal tutor, changes address, enrols on a different course, is reassigned to a personal tutor, and graduates.
(iii) A student enrols, is assigned to a personal tutor, and fails the course.
(iv) A student enrols, is assigned to a personal tutor, changes address, changes address again, and changes address a third time.

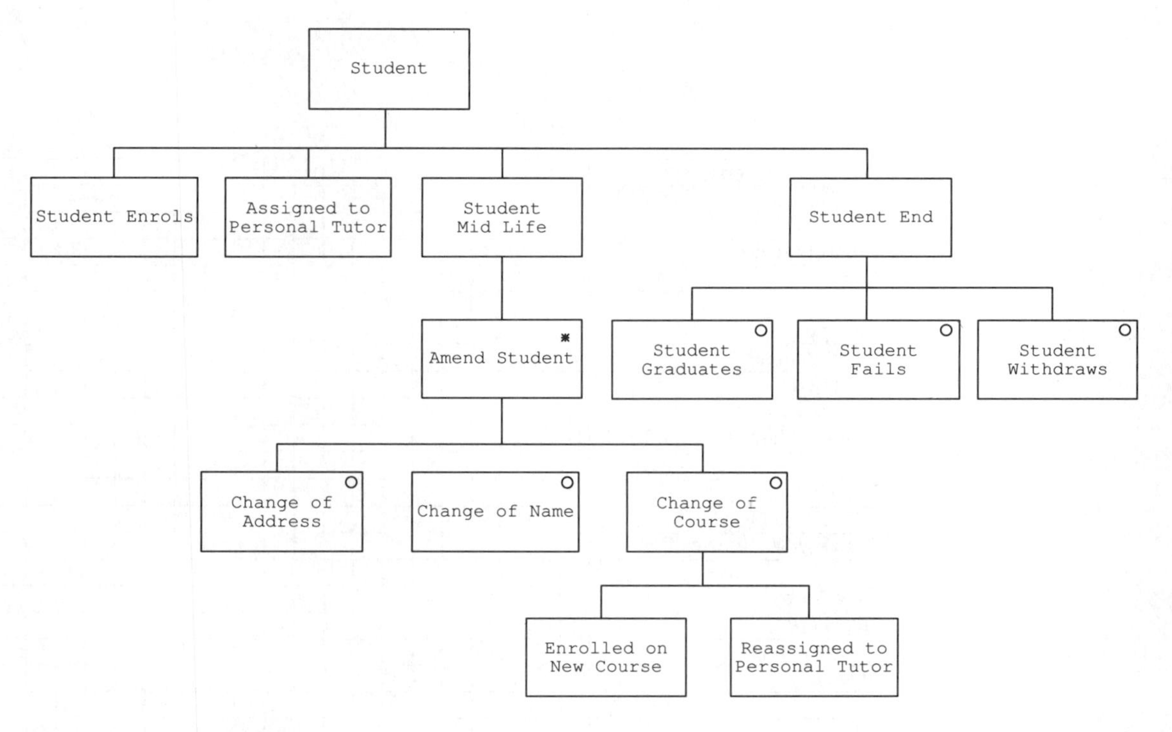

Figure 7.7 Exercise 7.2.

Approach to drawing an ELH

An ELH represents all the permitted sequences of events that may occur during the life of an occurrence of an entity in the information system. The first step in constructing an ELH for a given entity is therefore to identify all of the events that affect it. A technique that aids this step is the **event/entity matrix.**

The Event/Entity Matrix

The event/entity matrix (EEM) is a matrix that lists on one axis the entities from the E-R model and on the other the events that affect them. The latter are derived by examining the levelled set of required system DFDs and, in particular, the data flows that write to the data stores, since the latter are by definition creating, modifying or deleting data. The EEM shows which entities are affected by each event, and whether that effect is to create, modify or delete an entity occurrence.

Indicated by C, M or D, respectively.

Figure 7.8 contains an EEM for the estate agent case scenario. In constructing this EEM, several assumptions have been made:

1. The same attributes are to be stored for both potential buyers and buyers (name, address and telephone number), so they have been treated as one entity.

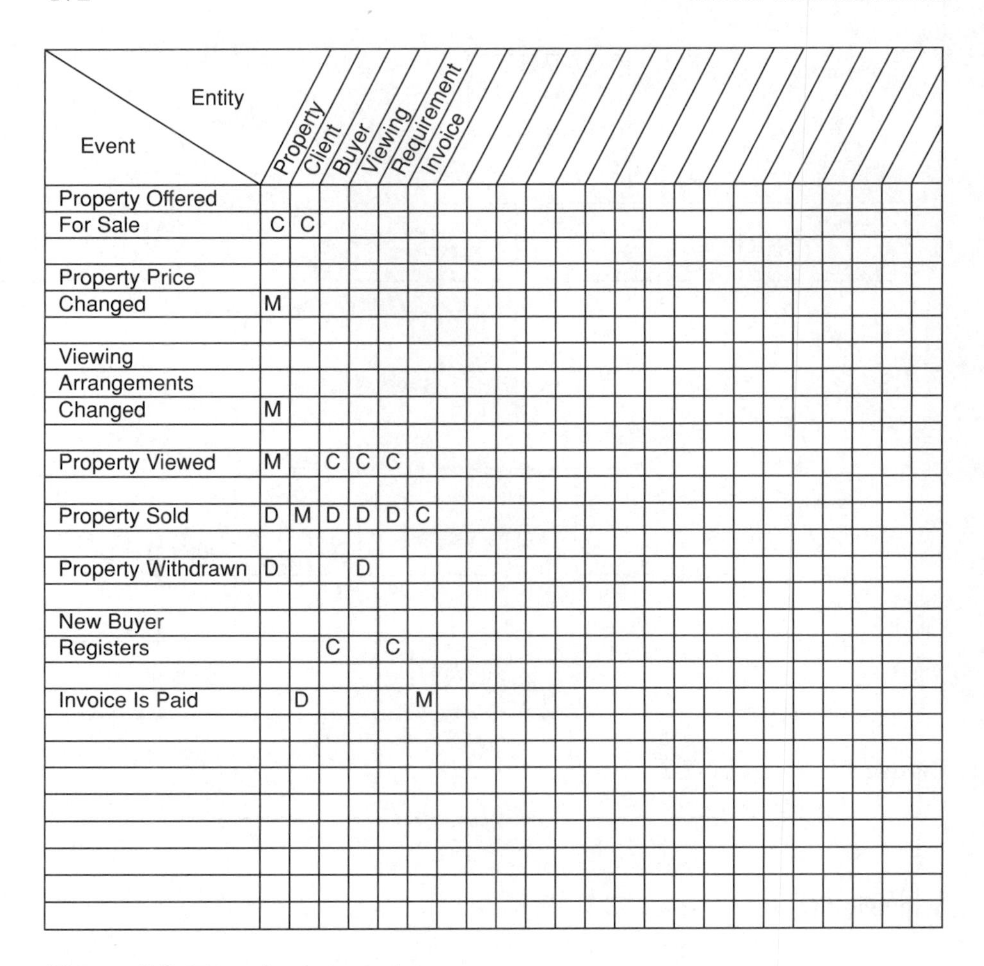

Event \ Entity	Property	Client	Buyer	Viewing	Requirement	Invoice
Property Offered For Sale	C	C				
Property Price Changed	M					
Viewing Arrangements Changed	M					
Property Viewed	M		C	C	C	
Property Sold	D	M	D	D	D	C
Property Withdrawn	D			D		
New Buyer Registers			C		C	
Invoice Is Paid		D				M

Figure 7.8 Event/entity matrix.

2. When a property is viewed by someone who is not already a potential buyer, the details of the person viewing are taken and stored as a potential buyer, together with their requirements.
3. When a property is sold, details of the buyer and their requirements are deleted. Details of the client are retained until the invoice has been paid, at which point they are deleted providing the client has no more properties to sell.
4. When a property is withdrawn, details of the client are retained as the estate agent likes to follow them up in attempting to attract business at a later date.

A complete EEM will contain at least a creation and a deletion event for each entity, and usually one or more modification events. In Figure 7.8 the entities `Property` and `Client` are created, modified and deleted by events. The entities `Buyer`, `Viewing` and `Requirement` are created and deleted but not modified while the entity `Invoice` is not deleted. This highlights some omissions from the DFD. It is conceivable that the attributes of `Viewing` will rarely be modified, but it is possible for a buyer to change their personal details or their requirements, so these events must be taken into account. It must also be determined when, or whether, invoices are deleted; they may be retained for some time for auditing purposes, but at some point must be removed from the information system either by being deleted altogether or by being archived into some form of backup storage for permanent retention.

Guidelines for drawing an ELH

1. Construct an EEM.

For each entity in turn:

2. Identify which of the events that affect it are part of a sequence, which are selections and which are iterations.
3. Start drawing the ELH by placing the creation events to the left and the deletion events to the right of the structure diagram.
4. Add the modification events, ensuring that a valid sequence of events is maintained.
5. Review the ELH, checking that the notation is correct and that a complete and permissible combination of events has been modelled. Return to step 2 and repeat all the steps if necessary.

Worked example 7.1

Using the specified guidelines, develop an ELH for the entity `Animal` from the Somerleyton Animal Park case scenario.

See the set of DFDs developed in Chapter 3, Worked example 3.2 and the E-R model developed in Chapter 6, Worked example 6.1.

Solution:

Guideline step 1: See the EEM in Figure 7.9.

This EEM includes the entity `Animal` and all of the entities with which `Animal` has a relationship. Relationships such as `Lives_in` which are represented by a table in Chapter 6, Worked example 6.1, are included because they also contain data that needs to be created, deleted and modified. On the event axis, the EEM contains all of the events that affect `Animal`.

In creating the EEM the following assumptions have been made:

Assumptions

1. When a new animal arrives, `Species` is modified because the relationship `Is_species` has to be established. If the animal belongs to a species for which no record is held, a new occurrence of `Species` will need to be created.
2. When the animal food requirements are established, `Animal` and `Foodtype` are modified because the relationships with `Animal Food` have

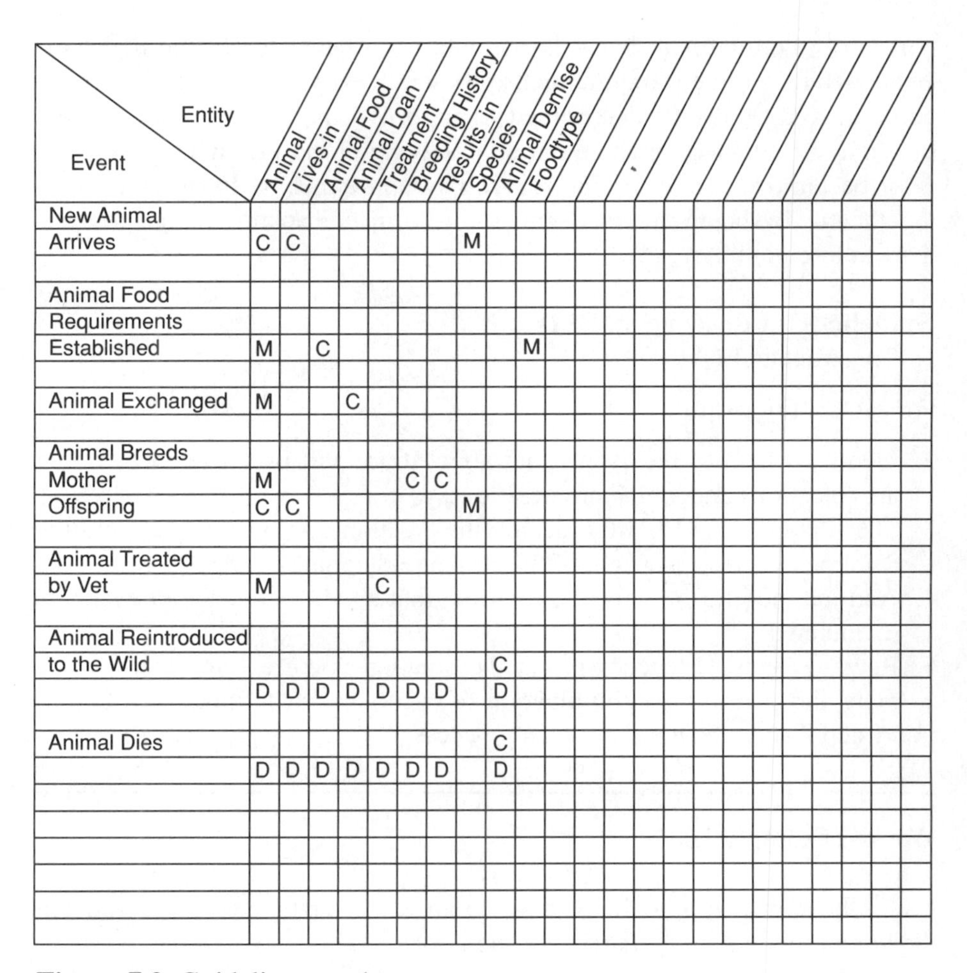

Event \ Entity	Animal	Lives-in	Animal Food	Animal Loan	Treatment	Breeding History	Results_in	Species	Animal Demise	Foodtype
New Animal Arrives	C	C						M		
Animal Food Requirements Established	M		C							M
Animal Exchanged	M			C						
Animal Breeds										
Mother	M					C	C			
Offspring	C	C						M		
Animal Treated by Vet	M				C					
Animal Reintroduced to the Wild									C	
	D	D	D	D	D	D	D		D	
Animal Dies									C	
	D	D	D	D	D	D	D		D	

Figure 7.9 Guideline step 1.

to be established. If the animal requires a type of food for which there is currently no record, a new occurrence of `Foodtype` will need to be created.

3. When an animal is exchanged, `Animal` is modified because the relationship with `Animal Loan` has to be established.
4. When an animal breeds, `Animal` is affected in two ways; the occurrence for the mother animal will be modified and, if live offspring result, new occurrences for the young animal/s will be created.
5. When an animal is treated by the veterinary surgeon, `Animal` is modified because the relationship with `Treatment` needs to be established.
6. When an animal dies or is reintroduced to the wild, an occurrence of `Animal Demise` is created, then all of the data relating to that animal, including `Animal Demise`, is archived (see Chapter 3, Figure 3.13c), that is, it is deleted from the live system.

Within SSADM the entity is said in this case to be assuming different roles. See the SSADM Reference Manual (1990).

See Chapter 8 for a discussion of the issues highlighted by the cross-checking of the three views of an information system.

The EEM highlights some omissions; for example, `Species` has no deletion events. It is possible that species data will never be deleted as a comprehensive collection of data is required, but this should be checked with the user.

It can be seen from Figure 7.9 that some ELHs are simple, consisting of no more than one or two creation and deletion events. Others, such as that for `Animal`, are more complex.

Guideline step 2: The following events form a sequence:
- Arrival of new animal;
- Animal food requirements established;
- Animal dies/Reintroduction to wild.

Assumptions (continued)

7. An animal's food requirements are established before it dies.

The following events are a selection:
- Animal dies; or
- Reintroduction to wild.

Each of these events will happen just once.

Assumptions (continued)

8. Once an animal has been reintroduced to the wild it is not recaptured for Somerleyton Animal Park.

The following events are iterations:
- Exchange of animal;
- Animal breeds;
- Animal treated by vet.

An occurrence of animal could be affected by none, by any or by all of them and they may happen in any order, any number of times.

Assumptions (continued)

9. A full veterinary treatment and breeding history is maintained for an animal that is on loan from another animal park.
10. A full veterinary treatment and breeding history is maintained for Somerleyton animals that are on loan elsewhere; the other park sends details to Somerleyton so that the latter's records are complete.

Guideline step 3: A new occurrence of `Animal` is created when a new animal arrives at the park or when an existing animal successfully breeds. This gives another selection for the ELH.

Assumptions (continued)

11. `New Animal Arrives` includes both the permanent acquisition of an animal and the arrival of an animal on loan from another park.

An occurrence of `Animal` is deleted when an animal dies or when it is reintroduced to the wild. However, what about animals on loan to Somerleyton, that are returned to their original park? These records need to be deleted when the animal is returned, giving a third selection for the ELH.

Assumptions (continued)

12. When an animal on loan to Somerleyton is returned to its original park, or passed on to another one, its record is archived.

Figure 7.10 shows the ELH for `Animal` after guideline step 3.

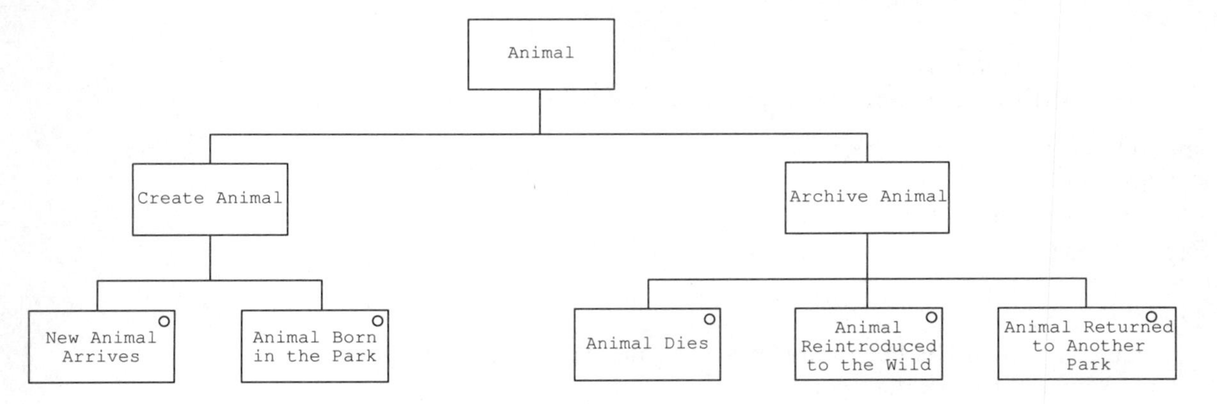

Figure 7.10 Guideline step 3.

Guideline step 4: The remaining events that affect `Animal` are:

- Animal food requirements established;
- Exchange of animal;
- Animal breeds;
- Animal treated by vet.

For the moment, it will be assumed that an animal's food requirements are established as soon as it arrives at the park.

Assumptions (continued)

13. An animal's food requirements are established before it is exchanged, breeds or is treated by the veterinary surgeon.

It was noted under guideline step (2) that an occurrence of `Animal` could be affected by none, by any or by all of the remaining three events, and that they may happen in any order, any number of times. This can be represented on the ELH by showing them as an iteration of a selection.

Figure 7.11 shows the complete ELH for `Animal`. An animal must either arrive at the park or be born there, then have its food requirements established, and then either die, be reintroduced to the wild or be returned to another park. As the iteration asterisk denotes 'zero or more times' the

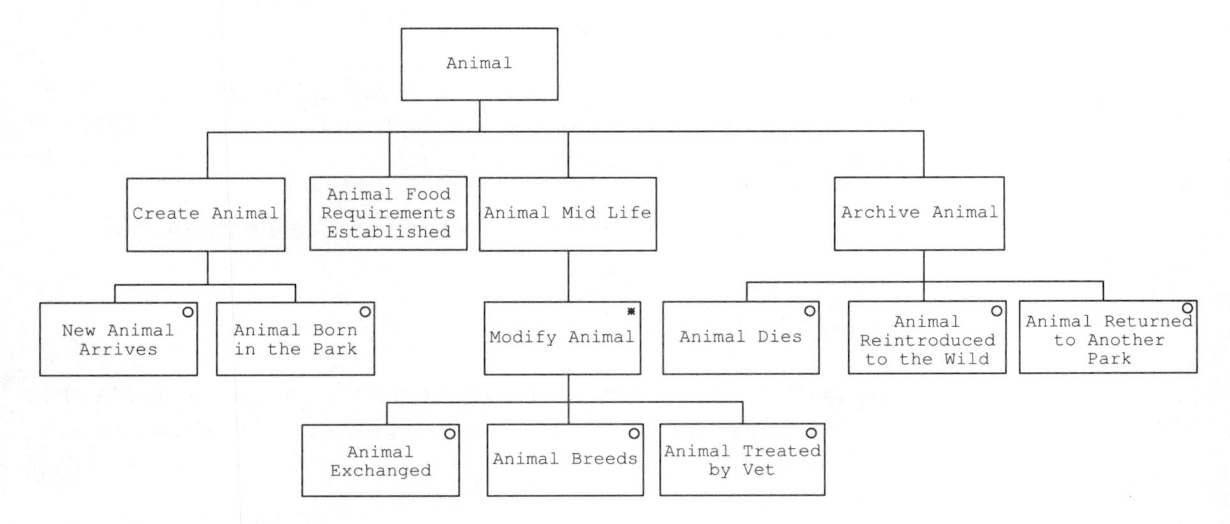

Figure 7.11 Guideline step 4.

model supports the case where an animal is never exchanged, never breeds and is never treated by the veterinary surgeon, as well as the case where an animal experiences one or more of these events several times.

Assumptions (continued)

14. `Animal Exchanged` refers to a Somerleyton animal being loaned to another park; the loan of an animal to Somerleyton from elsewhere has been discussed in guideline step (3) above.

Guideline step 5: The notation of Figure 7.11 is correct. The ELH is consistent with the previous models of the case scenario, and with the further assumptions made in this worked example.

In developing this ELH, some events have been introduced that were not on the EEM in Figure 7.9, namely `Animal Born in the Park` and `Animal Returned to Another Park`. This implies that the DFDs may be incomplete, and the systems analyst should at this point return to the previous models to recheck them. The EEM also needs to be amended with the addition of these events together with an indication of which entities are affected by them.

See Chapter 8.

Exercise 7.3

On further checking with the user it has been discovered that an animal's food requirements can be established at any time, and that they can subsequently be modified, sometimes several times. Moreover, an animal may be moved to another enclosure at any time, and this may also happen more than once. Modify the ELH in Figure 7.11 so that it supports these events.

Exercise 7.4

Draw an EEM for the Albany Hotel case scenario. Incorporate all of the entities from the solution to Chapter 6, Exercise 6.9, and all of the events that affect them from the DFDs in the solutions to Chapter 3, Exercises 3.3 and 3.4.

Check your answer to Exercise 7.4 before attempting Exercise 7.5.

Exercise 7.5

The handling of these events requires further notation, which is introduced below.

Using the EEM developed in Exercise 7.4 and the specified guideline steps (2) to (5), develop an ELH for the entity `Booking` from the Albany case scenario. However at this stage *omit the events* `Booking Changed` and `Booking Cancelled`.

Further notation

There are two further constructions that can be used in an ELH if the notation introduced so far cannot adequately represent a permissible combination of events. These are the **parallel structure** and the **quit and resume.**

Parallel structure

If a sequence of events cannot be specified, because the events may happen in random order, a parallel structure can be used. For example, a student undergoing a period of work experience is allocated a work placement and is assigned to an industrial tutor. But some students are allocated the work placement first, and others are assigned to an industrial tutor before being allocated a placement. Either event may happen first to a given student. This may be represented by using a parallel structure as in the ELH fragment in Figure 7.12. This denotes that both of the events, `Assigned to Industrial Tutor` and `Allocated a Placement` will happen, but either may happen first.

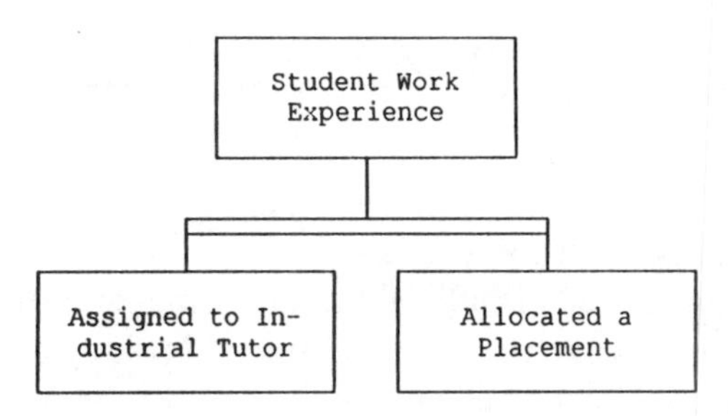

Figure 7.12 Parallel structure.

Quit and resume

The quit and resume construct can be used if an entity occurrence reverts to an earlier stage of its life or jumps to a later stage. Take, for example, the ELH for `Student` in Figure 7.7. Supposing that, if a student changes course, it is regarded as a withdrawal from the course on which he or she is currently enrolled. A new occurrence of `Student` will be created when he or she enrols on the new course. The ELH can be redrawn using quit and resume as in Figure 7.13.

When the event at Q1 occurs, the life jumps to the event at R1. Further quit and resume constructs may be used on the same ELH using the labels Q2, R2 and so on.

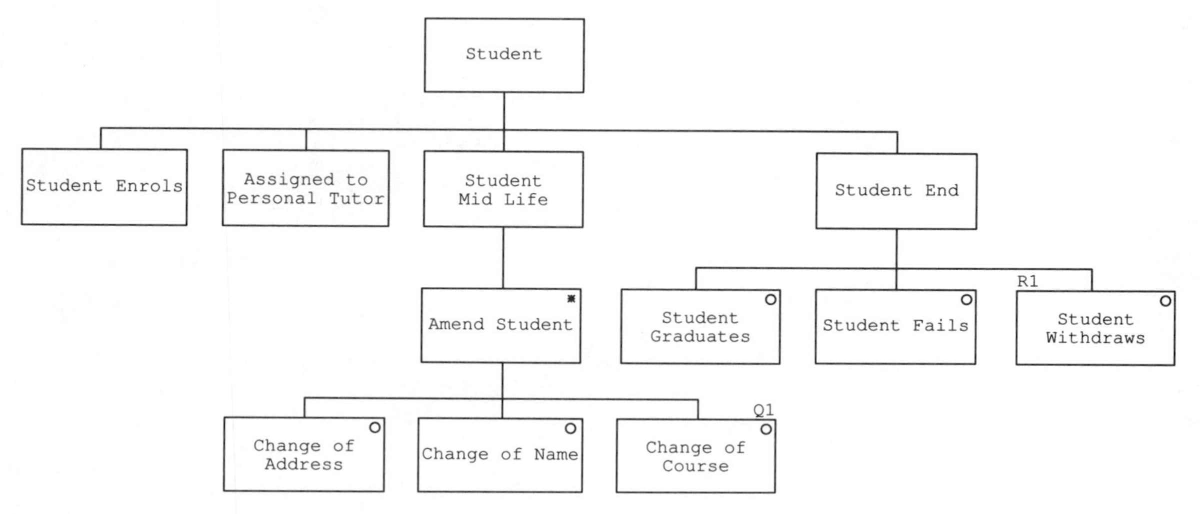

Figure 7.13 Quit and resume.

A special kind of quit is the **quit from anywhere** on the ELH to a common resume box, drawn separately from the main ELH structure. It is used to deal with random or unusual events. For example, with reference to Somerleyton Animal Park, an animal could at any time contract a notifiable disease, as a result of which the appropriate government department is notified and the animal is destroyed. Figure 7.14 illustrates how this exceptional event can be catered for using quit from anywhere.

Quit and resume should only be used to handle abnormal or error conditions. If the same meaning can be conveyed with alternative notation, then that should be used in preference.

Exercise 7.6

Changes can be made to a booking, or it can be cancelled, any time after a provisional booking has been made and before the guest is resident. Modify the `Booking` ELH in Figure E7.5b so that it supports these events.

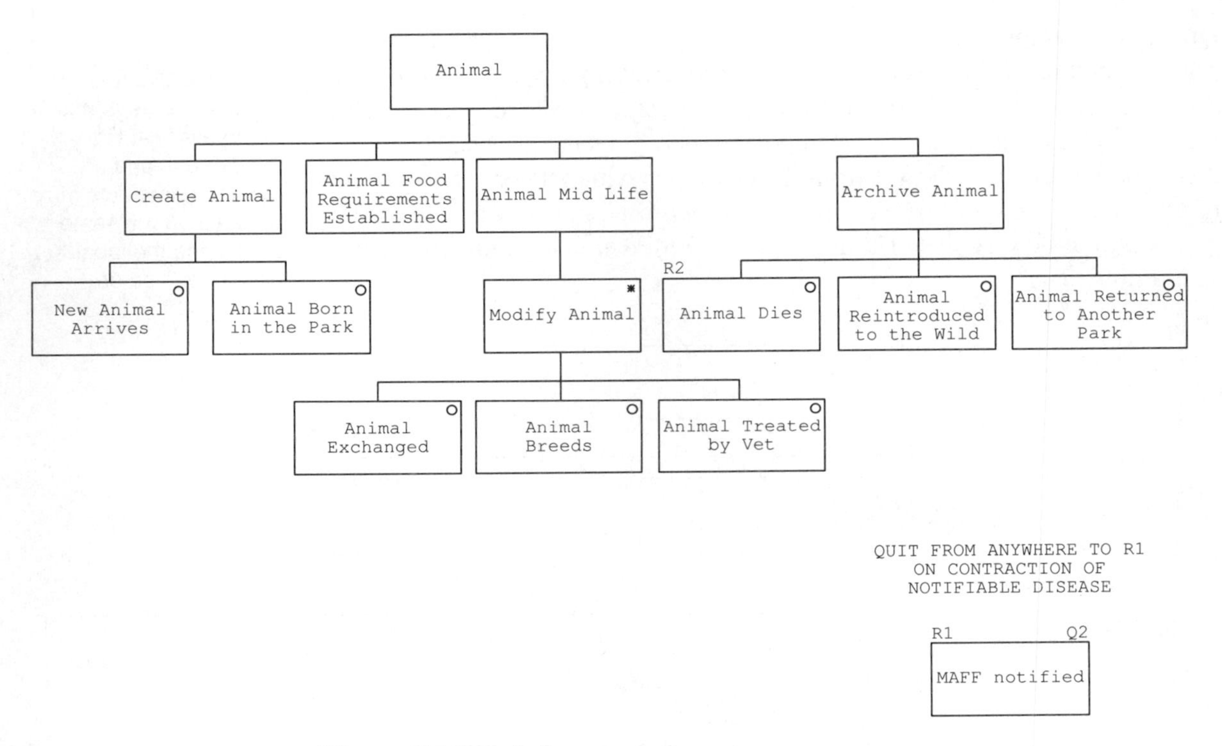

Figure 7.14 Quit from anywhere.

State indicators

The ELH is completed at the analysis stage by the inclusion of **state indicators**. These are written under the boxes which are leaves and show for each leaf box two values: the 'valid prior' value and the 'set to' value. That is, the value or values that indicate the state an entity occurrence must be in for it to be valid for the event to take place, and the value that indicates the state that the entity occurrence is in after the event has taken place. Figure 7.15 shows the `Animal` ELH with state indicators.

The purpose of the state indicators is to show what state an entity occurrence is in at a given point in time, and thus to determine whether it is valid for an event to take place. It would not be valid, for example, to attempt to modify an occurrence of `Animal` that is in state 1. They are used in the systems design and programming stages of the systems development lifecycle to ensure that only the permitted sequence of events takes place for each entity occurrence.

The 'valid prior' value is null (–) before a creation event, and is set to null by a deletion event. The 'set to' states are numbered sequentially from the beginning to the end of the ELH. When an iteration comprises a selection, as with 'Modify animal', the 'valid prior' values include the 'set to' values of all of the other selections. This is because any of the selection

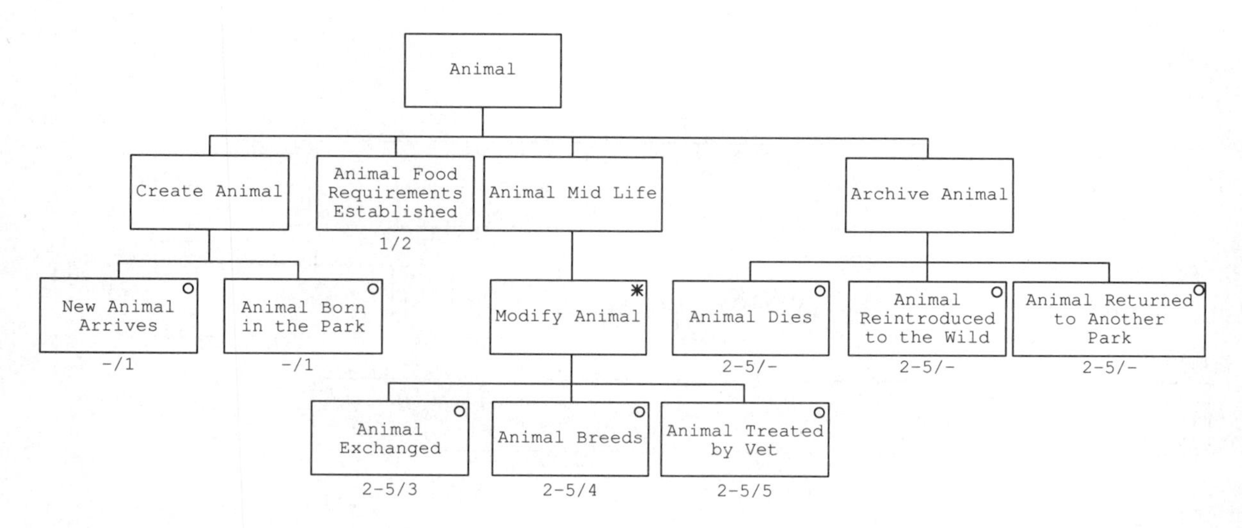

Figure 7.15 `Animal` ELH with state indicators.

events may happen, any number of times, so an entity occurrence can be in any of these states each time through the iteration.

Exercise 7.7

The 'set to' values for the `Modify Animal` events are 3, 4 and 5, so why do the 'valid prior' values for the three deletion events include state 2?

Exercise 7.8

Add state indicators to the `Booking` ELH developed in Exercise 7.5.

See Figure E7.5b.

Where a parallel structure has been used, one leg of the structure updates the main state indicator in the normal way. The second leg leaves the main state indicator unchanged and a subsidiary state indicator is introduced. The next effect sets the subsidiary state indicator back to null. The use of state indicators with a parallel structure is illustrated in Figure 7.16, where they have been applied to the solution to Exercise 7.6.

See the SSADM Reference Manual (1990) for further details.

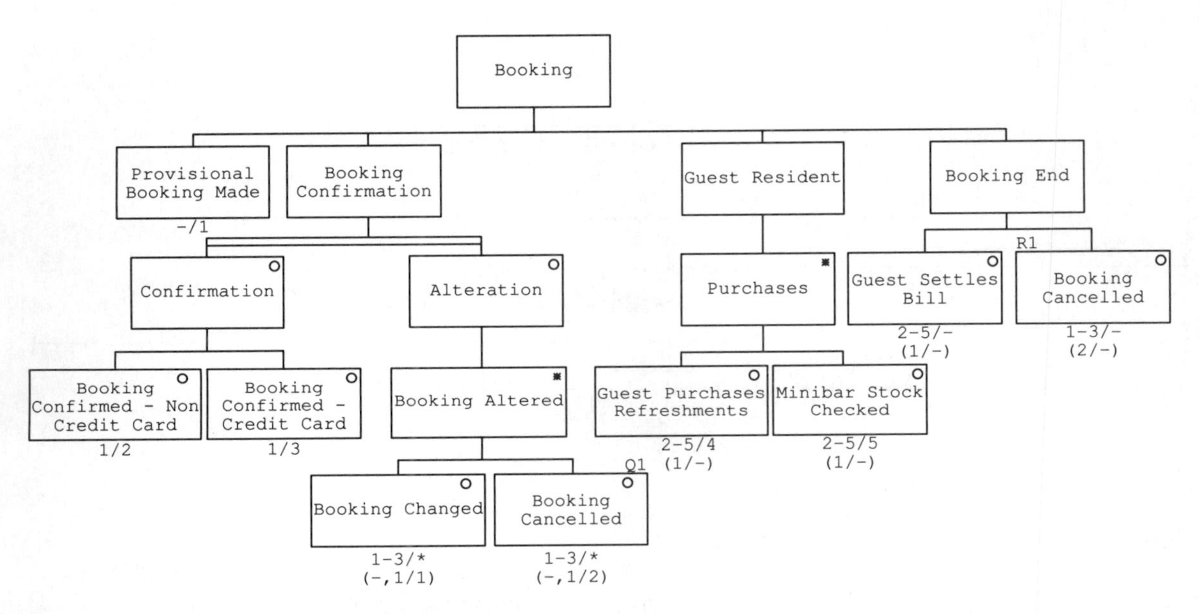

Figure 7.16 State indicators with a parallel structure.

Summary of ELHs

For example Skidmore, Farmer and Mills (1992), Ashworth and Slater (1993) or Downs, Clare and Coe (1992).

The ELH is a technique used in SSADM. Further discussion and examples can be found in the SSADM Reference Manual (1990), and in texts devoted to this systems development method.

Within SSADM, 'operations' are added to the leaf boxes to indicate what processing is associated with each effect. As the addition of operations is considered to be a design rather than an analysis issue, it has been omitted from this text.

The completion of a set of ELHs provides the analyst with a third view of the information system. Throughout this chapter there have been references to the fact that the construction of an ELH has highlighted an error or omission from one of the previous models. The ELH provides a valuable cross-checking mechanism that can help to ensure that the models produced during the systems analysis phase of a project are complete and consistent. A discussion of the issues involved in cross-checking the three views of the information system forms the subject of the next chapter.

Chapter 8

Relationship between the Three System Views

OBJECTIVES

In this chapter you will learn:

- how to cross-check a set of levelled data flow diagrams (DFDs) with an entity relationship (E-R) model;
- how to cross-check an E-R model with one or more entity life histories (ELHs);
- how to cross-check a set of levelled DFDs with one or more ELHs;
- how to use the data dictionary (DD) in cross-checking the other models.

Introduction

Previous chapters have concentrated on developing the three views of a system; process, data and time. These were modelled using a set of levelled DFDs, an E-R model and a set of ELHs, respectively. As each model is drawn, consistency and completeness checks may be performed. This chapter explains how the different models interrelate, and can be used to cross-check each other to increase the quality of the systems analysis, and also how to make use of the DD which underpins all of the three other models. The relationships between the three views of a system are shown in Figure 8.1. The case scenarios studied in earlier chapters are used to explain the issues involved, and to offer guidance on actions to be taken.

Correspondence between a set of levelled DFDs and an E-R model

As stated in earlier chapters, DFDs depict the transformation of data, and show data in motion and data at rest. Data in motion is represented by the data flows and data at rest by the data stores. The contents of each are

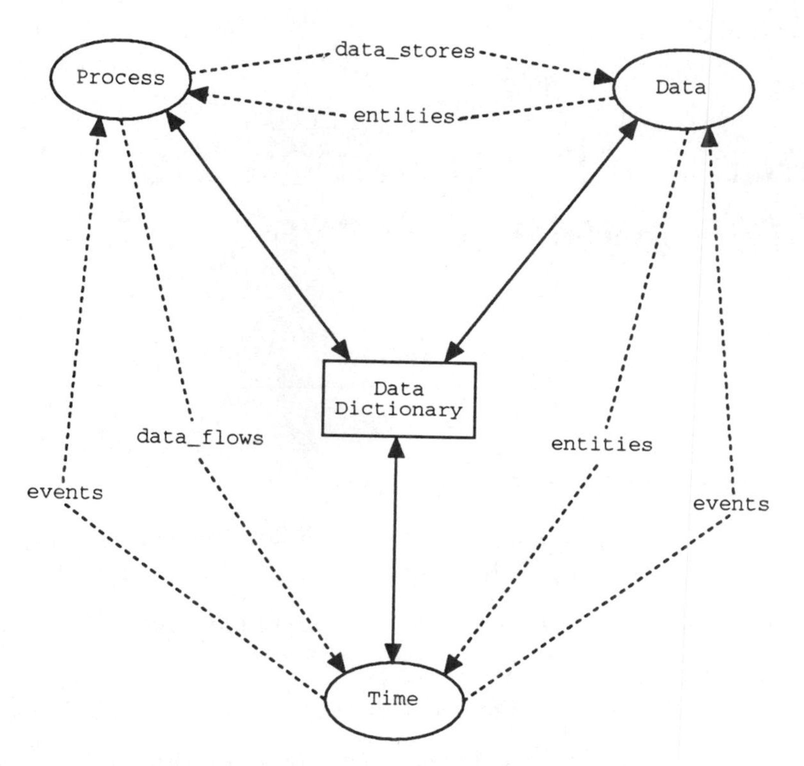

Figure 8.1 Relationship between the three views of a system.

included in the DD. An E-R model provides a logical view of the data required to support the information system under development and comprises an E-R diagram and a set of well-normalized tables. Each data store on a DFD should correspond to one or more entities in the associated E-R model.

Yourdon (1989) advocates a 1:1 correspondence which means that the DFDs become very cluttered.

Each entity in the E-R model should be held completely within one data store in the DFD. This may not be the case in the current physical DFD, where there may be duplication and redundancy of data, but this should be rectified during logicalization. Transient data stores, where data is held temporarily, may or may not have associated entities. Corresponding data store and entity names may or may not be the same. Examination of the DD contents for the data store and the tables for the entities will assist in the establishment of correspondence. It is not always easy to identify the correspondences between data stores and entities, but if entity occurrences from different entities are created at the same time and/or are part of the major inputs and outputs of the system, they are likely to belong to the same grouping. Data in motion is represented on DFDs by data flows. The data items on the data flows passing in and out of data stores will obviously be held in the data store, and therefore should be held within the associated entities. In our examples, the contents of the data store are documented in the DD, the contents of the entities are shown in well-normalized tables. It is therefore possible to check that the entries are consistent. If the DFDs and the E-R model are prepared in parallel, it is possible to validate the models as they are developed and to include one set of complete and consistent entries in the DD. The correspondence of the data stores on DFDs and entities in an E-R model may be shown in tabular form. An example of this is shown in Worked example 8.1.

Worked example 8.1

For the Somerleyton Animal Park case scenario:

(a) Prepare in tabular form the correspondences between the set of DFDs for the required (proposed) system, shown in Chapter 3, Figures 3.13 a–j, and the E-R model developed in Chapter 6, Worked example 6.1.
(b) For the data stores whose data elements have been defined in the DD in Chapter 4, Worked example 4.1, check that the data elements in the data stores are consistent with the attributes in the well-normalized tables listed in Chapter 6, Worked example 6.1.
(c) Comment on the issues, identified in (a) and (b), that need to be resolved to ensure that the DFDs and the E-R model are consistent and complete.

Solution:

(a)

Data stores	*Entities*
`Animal Archives`	`Animal Demise`
`Animal Foods`	`Animal Food`
	`Foodtype`
`Animals`	`Animal`
	`Treatment`
	`Breeding History`
	`Results_in`
	`Animal Loan`
	`Lives_in`
`Food Type Stock`	`Animal Food Stock`
`Orders`	`Order`
	`Orderline`
	`Delivery`
	`Delivery Item`
	`Corresponds_to`
	`Invoice`

The entity `Foodtype` is included here because of its obligatory relationship with the entity `Animal Food.`

The data store `Orders` incorporates all the details for an order from the time that it is raised until the time that payment is received. Initially, all the entities associated with an order through to payment are listed. This correspondence will be reviewed on examination of the data store elements and entity attributes in part (b) of this example.

Party Bookings	Visit
	Visit Species
Species	Species
Suppliers	Supplier
	Foodtype Supplier
	Area
	Enclosure
	Keeper

The entities Area, Enclosure and Keeper appear to have no corresponding data stores.

(b)

The table produced in part (a) will be used as a reference in this example.

For the following data stores and associated entities, the data stores and attributes are consistent:

Data stores	*Entities*
Animal Archives	Animal Demise
Food Type Stock	Animal Food Stock
Party Bookings	Visit
	Visit Species
Species	Species
Suppliers	Supplier
	Foodtype Supplier

The data store Animal Foods contains a data element area_no which does not appear in either the entity Animal Food or Foodtype.

The data store Animals corresponds to the entities listed in the table in part (a) of the example with the following exceptions:

1. The data store Animals contains an additional data element area_no.
2. The entity Animal has an additional attribute special_notes.
3. The entity Results_in has an attribute offspring_animal_no.

The data store Orders corresponds to the entities listed in the table in part (a) of the example with the following exceptions:

1. The entities contain the attributes quantity_ordered and quantity_delivered and maintain them separately. The data store element foodtype_requirement will contain the value of the quantity ordered or the quantity delivered, depending upon the status of the order.
2. The entity Delivery has an additonal attribute delivery_date.
3. The data store Orders does not contain the attributes specific to the entity Invoice.

The attributes within the entities `Area` and `Keeper` are not found within data stores. The data store `Animal` does contain data elements, indicating which area and enclosure house a particular animal, but does not have a separate list as in the entity `Enclosure`.

(c)

Closer examination of the events to be supported by the system will have to be undertaken.

The inconsistencies where there appear to be additional data elements or attributes in the data stores and entities have to be resolved. Appropriate changes should then be made to the models to ensure their consistency. Examples such as the data store `Animals` not containing a data element for the special notes should be relatively simple to resolve. Some of the other issues are more fundamental. Investigation of the attribute `offspring_animal_no` in the entity `Results_in` determined that an entity occurrence for the offspring animal is created in the entity `Animal`; within the DFD, the breeding data for the parents are recorded, but does the data flow `new_arrival` cover offspring as well as animals that are acquired by the park? Within the processing of orders a decision will need to be made on whether both the quantity ordered and quantity delivered are required, and whether the invoice details should be held. At present, in both cases the processing is supported on the DFD but the data is not held. Do we need to keep the details about the keepers to support the events proposed within the system? The maintenance of the food requirements for the animals appears to have been interpreted slightly differently between the two models – which is correct? The maintenance of the data regarding areas and enclosure appears to be more efficient in the E-R model – an argument in favour of normalization? Some of these issues will involve communication with the user to obtain clarification. Some are unlikely to arise if the same person develops both models, or if a CASE tool is used which encourages consistency.

Exercise 8.1

For the Albany Hotel case scenario:

(a) Prepare in tabular form the correspondences between the set of DFDs for the required (proposed) system, shown in Chapter 3, Figures 3.4a–c and Figures 3.3d–h, and the E-R model developed in Chapter 6, Exercise 6.9.
(b) For the data stores whose data elements have been defined in the DD in Chapter 4, Exercise 4.1, check that the data elements in the data stores

are consistent with the attributes in the well-normalized tables listed in Chapter 6, Exercise 6.9.

(c) Comment on the issues, identified in the two previous points, that need to be resolved to ensure that the DFDs and the E-R model are consistent and complete.

Correspondence between an E-R model and one or more ELHs

An ELH may be drawn for each entity in an E-R model. Each entity occurrence and the data items within it must be created, deleted and possibly amended by the events in the ELH. The event/entity matrix (EEM) shows the relationship between the E-R diagram and ELHs. The vertical view for each column displays the ELH view for the entity shown. As stated in Chapter 7, Worked example 7.1, the EEM highlights some omissions. Each entity must have at least one creation event and one deletion event. The EEM in Chapter 7, Figure 7.9 shows that the entity `Species` does not have a deletion event. Does this mean that the data is to be retained forever? This must be clarified with the user. The E-R diagram should also be checked against the EEM and ELHs to ensure that the events support the membership classes specified on the E-R diagram.

At the design stage the cross-checking between these two models will be undertaken in more detail. At that time the cross-checking may be performed by the use of a **transaction attribute grid**, well-documented in Howe (1989) and Skidmore and Wroe (1990). A single event may affect more than one entity. The EEM and the E-R model may be validated to ensure that the navigation path for an event is supported. This navigation path may be checked at the design stage using a **logical access map**, well-documented in Skidmore and Wroe (1990).

This cross-checking is undertaken in SSADM by the use of an **effect correspondence diagram** – see SSADM Reference Manuals (1990).

Correspondence between a set of levelled DFDs and one or more ELHs

As stated in Chapter 7, an ELH models all the possible events that can affect an occurrence of the entity focussing on the creation of an occurrence (birth event), the deletion of an occurrence (death event), and probably the modification of an entity occurrence (life event). These birth, life and death events on the ELH should correspond to data flows which update data stores on the DFDs. The presence of the data flows and the consistency of their contents may be checked via the DD. If an event is not supported by one or more data flows on the DFD, it may be that the data store is updated by an associated subsystem, in which case the event may be considered to be supported.

The cross-checking required here may be done in conjunction with that between the E-R model and ELHs. The events have been identified. The correspondence between the data stores and entities has been determined. The DD may be used to check that all the data elements within the data store are maintained. One of the steps when drawing DFDs was to check that a data store was not either read-only or write-only, and that items may be created and deleted. The cross-checking process between the ELHs and the DFDs should highlight any errors not found at that time. Each data store must either be maintained by the system under development or specifically by another associated subsystem.

Worked example 8.2

For the Somerleyton Animal Park case scenario: Examine the EEM and ELHs developed in Chapter 7, Worked example 7.1, and cross-check the events with the data flows on the set of DFDs for the proposed system shown in Chapter 3, Figures 3.13a–j to determine whether or not the events are supported.

Solution:

Events on EEM/ELHs	*Data flows on DFDs*
`New Animal Arrives`	`new_arrival`
`Animal Born in the Park`	`new_arrival?`
`Animal Exchanged`	`exchange_data`
`Animal Breeds`	`breeding_data`
`Animal Treated by Vet`	`treatment_data`
`Animal Reintroduced to the Wild`	`reintroduction_to_wild`
`Animal Dies`	`death`
`Animal Food Requirements Established`	`animal_food_requirement`
`Animal Returned to Another Park`	`exchange_data`

The modification of the entity `Species` denotes the establishment of the relationship between the entities `Animal` and `Species`. Within the DFD the data element `species_no` will be created within the data store `Animals`. See Worked example 8.1c.

See Worked example 8.1c for comment on the different interpretation of food requirements between the views.

A complete DD could be used to check that all the data elements within the data flows used to update the data stores are equivalent to the attributes within the events that update the entities.

Examination of the ELHs and DFDs reveals the following situation for the data stores:

- Animal Archives – write-only;
- Species – no deletion;
- Orders – no deletion of back orders;
- Suppliers – read-only; and
- Animal Foods and Food Type Stock – identified for further examination in Worked example 8.1.

The DFDs and DD need to be checked closely to ensure that they support the exchange of animals both to and from the Animal Park. For example, Chapter 7, Worked example 7.1 states in Assumption (9) that when an animal on loan to Somerleyton is returned to its original park, or passed on to another one, its record is archived. This is not supported on the DFD.

All the issues identified during the cross-checking process have to be resolved and changes made to the appropriate models.

Exercise 8.2

For the Albany Hotel case scenario:

(a) Examine the EEM and ELHs developed in Chapter 7, Exercises 7.4–7.8, and cross-check the events with the data flows on the set of DFDs for the proposed system shown in Chapter 3, Figures 3.4a–c and Figures 3.3d–h to determine whether or not the events are supported.
(b) Examine the data stores and identify where further clarification for their maintenance is required.

Correspondences between the DD and the other models

As has been stated, a number of the correspondences between the models are made through references to the DD. There are a number of further checks that may be made.

In a DFD, all data flows, data stores and their contents must be specified within the DD or they are considered to be undefined. There must be a **process specification** in the DD for all functional primitives in the DFD. All the inputs and outputs to the process specification must be defined in the DD, and shown on the DFD. All data flows, data stores and data elements referenced in the process specification must be defined in the DD. The process specifications should also contain the logical steps that would create and delete an occurrence of each entity in the E-R model.

Many of the specified checks demonstrate the checking of another model with the DD. Of course, the converse checks should also be true. For example, for each process specification in the DD there must be a corresponding functional primitive in the DFD.

As can be seen the DD itself, though it is not one of the three system views, is an important model underpinning all the others.

Use of CASE tools and prototyping

For fuller coverage on CASE tools and prototyping, see Martin and McClure (1988), McClure (1989), Spurr and Layzell (1992) and Vonk (1990).

As stated in Chapter 4, a CASE tool facilitates systems development. The cross-checking described in this chapter may be performed as the models are developed and balanced.

Entries for the component parts of the models (for example data flows) are entered into the DD as the models are created. Entries are made for the logical steps within process specifications. The CASE tool will check that the data flows, stores and elements referenced within the process specifications are referenced elsewhere in the DD. A model, once developed, may then be balanced. Any errors are highlighted by the CASE tool, then amendments may be made. The modification of one diagram has an effect on the others. In the exercises in earlier chapters you will have found that it is difficult to detect all the consequences of any changes you wish to make. A CASE tool will do this automatically. Most CASE tools have a facility for report generation from the DD, for example on the usage of a data element throughout the models. Reports of this nature are useful during the initial system development and during maintenance. The use of CASE tools also provides good documentation for the system which makes maintenance easier.

Prototyping is often used to model the user interface and system functionality during systems development. It is possible to develop a system making the best uses of prototyping without compromising on the use of the recognized modelling techniques described in this text. It would be possible to develop a top level DFD for the proposed system which would be used to help identify the areas to be used for prototyping. Prototypes could then be used to assist the requirements definition by modelling the user interfaces and specified areas of functionality. The authors suggest that E-R modelling should take place alongside prototyping, as prototyping gives no support to the modelling of data and the associated data structures required. ELHs can then be drawn for those entities considered to be complex.

Chapter 9

Approach to Case Scenario Assignments and Examinations

OBJECTIVES

In this chapter you will learn:

- how to approach a systems analysis assignment or examination.

Introduction

Students learning about systems analysis will find themselves having to complete assessed work. Systems analysis assignments and examinations typically use a case scenario similar to those that have been used throughout this text, and require the student to demonstrate their grasp of systems analysis techniques by applying them to this scenario.

This chapter offers some suggestions on the approach to take when faced with this task, and provides a further exercise with solutions for the student to practise the techniques that have been covered in this text.

Approach to an assessment

The case scenario

For example, see Chapter 1, the Somerleyton Animal Park case scenario, where the `Species Card` and the `Animal Record` contain duplicate information.

The case scenario should of course be read very carefully, several times. In particular, examine closely any sample documents that are provided. They have been included for a reason; they will contain clues about the data that needs to be stored and processed by the information system. They may also illustrate problems with the existing system.

The tasks

Study carefully the tasks you have been set to ensure that you understand exactly what is required. For example, if you are required to provide a process view of the system, have you been asked for one top level data flow diagram (DFD) or for a set of levelled DFDs? for a context diagram as well? for a current physical view, for a current logical view or for a required logical view? Have you been asked to model the information system for the whole organization that is described in the case scenario, or to model just part of it?

If the assessment is a group assignment the tasks need to be divided between the members of the group. It may be appropriate for different members to take on the modelling of the three different views, but allow time for the views to be cross-checked after each draft to ensure their consistency and completeness.

See Chapter 8.

Assumptions

It is difficult for systems analysis assessments to avoid being artificial. The 'real world' of a systems development project can be simulated within group assignments where a group of students form a systems development team and other individuals (usually teaching staff) play the role of clients and/or users of the information system in question. In this situation there are 'users' who can be interviewed and asked to clarify any ambiguities in or omissions from the written case scenario. Even so, the opportunities for discussion with the 'users' may be limited.

Other assignments, and all examinations, are likely to be carried out without the benefit of a user to interview. In these situations it is essential that you make and state whatever assumptions are necessary to support the models that you produce.

This is the procedure that has been adopted throughout this text (see Chapter 2).

There is rarely one 'right' answer; alternative solutions are often acceptable providing that they are supported by reasonable assumptions that are consistent with the case scenario.

Examination preparation

A case scenario can be quite a lengthy document, and for this reason part of it is often released in advance of an examination with further, perhaps more detailed, information being provided with the examination paper. This procedure is adopted so that the student has the opportunity to read and absorb the case scenario at leisure, away from the time constraints of the examination. The examination time can then be used more effectively by the candidate to demonstrate the ability to apply techniques rather than the ability to read a large document in a short space of time.

If you are issued with a case scenario in advance of the systems analysis examination to which it applies, there are several things that you might do to prepare yourself in the best possible way for the examination:

1. Read the scenario through carefully, several times, to ensure that you are thoroughly familiar with it. It is not the purpose of examiners to trick or mislead you; most details are likely to be relevant, although there might be some padding with superfluous information.
2. Sketch a top level DFD and a context diagram. Sketch a current physical DFD first and then logicalize it; this will ensure that you are prepared for

whatever view you may be required to produce in the examination. Consider whether there is enough information to explode any of the processes to a further level of detail, and sketch these DFDs if possible. Make a note of the areas where information is lacking; it may be provided with the examination paper, and you will be in a better position to absorb the extra detail if you have already given some thought as to what it might be. Look for references to the requirements for any new system which would need to be added to the current logical DFDs to produce a set of required logical DFDs.

3. Sketch an entity-relationship (E-R) diagram, selecting an identifier for each entity. Often the detail that is required to produce a complete E-R model is lacking in the part of the scenario released in advance; sample documents containing the data items necessary for detailed data analysis may be presented with the examination paper. Make a note of any key documents that are referred to, and look out for more information about these when you enter the examination. Meantime, make a list of as many attributes as you can find and allocate them to the entities. Even the production of an outline, incomplete, preliminary E-R model will have been a worthwhile exercise, providing a valuable basis for the work to be completed in the examination itself.

Even if the models you prepare in advance are not all asked for in the examination questions, your preparation will not have been in vain. The application of each technique provides valuable practice, and ensures that you are thoroughly familiar with the case scenario; this will save you time in the examination.

Use of guidelines

See Chapters 2, 3, 6 and 7.

When developing a set of DFDs, data dictionary (DD) entries, an E-R model or one or more entity life histories (ELHs), remember to use the guidelines introduced in earlier chapters.

The three groups of guidelines for DFDs indicate how to develop a current physical top-level DFD, how to convert the latter to a current logical DFD and how to develop a levelled set of DFDs. These guidelines may need to be adapted; for example, you may be required to develop a top-level current logical DFD which will require the combination of the guidelines presented in Chapter 2. The guidelines are intended to be guiding principles, not inflexible rules.

Exercise 9.1 will require the adaptation of the DFD guidelines.

Exercise 9.1

The timing associated with each question indicates the amount of time likely to be allowed for its completion within an examination. It is assumed that the GreenField case scenario, without Figures 9.1, 9.2 and 9.3, would have been released to the candidates in advance. The times have been included to enable you to practise under examination conditions.

For the GreenField Garden Centre case scenario (see below):

(a) Develop a top level DFD for the required system.
(b) Explode the process dealing with the maintenance of greenhouses to a further level of detail.

(Questions 1 and 2 – 50 minutes)

(c) From your answer to question (b), choose one of the functional primitives from the exploded DFD:

(i) write a data dictionary entry for two flows that either enter or leave that process.
(ii) write the process specification for that process.

(20 minutes)

(d) Develop an entity-relationship (E-R) model, consisting of an E-R diagram and a set of well-normalized tables.

(45 minutes)

(e) Draw an entity life history (ELH) for the entity type `Greenhouse Plant`. Your solution should include state indicators.

(30 minutes)

GreenField Garden Centre

Background

GreenField Garden Centre started from very humble origins some 10 years ago, when Hilary Green began to grow and supply herb plants to shops and other outlets. Her customers included health food shops and one or two garden centres, and she also set up a stall at fetes and craft fairs throughout each spring and summer.

After three or four years, Hilary's Herbs had begun to expand, with more and more outlets interested in the herb plants. Then the owners of a garden centre that had been one of Hilary's first customers retired and put the business up for sale. Hilary and a friend, Alec Field, raised sufficient capital and went into partnership as the new owners of the centre, which they renamed.

The GreenField business had become somewhat run down before Alec and Hilary took it over. The owners had been in ill health for some time before they finally retired and had put little energy into ensuring the variety or quality of their products. The new owners spent a couple of years improving the plant stock. With Hilary's Herbs as the core of the business, emphasis was placed on the supply of such products as organically-grown vegetable seedlings as well as ornamental garden plants. This approach worked well, and as the business started to pick up, at first slowly and then more rapidly, they continually increased the range of their products. The previous owners had not bothered to sell more than a few garden implements and compost in the small shop attached to the greenhouses. Alec and Hilary changed all that, greatly expanding the shop area and stocking it with a wide variety of goods such as books, fertilisers, gardening implements, garden furniture and barbecue equipment.

At first Hilary and Alec ran the business on their own, but they now employ several gardeners, three full-time staff to run the shop and to help with the office work, and extra help at the weekend. The past 18 months have seen a period of particularly rapid expansion. The current manual records have become cumbersome to maintain, and it is impossible to extract from them the information necessary to plan the future direction of the business. Realizing that not only further expansion but also the continued success of GreenField Garden Centre is in jeopardy unless something is done about this, Hilary and Alec called in a firm of consultants to advise on a computerized information system.

Greenhouses and Plants

The seedlings, plants and shrubs that are the backbone of the business are grown in several large greenhouses. The temperature and humidity in these are carefully monitored each day; each value is controlled so that it stays between an upper and a lower limit. These desirable limits vary among the greenhouses so that seedlings can be grown in the most appropriate conditions. Extra greenhouses have been added as the business has grown and there is still sufficient space for further expansion. From time to time the temperature and humidity limits of an individual greenhouse are altered. Each greenhouse contains a number of different plants; each type of plant can be grown in more than one greenhouse. Each of the greenhouses is under the charge of one of the more experienced gardeners. This person is responsible for completing the greenhouse record sheet (see Figure 9.1). Each week the plants that are ready for sale are removed and put on display for customers. This might be in the open-air part of the garden centre, but delicate plants are put into one of the two greenhouses that display goods for sale (the public is not allowed into the growing greenhouses).

Each day the supervising gardeners tell the others what work is to be done in the greenhouses for which he or she is responsible. The temperature and humidity are marked up on the greenhouse record sheet once each day. Values beyond the required limits are indicated by an asterisk on the sheet and action is taken by the supervisor. When the greenhouse plants have been watered, this is also indicated on the sheet. The supervising gardeners know from experience the volume and frequency of watering required for each type of plant. As new trays of seeds are sown and placed in the greenhouses a record is kept of the type of plant, when the seed was sown, how many trays were sown (all plants are grown in standard-sized trays) and in which greenhouses, how often the seedlings should be watered and when

Greenhouse Record Sheet

Greenhouse Number: 6

Filled in by: KM

Date	Temperature (min 15 max 20)	Humidity (min 60 max 90)	Watering completed	Notes
16-5-94	22 *	85%	✓	
17-5-94	19	80%	✓	
18-5-94	20	85%	✓	
19-5-94	23 *	85%		Temp. high for second time this week.

Figure 9.1 Greenhouse record sheet.

they should be ready for sale. Plants that sell well are sown several times throughout the year so that supplies are always available. It is up to the experienced gardeners to decide which greenhouses are suitable for growing particular plants.

The plant information is recorded on a card (see Figure 9.2), which is filed in the office in a card file, arranged alphabetically by the plant name. The gardeners complete these and pass them on to the office staff for filing.

Plant Record Card			
LUPINUS	(Lupin)		
Date Sown	No. of Trays	Greenhouse Nos.	Date Ready
31-3-94	10	5	28-4-94
18-4-94	12	4, 5	16-5-94
20-5-94	10	4	17-6-94

Figure 9.2 Plant record card.

For plants for which there is already a card in the file the staff transfer the details of the latest planting onto the existing card. Hilary started this system for her herbs and extended it when she and Alec took over GreenField. While she was the only one maintaining the cards the system worked well. However, the office staff who now have to help maintain the file sometimes use the Latin and sometimes the common name to file the cards. A system of cross-references was started, but under the pressure of work it has not been consistently maintained. Thus the card for a particular plant often cannot be located (the staff may not know the Latin name) and a duplicate card is made out. This means that frequently it is impossible to find out important information such as how many batches of seeds have been sown for a given plant, or when the next batch will be ready for sale. Sometimes the date a batch is due to be ready for sale has to be amended, and the gardeners do not always remember to tell the office staff that this has happened.

As the number of greenhouses and gardeners increased the latter found it more convenient to fill out a 'growing history' sheet (see Figure 9.3), which is kept in the greenhouse on the same clipboard as the greenhouse record sheet. The gardeners' experience will determine when the seedlings are ready for sale, and the date they are transferred to the public areas for sale is recorded. Under pressure of work the gardeners often omit to fill out a plant record card for the office staff as well. All the trays planted at the same time are removed together. Very occasionally the plants die, for example as a result of blight, and the contents of all of the trays that were planted at the

Growing History

Greenhouse Number: 5

Plant	Date Sown	No. of Trays	Removed for Sale
Lupin	31-3-94	10	29-4-94
Lobelia	31-3-94	25	14-5-94

Figure 9.3 Growing history.

same time are destroyed to ensure that the problem has been eradicated.

The growing history sheets are kept, with the intention of using them to derive information such as the totals of plants grown. But on the rare occasions Alec tried to do this, he was discouraged to find the data relating to one type of plant scattered among several sheets, and it did not tally with the corresponding plant record card.

Some plants, such as the more exotic varieties and trees, are bought in rather than grown from seed. The names, addresses and telephone numbers of suppliers and their price lists are kept in a folder.

Shop

Customers bring the plants they have chosen to the till in the shop for payment. Currently, the price of each plant is derived in a rather arbitrary fashion by Alec. He would like to have more information on which to base his prices, such as the cost of materials (for example, seed and compost) and the length of time it takes for seedlings to reach the stage where they are ready for sale.

The shop also stocks a wide variety of related products. A shop manager, under Hilary's supervision, is responsible for ordering stock, using the supplier catalogues. All orders, including orders for the plants that are bought in, are written out on a standard two-part GreenField order form. The top copy is sent to the supplier and the carbon copy is placed in an orders file. When goods and invoices are received the shop staff unpack them, check them against the copy of the order form and pass any invoices on to Hilary or Alec to authorize payment. Any mismatch between what has been ordered and what has been supplied is queried with the supplier. The old order forms are filed in an 'orders fulfilled' file where they are kept for two years.

Requirements

The requirements of the new system include:

1. To record data on plants, including their common and Latin names and watering instructions.
2. To maintain accurate records of the plants grown, including the date the seeds were sown, the number of trays, in which greenhouse, the date they were estimated to be due to be ready for sale and the date they were actually removed for sale. These records are to be kept for one year.
3. To maintain accurate records of the greenhouses, including the minimum and maximum values of temperature and humidity permitted in each one, the actual readings taken each day, whether watering has been carried out each day and any special notes made each day.
4. To maintain details of suppliers of the products purchased for the shop and for use in the greenhouses.
5. To record data about the orders placed for products, including those for plants that are bought in, about the deliveries and invoices received and about the payments made.
6. To aid the pricing of plants, by providing data about the length of time each type of plant takes from sowing to being ready for sale, and also about the cost of materials purchased, such as seeds and compost.
7. To generate a variety of reports for Hilary and Alec, for example detailing the number of trays sown for specified plants over time, which

greenhouses have exceeded their daily temperature or humidity limits and the totals of orders and invoices dealt with each month.

Summary

This chapter has offered suggestions for how to approach a systems analysis assignment or examination. Any such exercise is inevitably an artificial one; there is no substitute for the experience to be gained from working on a live systems development project. Nevertheless, such exercises are an essential step in the mastery of the techniques described in this text. Once proficient in these techniques, the systems analyst will be well equipped to tackle the complex, challenging, but often rewarding and ever engrossing task of developing an information system.

Chapter 10

Solutions to Exercises

Chapter 2

Solution to Exercise 2.1

1. The process Make Decision has no identifier.
2. The data flow between processes (2) and (3) has no label.
3. The terminator UCAS appears on the DFD twice but no duplication mark is shown.
4. The data flow between data stores Applications and Rejections is not permissible – it must pass through a process.
5. A duplication mark should appear on all occurrences of the data store Applications.
6. The data flow between terminator UCAS and data store Applications is not permissible – it must pass through a process.
7. The data flow between the process Make Decision and the terminator UCAS has no label.

Notes:

1. The data flows between the process Invite To Interview and the data store Applications are acceptable. The unlabelled one indicates that the entire packet of data is passed to the process. However only the interview date and time are added to the application. Similarly, for the data flows between the process Make Decision and the data store Applications.
2. If a CASE tool is used, most of the syntax errors will not be allowed.

Solution to Exercise 2.2

Inputs and outputs	*Sources*	*Recipients*
property_details	Client	
requirements	Potential Buyer	
possible_properties		Potential Buyer
invoice		Client
payment	Client	
reminder		Client
sale_confirmed	Buyer	

Buyer could remain as potential buyer.

See Figure E2.2.

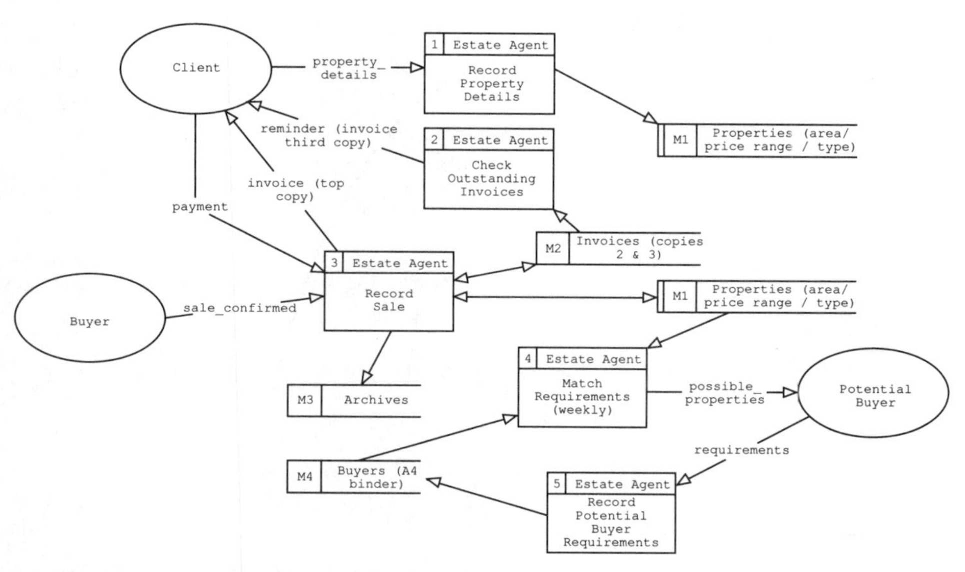

Figure E2.2 Solution to Exercise 2.2.

Assumptions

1. Clients will pay on receipt of the first reminder!
2. Maintenance of the Buyers file will be considered at a later date.

Solution to Exercise 2.3

See Figure E2.3a and b.

Solution to Exercise 2.4

See Figure E2.4.

Solution to Exercise 2.5

Guideline step 1: See Figure E2.5.

Guideline steps 2 to 4: Processes 1 and 2 will be combined – the data remains unaltered by process 1. The process `Despatch Goods` is changed to `Record Goods Issued` as it is less physical and concentrates on the transformation of data. Similarly the data store `Sales Ledger` becomes `Sales Accounts`, although some would argue that ledger has become an acceptable accounting term and does not imply 'large books and quill pens' – therefore this change is optional; the process `Print Invoice Set` becomes `Prepare Invoice Details`; and the data store `Credit List` becomes `Creditors`.

The processes `Prepare Invoice Details`, `Post Accounts` and `Prepare Statement` could also be combined.

See Chapter 3 ('Combining processes') for further discussion on this.

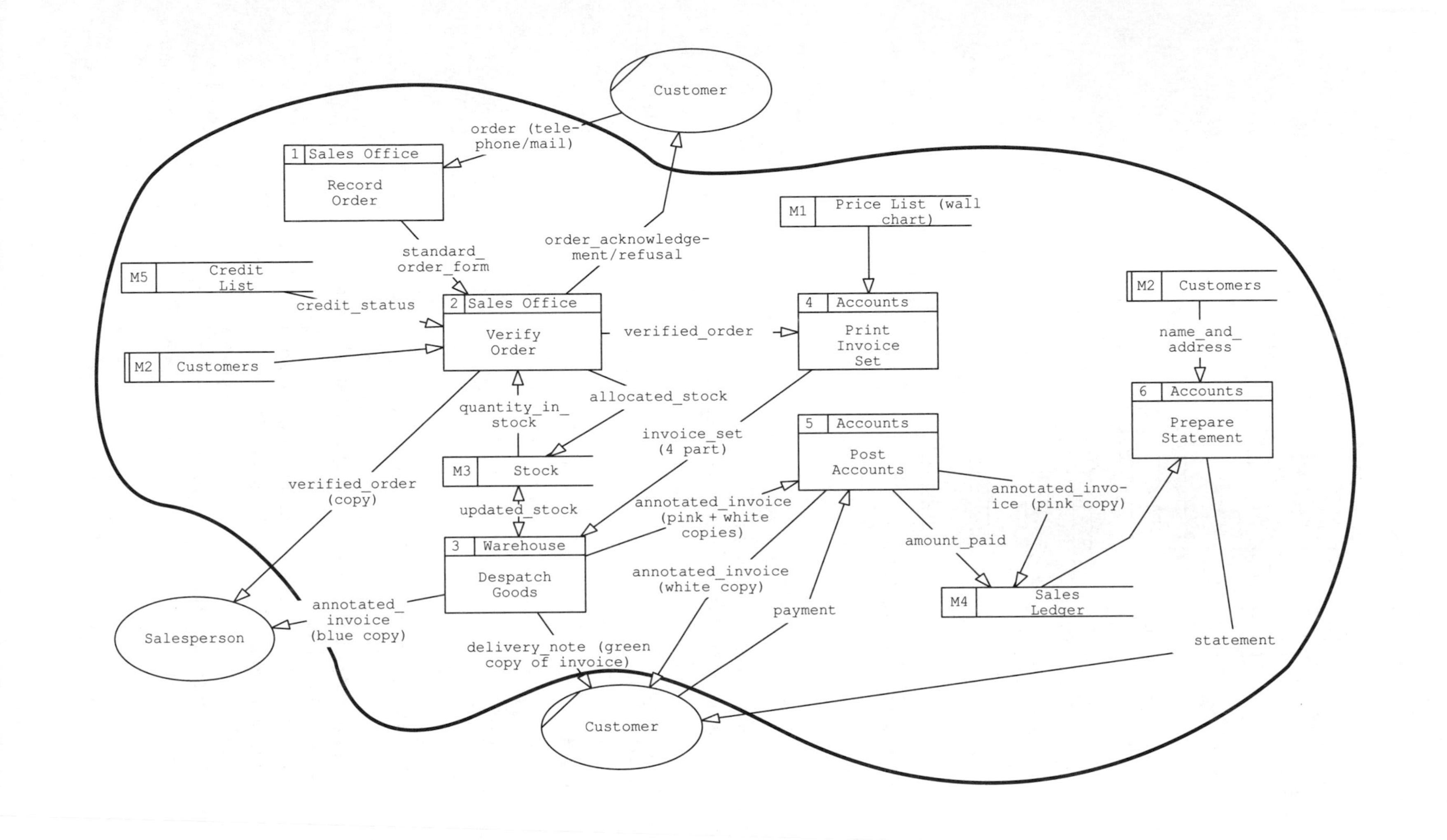

Figure E2.3 (a) Solution to Exercise 2.3.

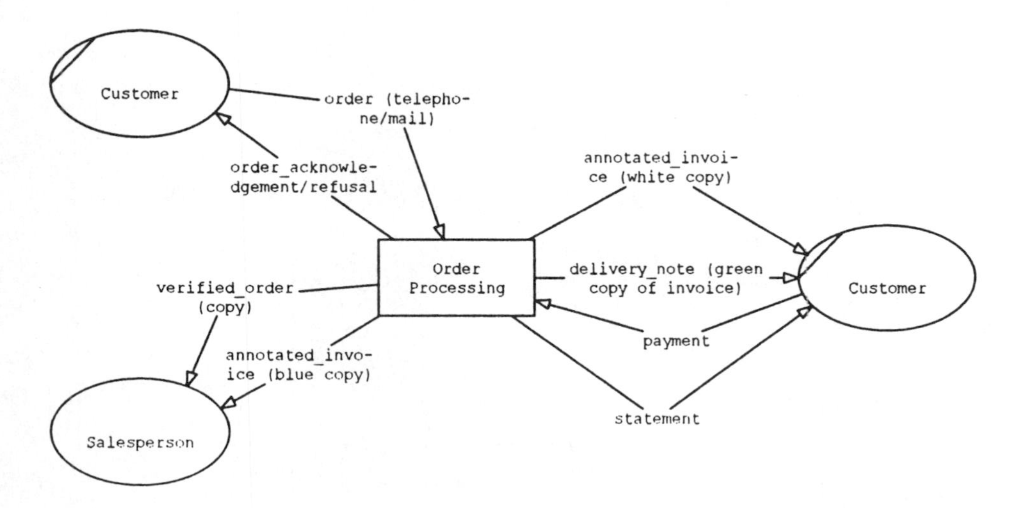

Figure E2.3 (b) Solution to Exercise 2.3.

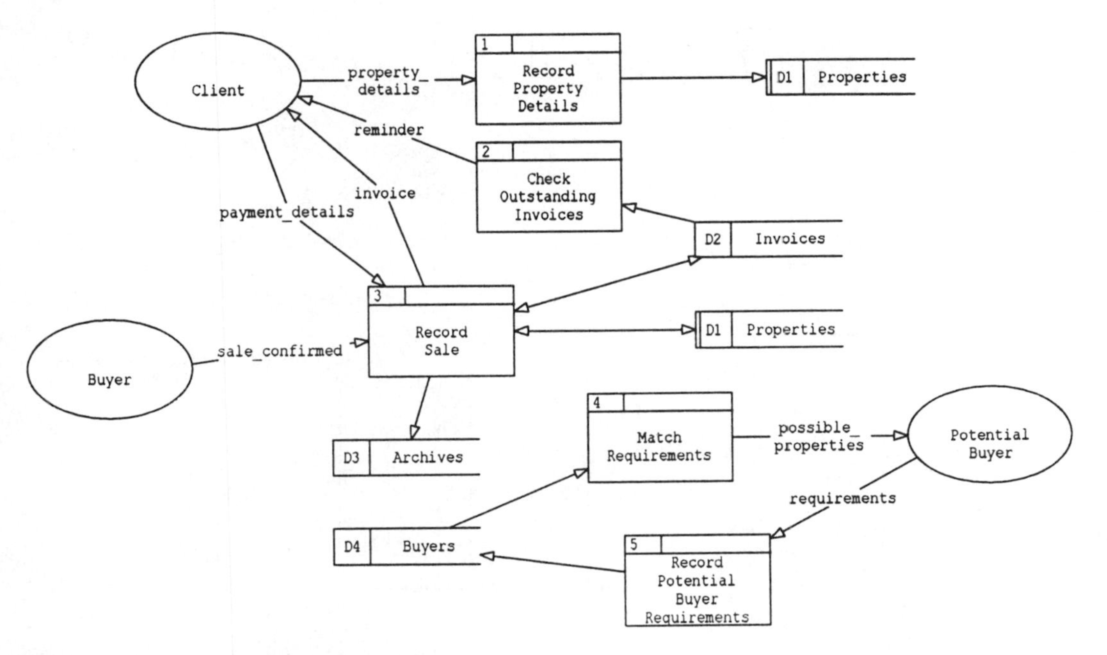

Figure E2.4 Solution to Exercise 2.4.

In this case, the changes required are the removal of the physical references to telephone/mail and copies of documents; and the word `details` has been added to `annotated_invoice` and `payment`.

Guideline steps 5 to 7: No further changes at present.
The context diagram should be checked and amended if necessary to reflect any changes made.

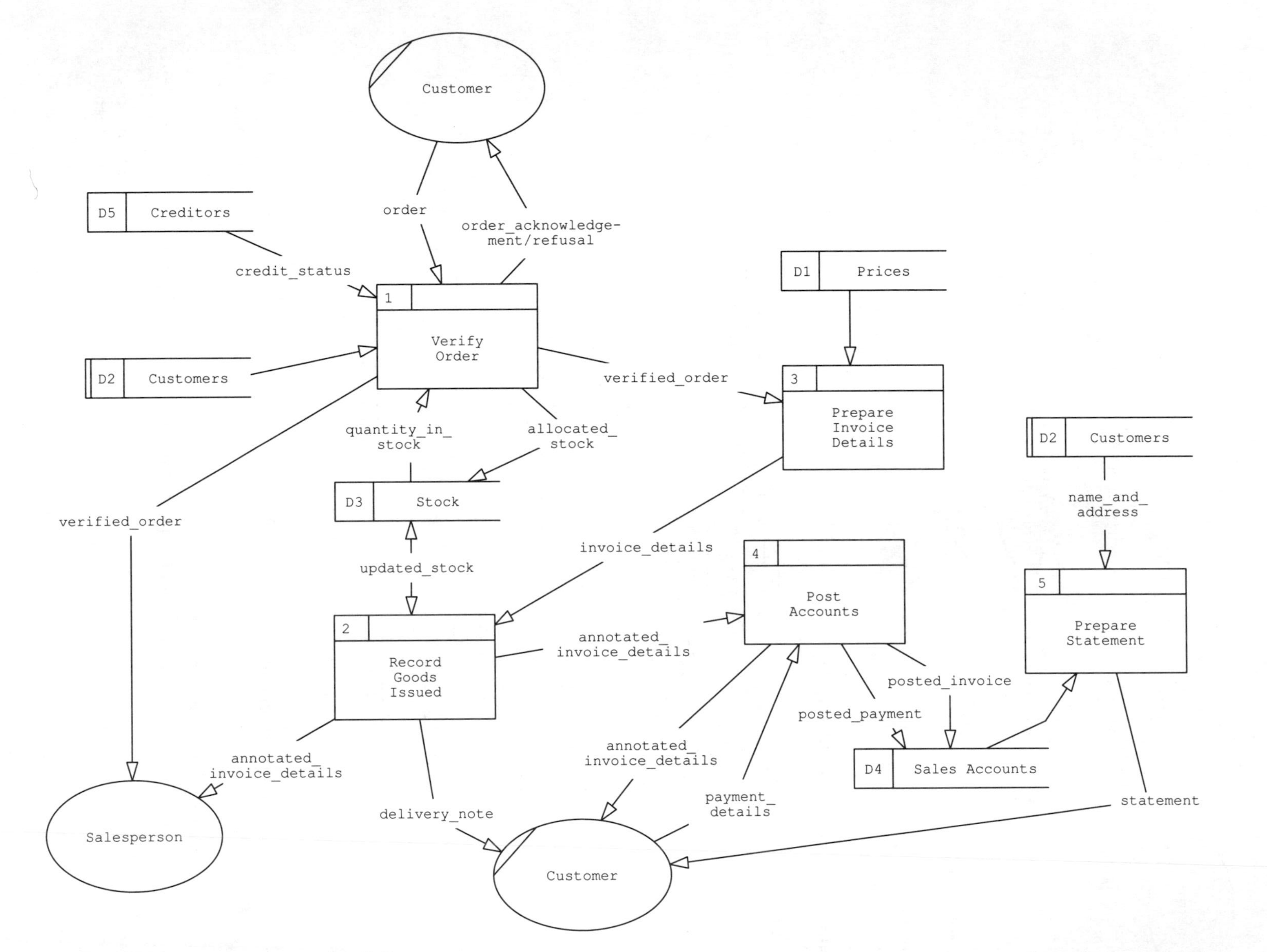

Figure E2.5 Solution to Exercise 2.5.

Solution to Exercise 2.6

(a) Current physical top level DFD

After guideline step 2.

Inputs and outputs	*Sources*	*Recipients*
telephone_enquiry	PotentialGuest/Guest	
enquiry_response		PotentialGuest/Guest
telephone_booking_request	Potential Guest/Guest	
non_availability		Potential Guest/Guest
verbal_provisional_booking		Potential Guest/Guest
booking_form/deposit_request	Potential Guest/Guest	
signed_booking_form	Potential Guest/Guest	
further_information_request		Potential Guest/Guest
further_information	Potential Guest/Guest	
letter_of_confirmation		Potential Guest/Guest
method_of_payment	Potential Guest/Guest	
alternative_method_of_payment_request		Potential Guest/Guest
alternative_method_of_payment	Potential Guest/Guest	
credit_details		Credit Card Company
credit_status	Credit Card Company	
signed_docket	Restaurant/Coffee Shop	
minibar_sales	Room Bursar	
account_request	Potential Guest/Guest	
guest_account		Potential Guest/Guest
settlement	Potential Guest/Guest	
receipt		Potential Guest/Guest

As stated earlier, you might need to return and repeat these and subsequent steps before all the inputs and outputs are identified. The solution given includes all the inputs and outputs.

Check your solution carefully with that given. There is more than one correct solution – doubtless different labels will be given to components, and/or the number of processes will differ.

Further details on the drawing of DFDs will be given in Chapter 3.

After Guideline step 5.

See Figure E2.6a1.

Assumptions

1. Other methods of payment are always provided.
2. Booking cancellations/changes are not included.

After Guideline step 9.

See Figure E2.6a2.

Note that the data store `Room Rates` is read-only. Further discussion will take place on read-only data stores in Chapter 8.

(b) Context diagram

See Figure E2.6b.

(c) Current logical top level DFD

See Figure E2.6c1.

See Figure E2.6c2 for revised context diagram.

For (b) and (c) the solutions given depend upon the solution to (a).

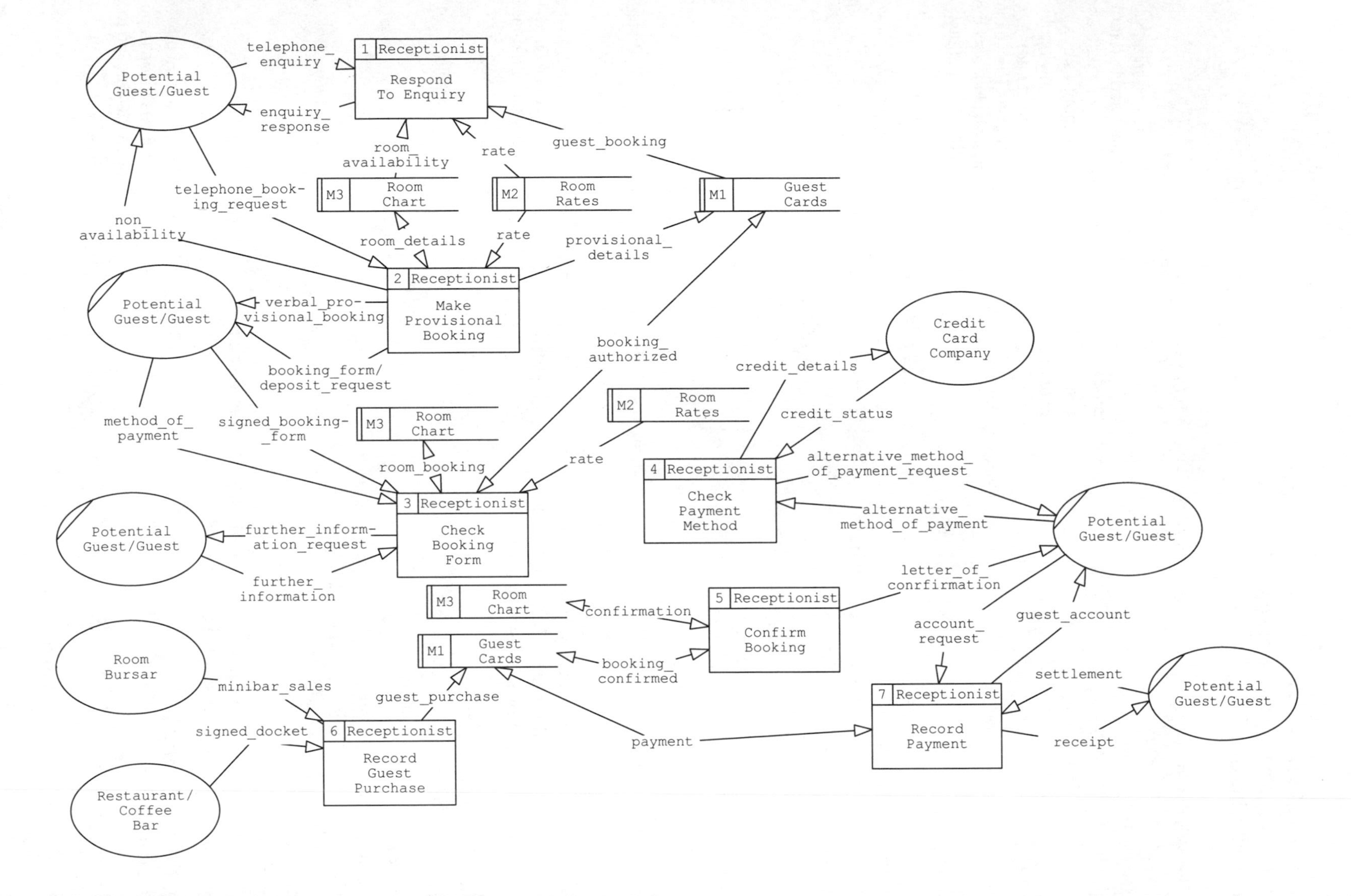

Figure E2.6a1 Solution to Exercise 2.6a after guideline step 5.

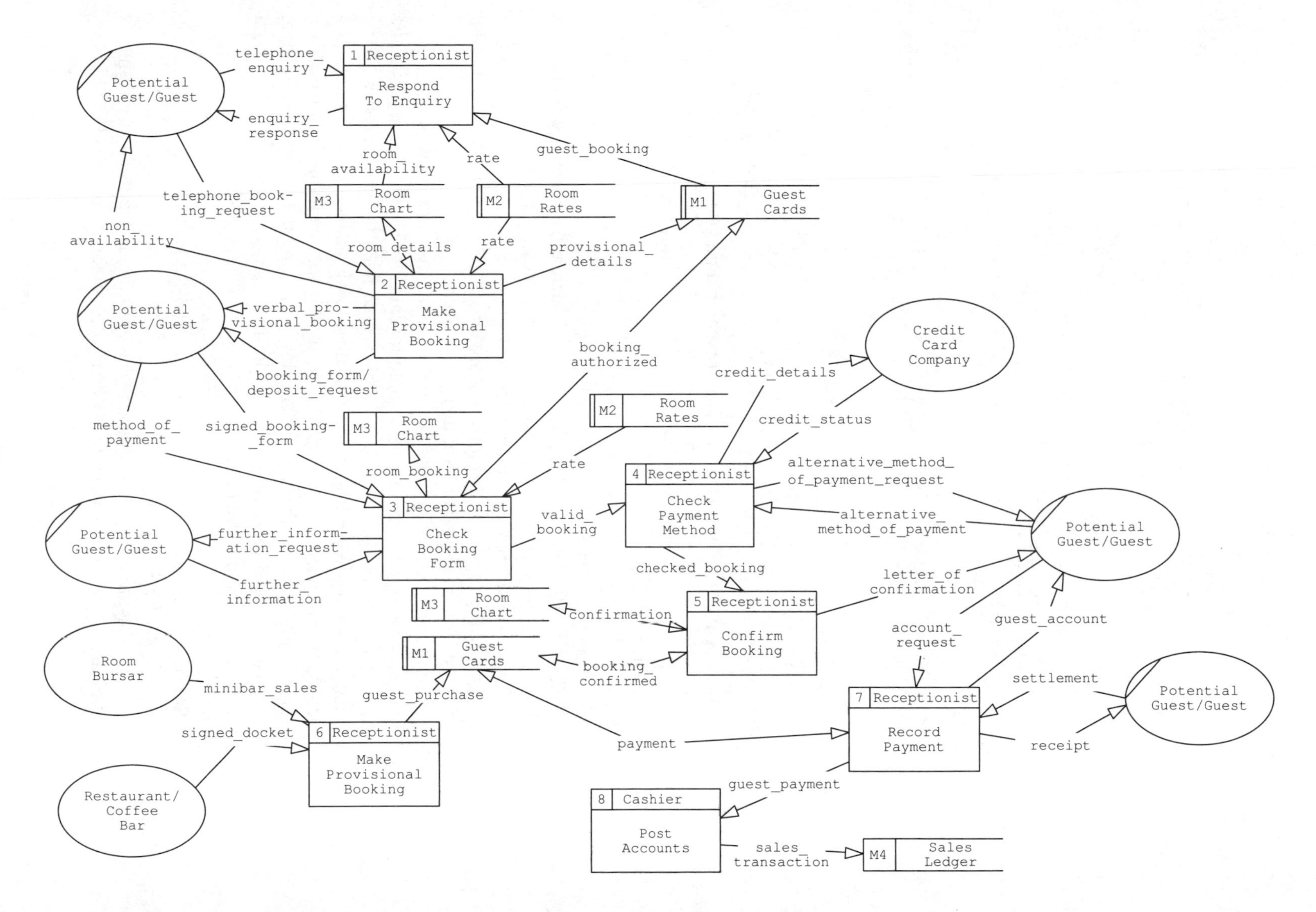

Figure E2.6a2 Solution to Exercise 2.6a after guideline step 9.

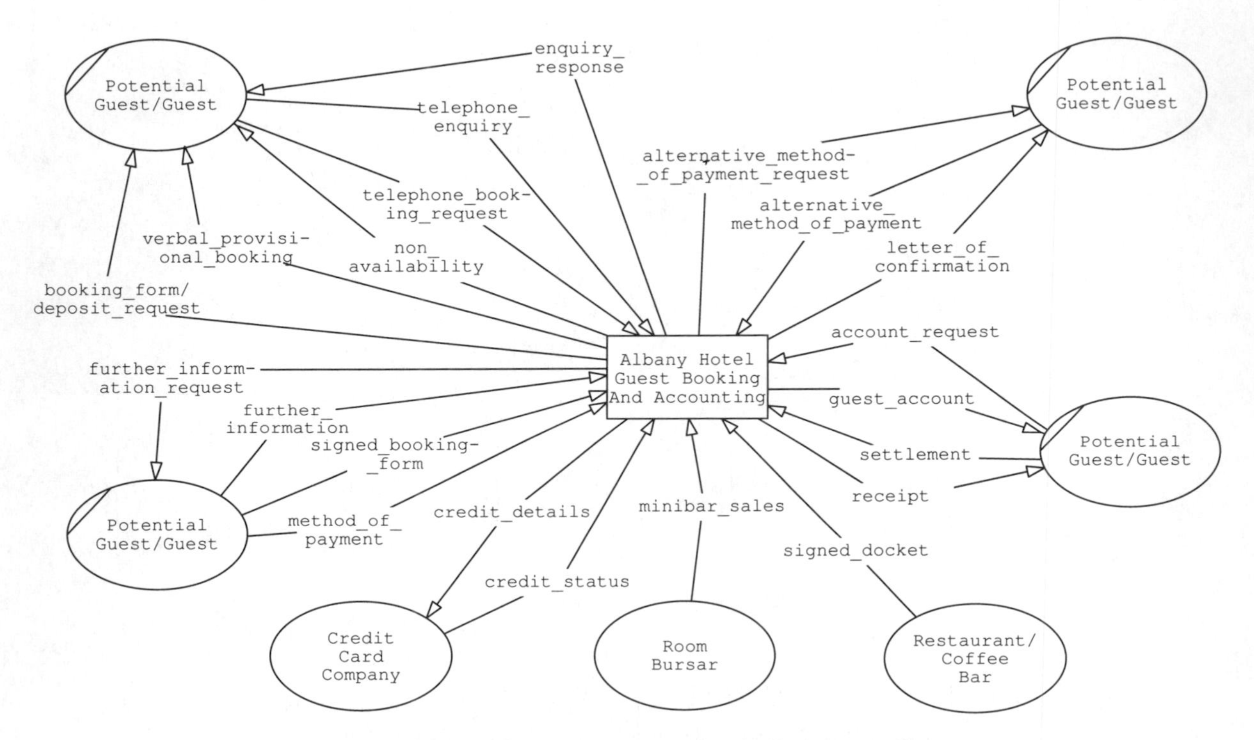

Figure E2.6b Solution to Exercise 2.6b context diagram.

Chapter 3

Solution to Exercise 3.1

Note: The data store Customer Accounts is outside the boundary frame as it is referenced by more than one higher level process.

The child diagram is numbered correctly.

On the higher level diagram, there are ten data flows entering and leaving the process; on the lower level there are eleven.

On further checking, on the higher level the data flow amount_paid seems as though it might be a generic data flow for deposit_paid and balance_paid on the lower level. This would need to be checked with the data dictionary to see if it is a case of dictionary balancing. If it is not, then changes will need to be made to ensure that the diagrams balance. Also, on the higher level there is a data flow balance, and on the lower level payment. Should these be the same? In this case, yes, and one of the data flow names should be changed.

Solution to Exercise 3.2

See Figure E3.2.

The data stores Customers and Stock are outside the boundary frame as they are referenced by more than one process at the higher level.

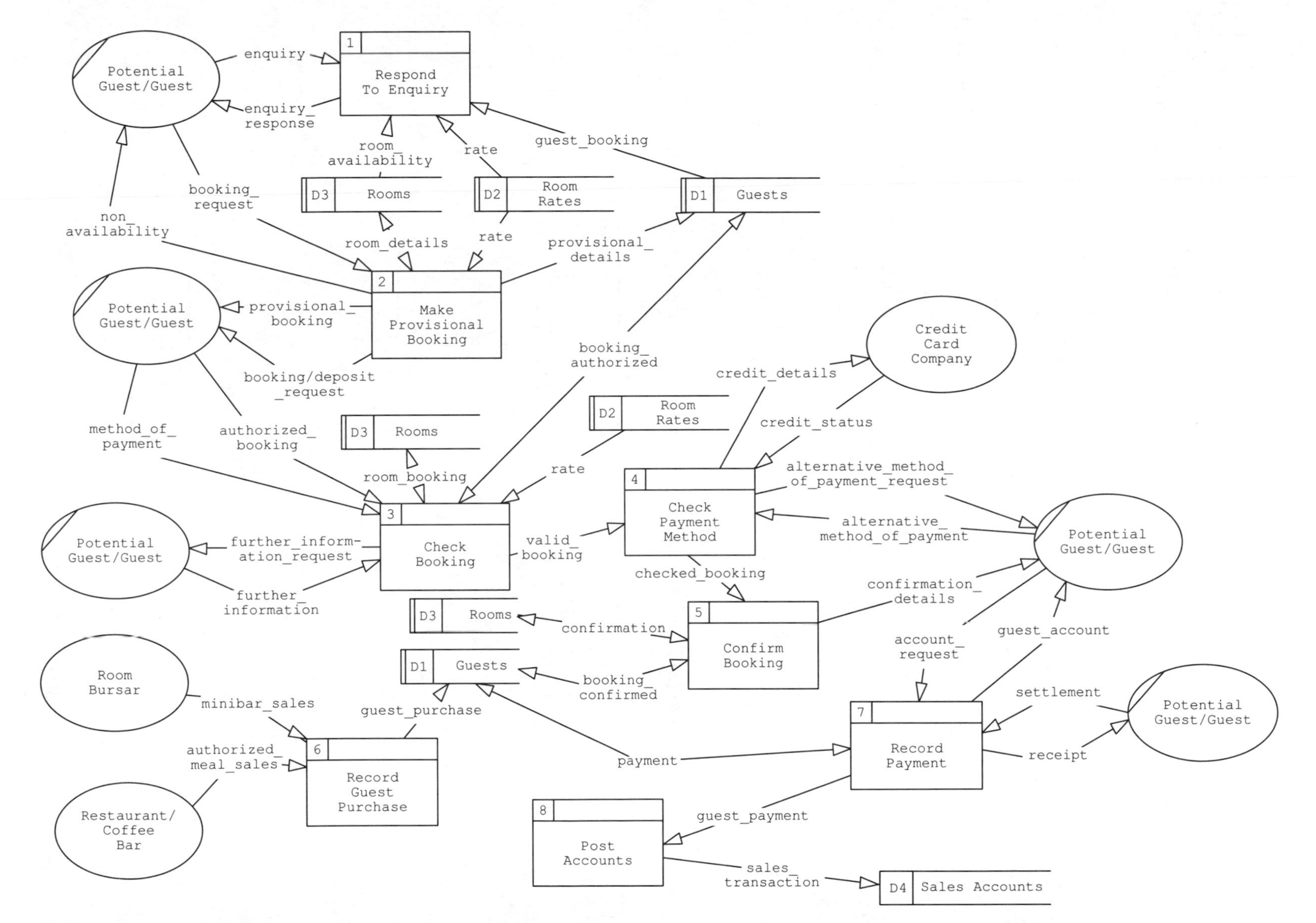

Figure E2.6c1 Solution to Exercise 2.6c current logical DFD.

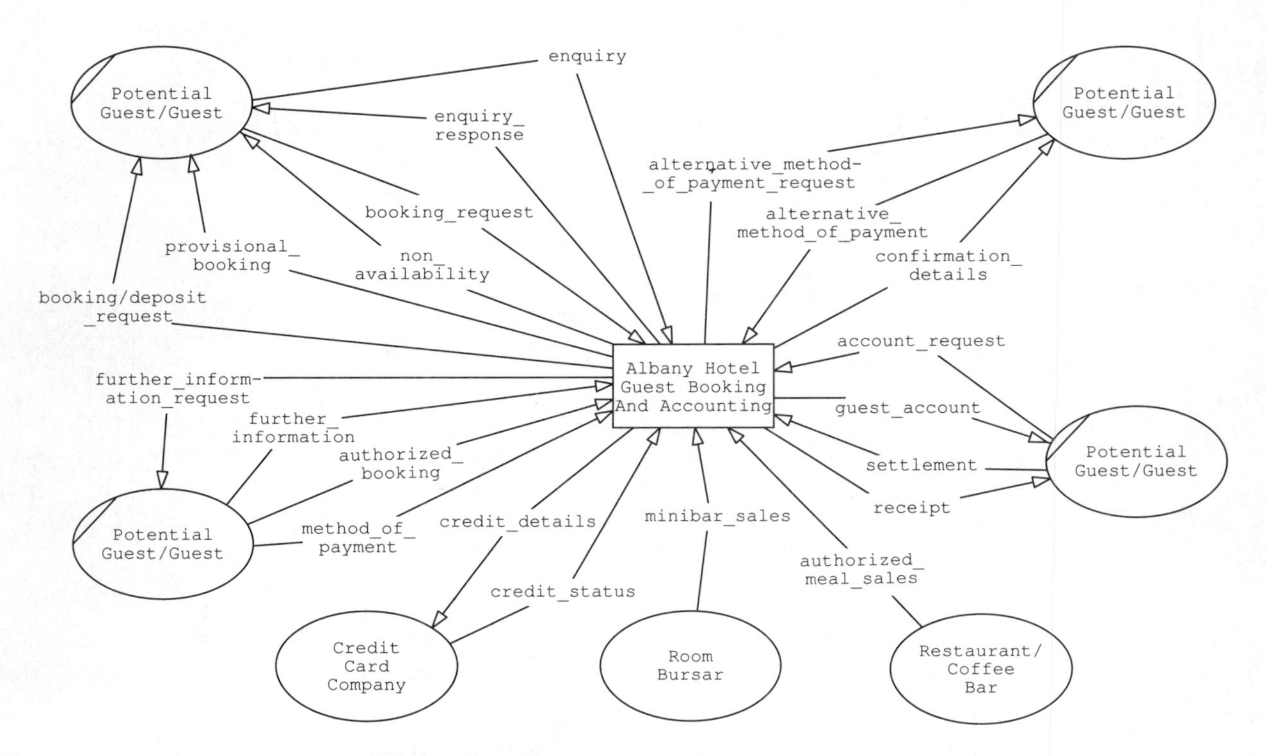

Figure E2.6c2 Solution to Exercise 2.6c context diagram.

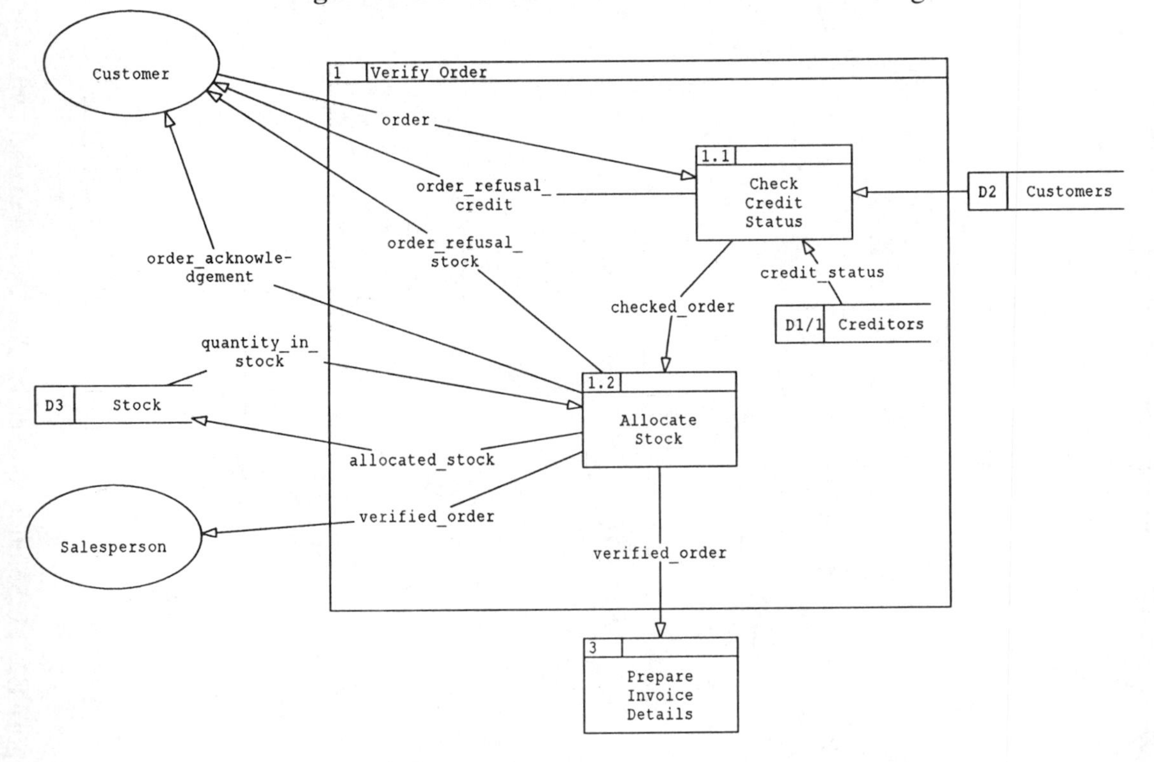

Figure E3.2 Solution to Exercise 3.2.

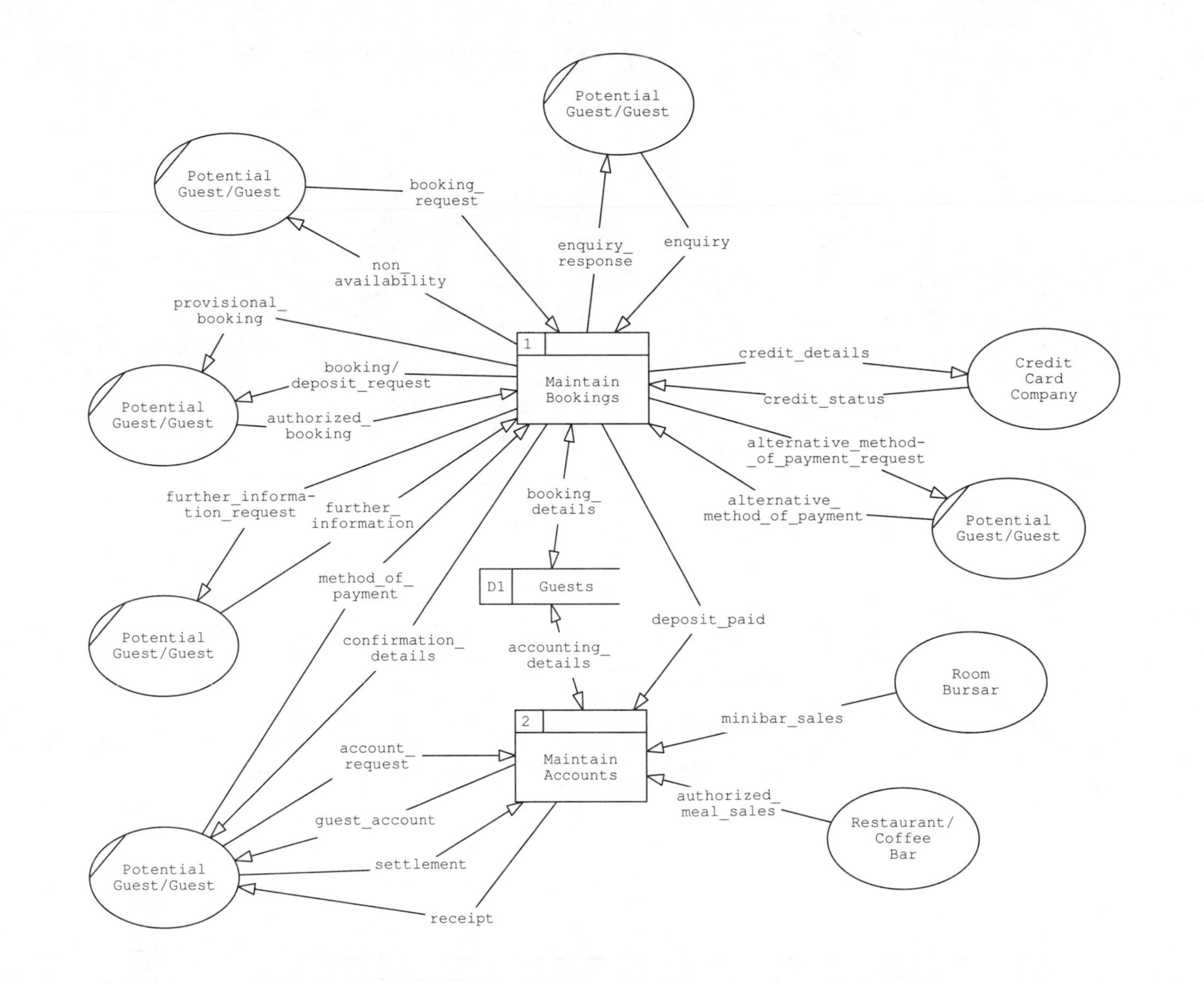

Figure E3.3 Solution to Exercise 3.3. (a) Top level diagram.

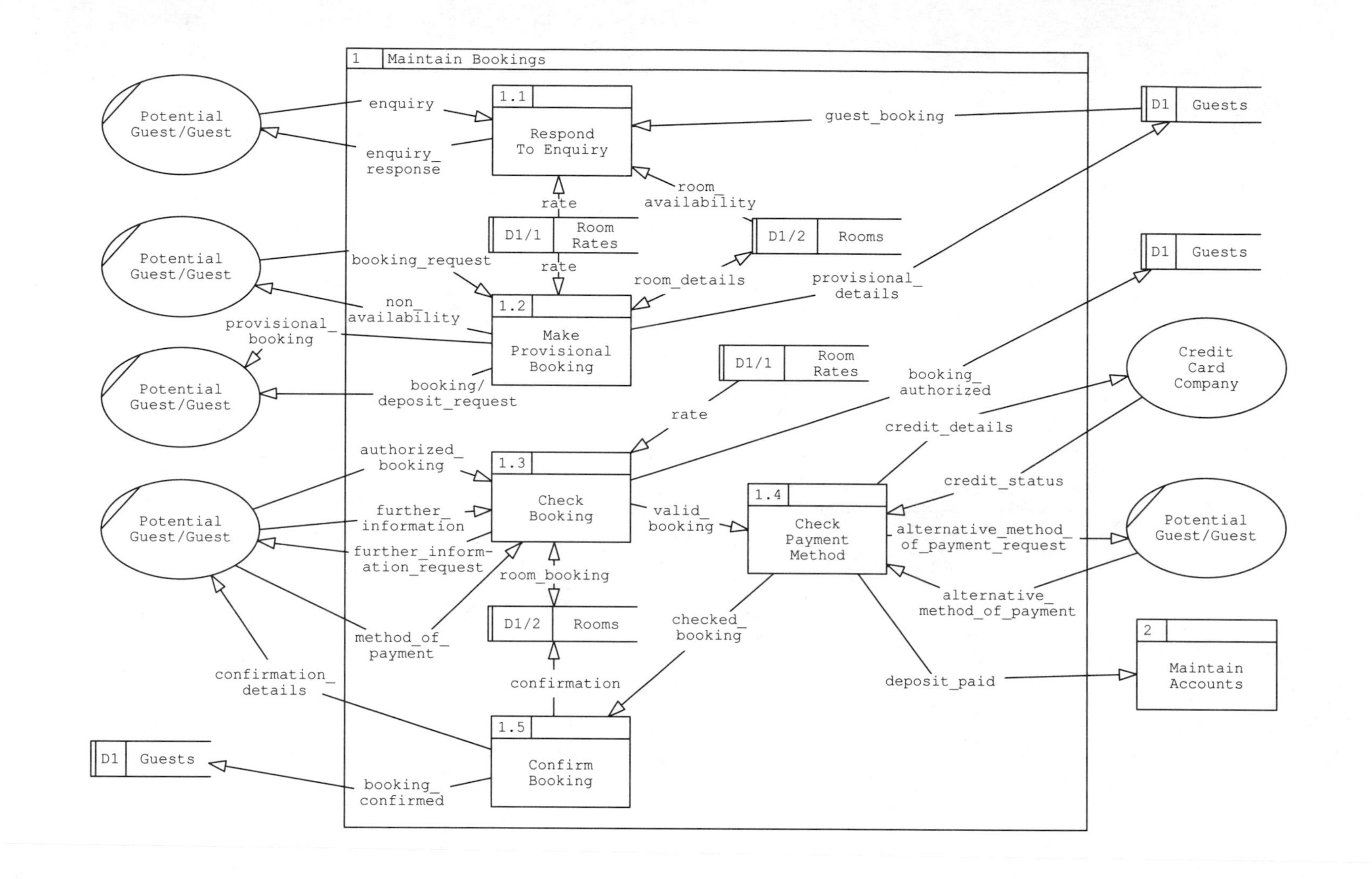

Figure E3.3 (b) Lower level diagram for process `Maintain Bookings`.

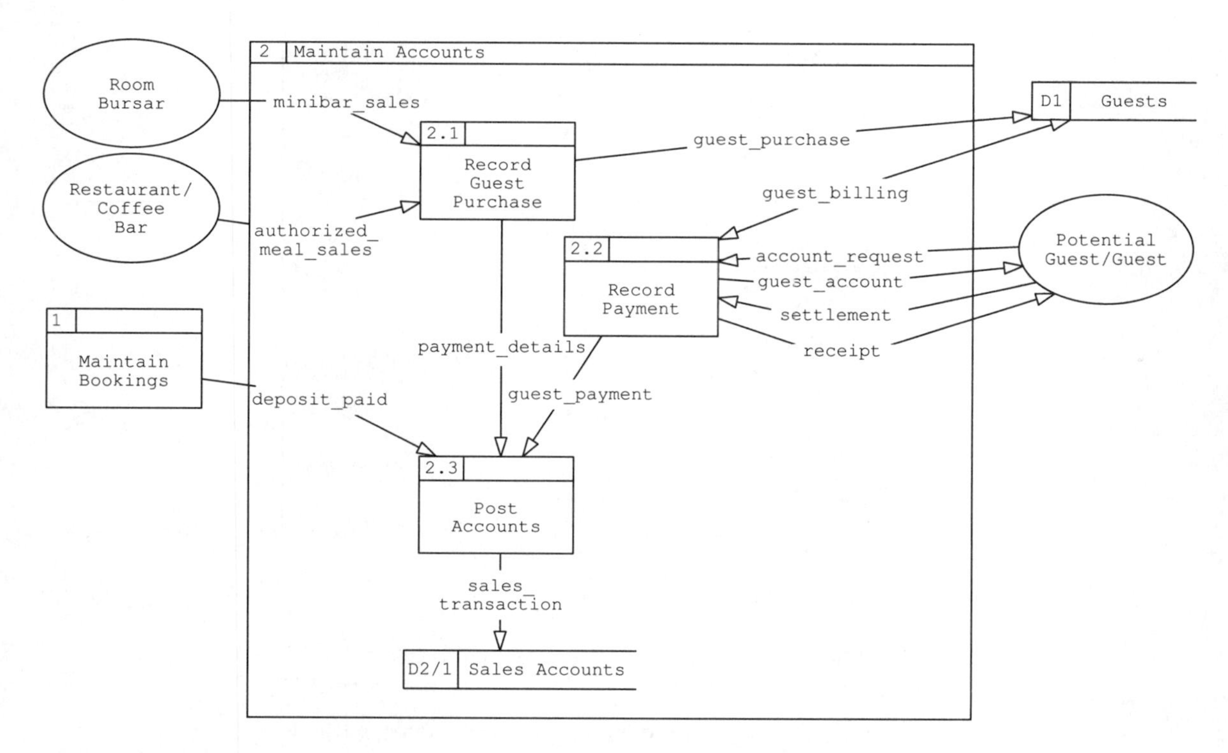

Figure E3.3 (c) Lower level diagram for process `Maintain Accounts`.

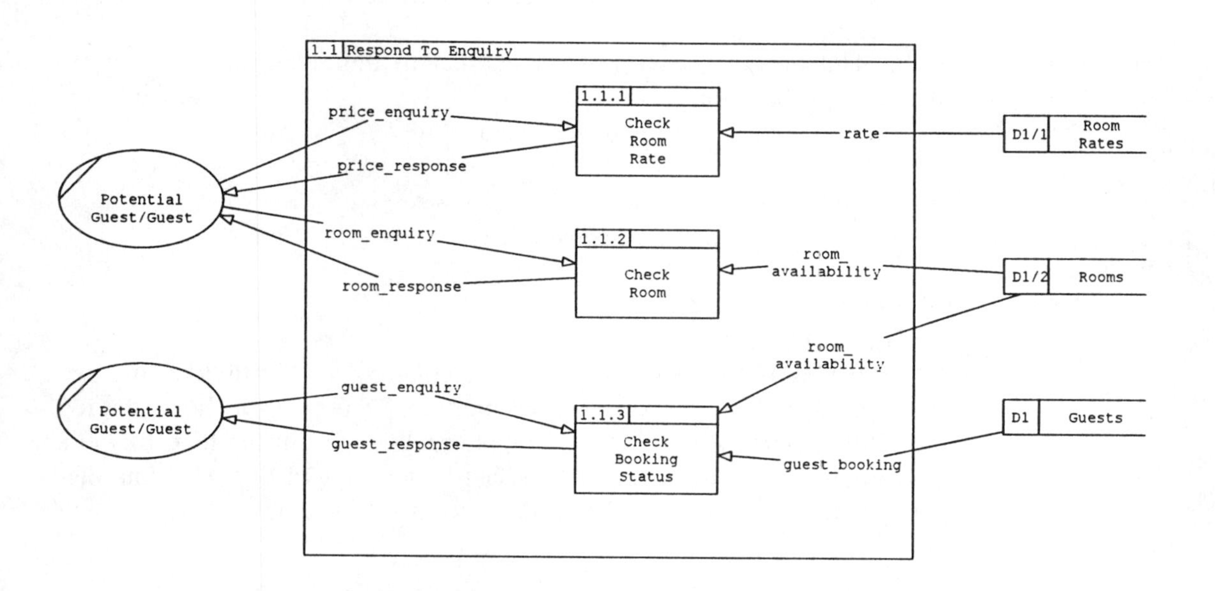

Figure E3.3 (d) Lower level diagram for process `Respond to Enquiry`.

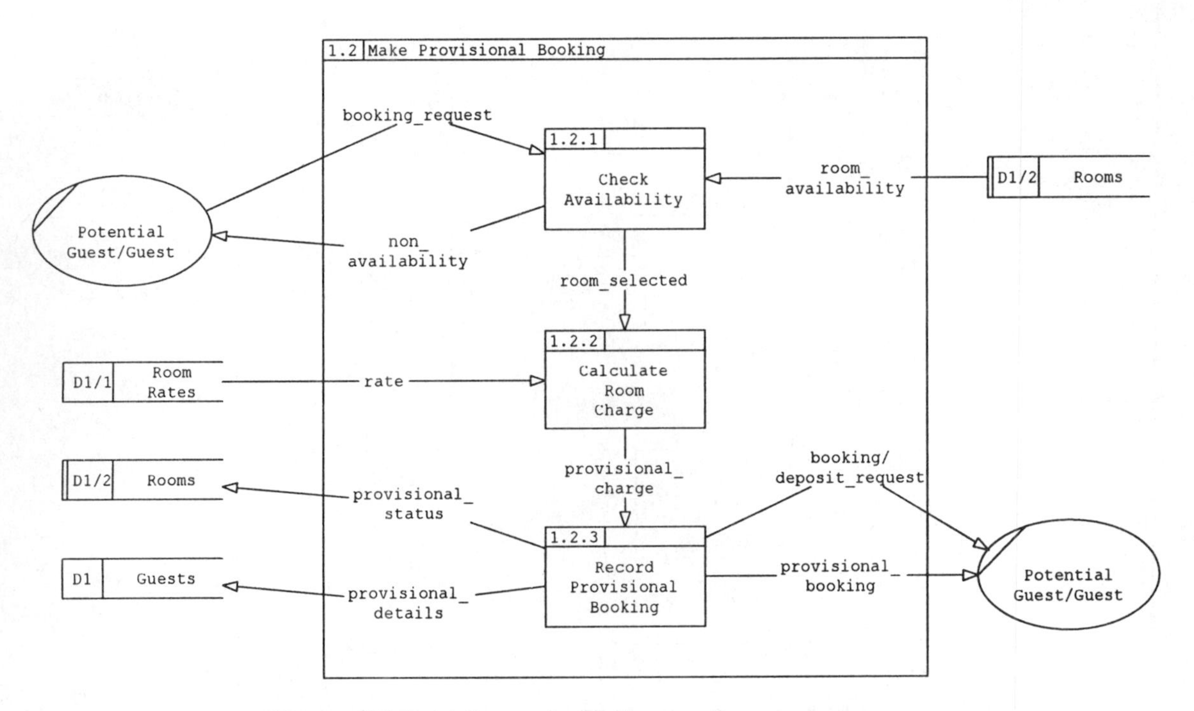

Figure E3.3 (e) Lower level diagram for process `Make Provisional Booking`.

The data store `Creditors` is inside the boundary frame as it is only referenced by process 1 at the higher level, and may be removed from the higher level diagram.

The data flow `checked_order` is internal to this level and is therefore unaffected by balancing.

The data content will be the same as for the data flow `verified-order` but the checked order has a different status when it is verified, and is given a different data flow name.

A data dictionary entry will be needed:

```
order_acknowledgement/refusal = [order_acknowledgement |
order_refusal_credit | order_refusal_stock]
```

Solution to Exercise 3.3

See Figures E3.3a–h.

The Albany Hotel guest booking and accounting system divides into two natural subsystems – booking and accounting. Figure E3.3a shows the top level DFD depicting these two subsystems. As stated in the text, this is a common approach for larger systems but is not compulsory. The data dictionary entry for the generic data flow `booking_details` is:

```
booking_details = [guest_booking | provisional_details |
                    booking-authorized | booking_confirmed]
```

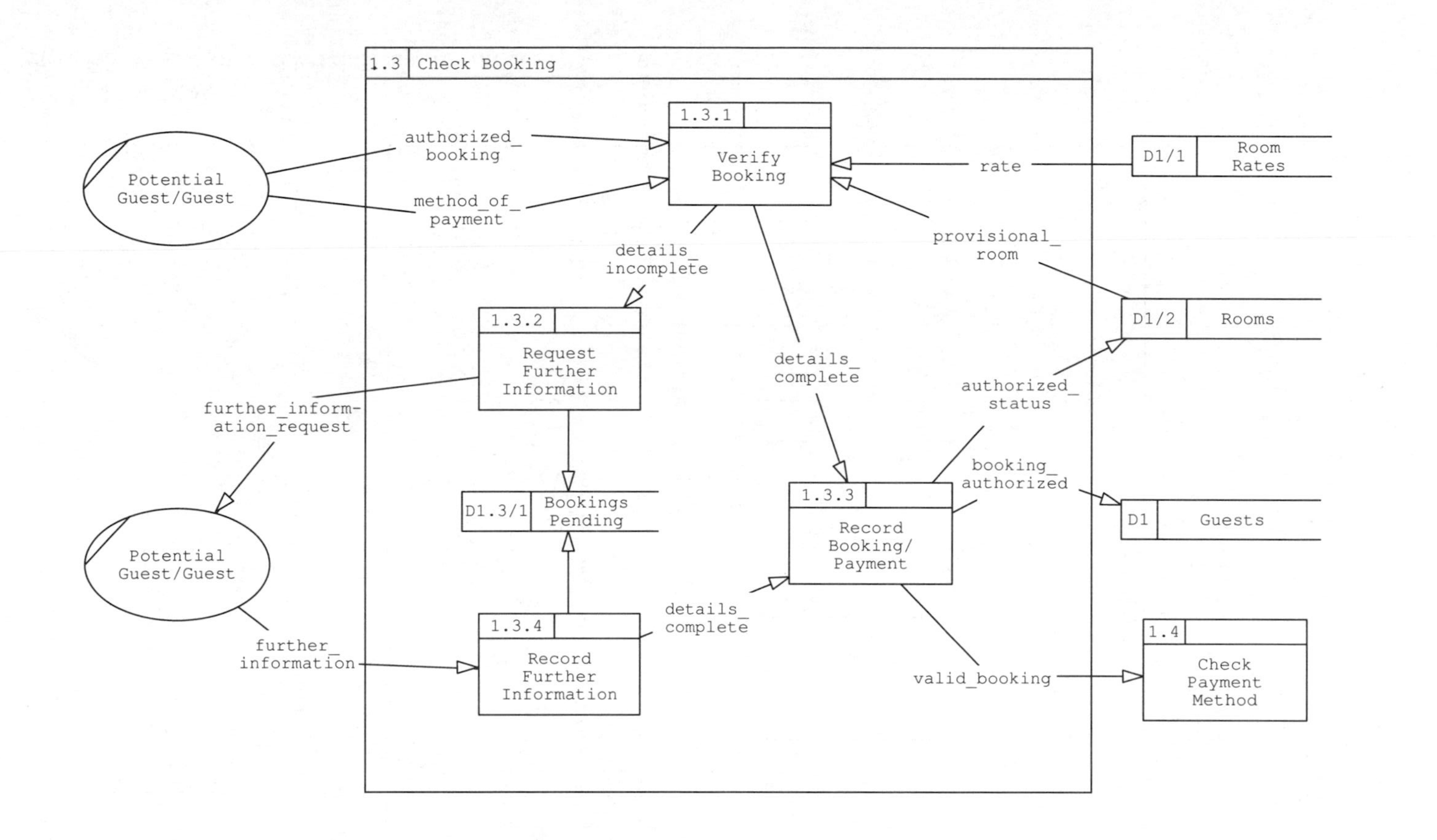

Figure E3.3 (f) Lower level diagram for process `Check Booking`.

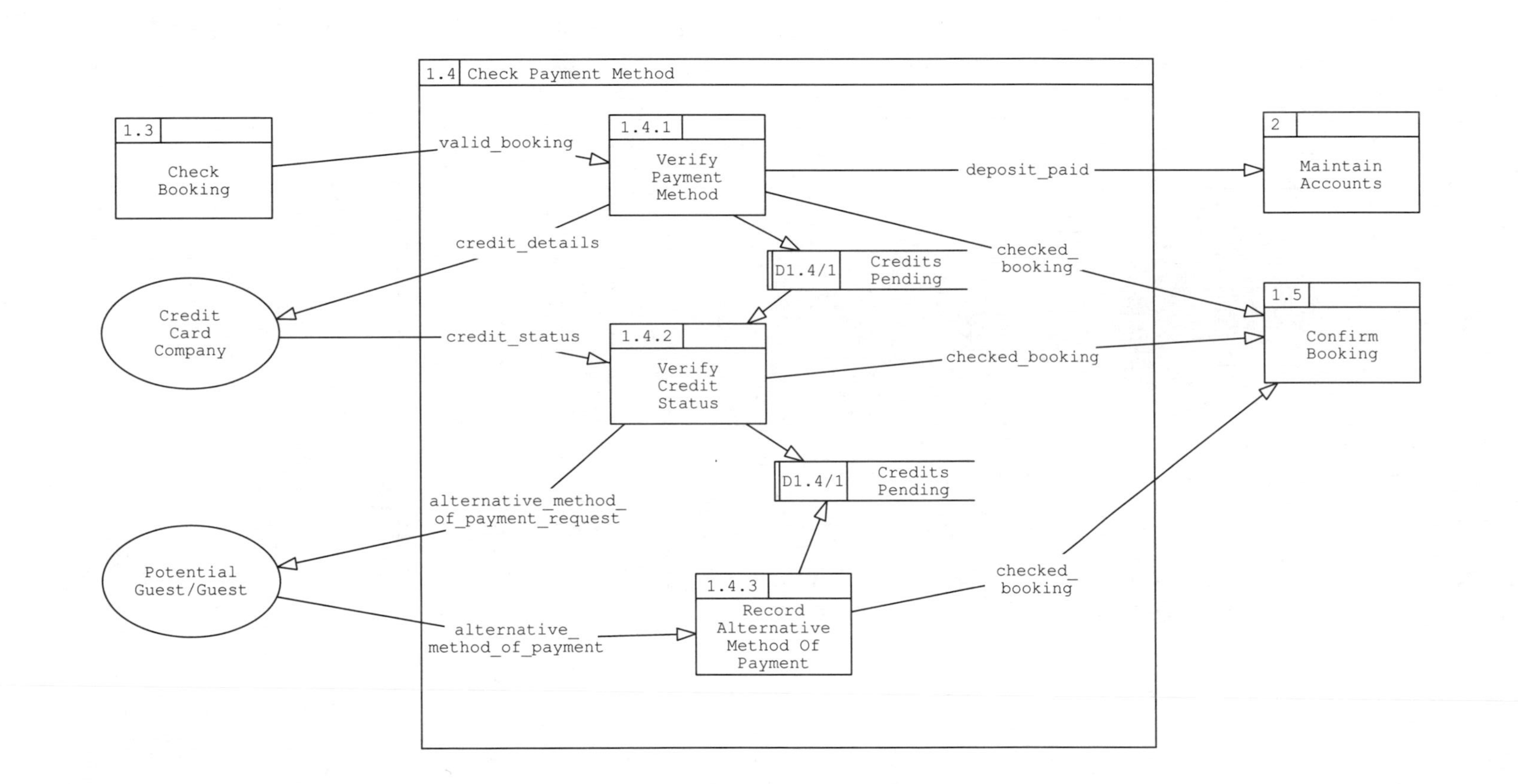

Figure E3.3 (g) Lower level diagram for process `Check Payment Method`.

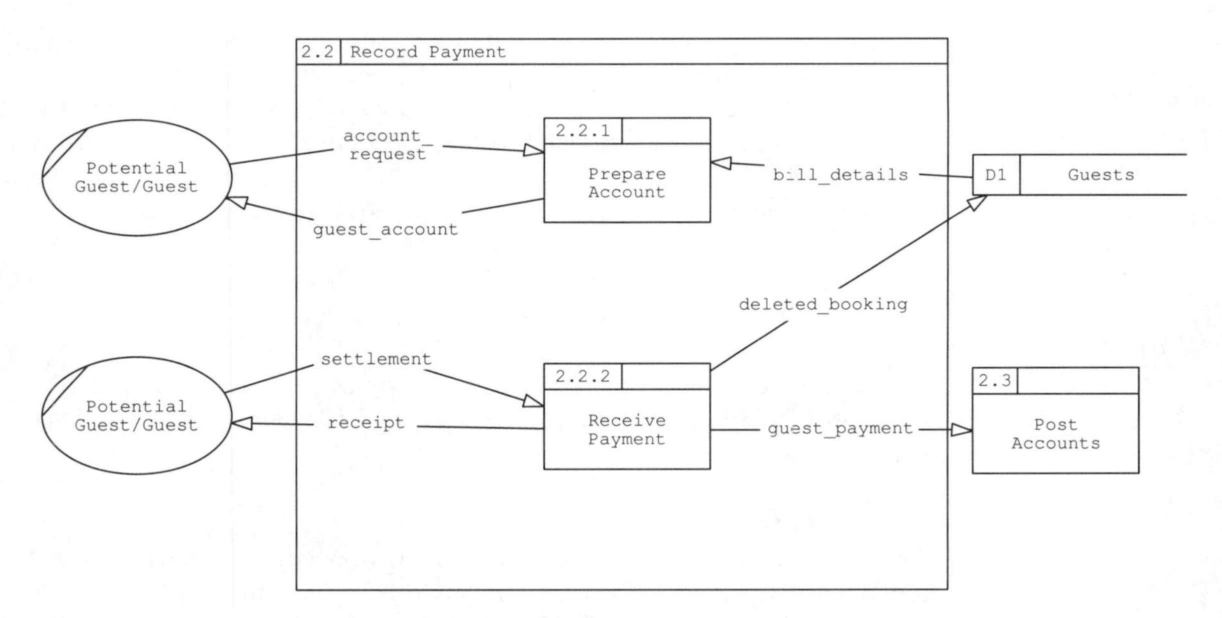

Figure E3.3 (h) Lower level diagram for process `Record Payment`.

A generic data flow `accounting_details` is defined within the data dictionary as:

```
accounting_details = [guest_purchase | guest_billing]
```

The data flow `guest_billing` is also defined as a generic data flow later in this example.

Figure E3.3b shows the decomposition of the process `Maintain Bookings`. Within this all the original processes are retained. The data stores `Rooms` and `Room Rates` are shown inside the frame boundary as they are local to this level. Processes 1.1–1.4 are considered complex enough to decompose. Process 1.5 `Confirm Booking` is not decomposed but has been modified to include two output dominant data flows `confirmation` and `booking_confirmed`.

The data dictionary entries for the generic data flows are:

The status of a room is assumed to be free, provisionally booked, authorized or confirmed. (See Chapter 4 for further details.)

```
enquiry = [price_enquiry | room_enquiry | guest_enquiry]
enquiry_response = [price_response | room_response | guest_response]
guest_billing = [bill_details | deleted_booking]
room_details = [room_availability | provisional_status]
room_booking = [provisional_room | authorized_status]
```

Figure E3.3c shows the decomposition of the process `Maintain Accounts`. The original three processes are retained, with only process 2.2 considered complex enough to decompose. The data store `Sales Accounts` is local to this level and shown inside the frame boundary. Figures E3.3d–g show the

It is assumed that adequate further information is provided.

decomposition of processes 1.1–1.4. The process `Respond to Enquiry` includes the different types of enquiry. The decomposition of the process `Check Booking` shows examples of output dominant data flows to the data stores `Rooms` and `Guests`; and the bookings pending data is examined before it is deleted.

The data dictionary entry for the data flow `method_of_payment` is:

```
method_of_payment =
 [cash/cheque_and_deposit | credit_and_card_no/expiry_date]
```

This allows for the alternative methods of payment offered to guests. The duplication of the data flow `details_complete` is intentional as the same data is carried in the same state.

Solution to Exercise 3.4

Requirements 2, 5 and 6 are supported in the current system. Within requirement 1, there is now an explicit requirement to support booking changes and cancellations. The booking changes and cancellations are part of the maintain bookings subsystem. Any change made to a booking between the provisional and authorized stage is covered by the existing system. However, changes may take place after the booking is confirmed. Cancellations may take place at any time. On the DFD, the additional processes are incorporated into the process `Maintain Bookings` and shown in Figure E3.4a.

You may have recognized that maintenance may be required for the data stores `Rooms` and `Room Rates`. Further discussion will take place in Chapter 8.

If you decide that the diagram for the process `Maintain Bookings` is too cluttered then it may be relevelled. For a booking change, it is assumed that any deposit made is held, the method of payment is the same, though the dates and type of booking may change, possibly necessitating a recalculation of charges. For a booking cancellation, the deposit is not returned. The DD entry `booking_details` defined in the Solution to Exercise 3.3 is redefined:

```
booking_details = [guest_booking | provisional_details |
booking_authorized | booking_confirmed | amended_charge |
guest_amendments | guest_cancellation]
```

In SSADM, the preparation of the Christmas Card List would be a minor retrieval, therefore entered into the requirements catalogue but not on to the DFDs. See SSADM Reference Manuals (1990) for full details.

Requirements 3 and 4 are additional, and need to be added to the required system DFDs. Both these requirements can be incorporated into the maintain accounts subsystem. Figure E3.4b shows the revised diagram for the process `Maintain Accounts`.

Guest records are archived when payment is received. The Christmas Card List is prepared using the data store `Guest Archives`. An additional process is added to delete the archived records after a period of 18 months. The DFDs need to be balanced and the top level diagram modified to incorporate the preparation of the Christmas Card List. See Figure E3.4c for

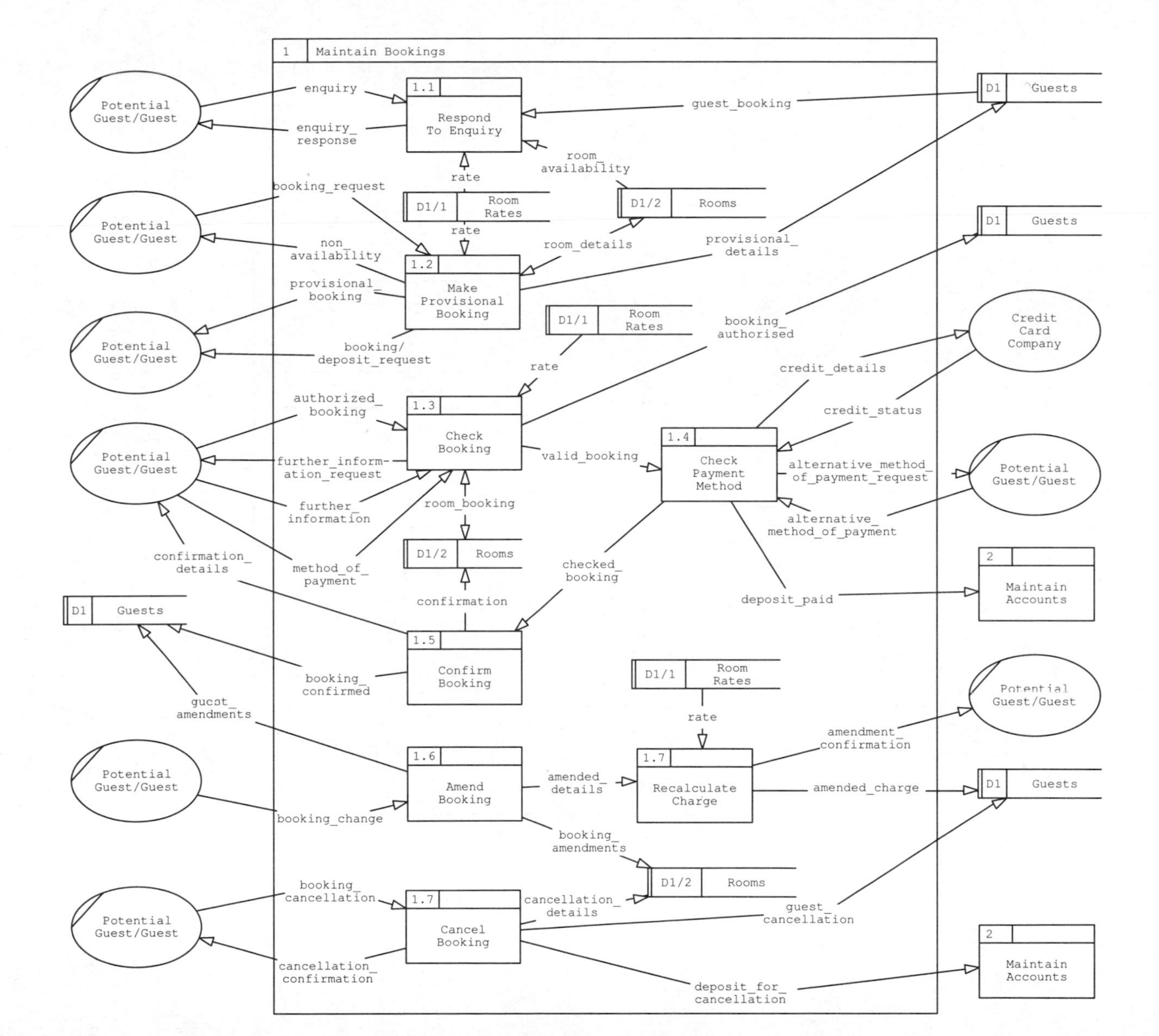

Figure E3.4a Solution to Exercise 3.4. Required lower level diagram for process `Maintain Bookings`.

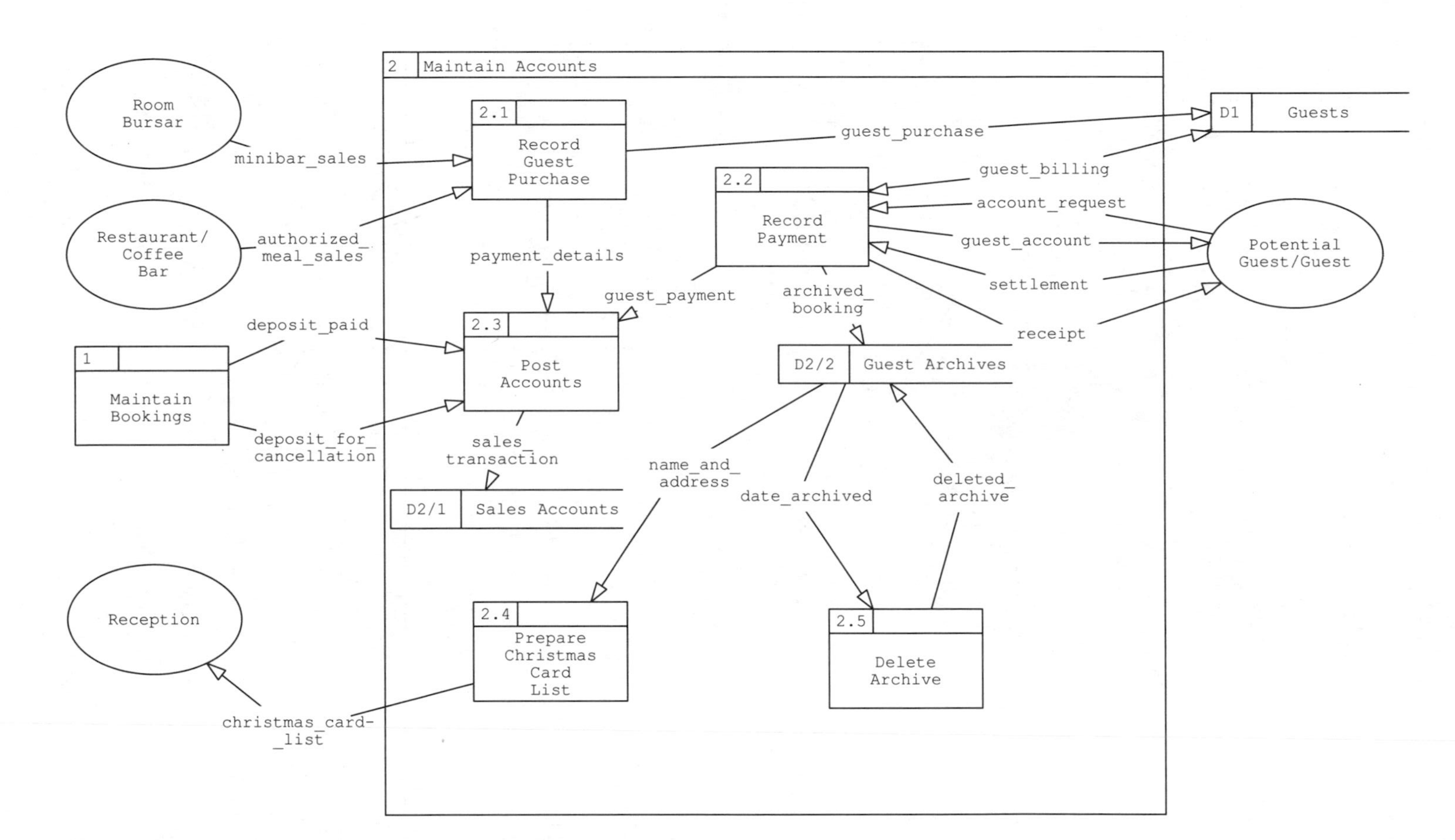

Figure E3.4b Required lower level diagram for process `Maintain Accounts`.

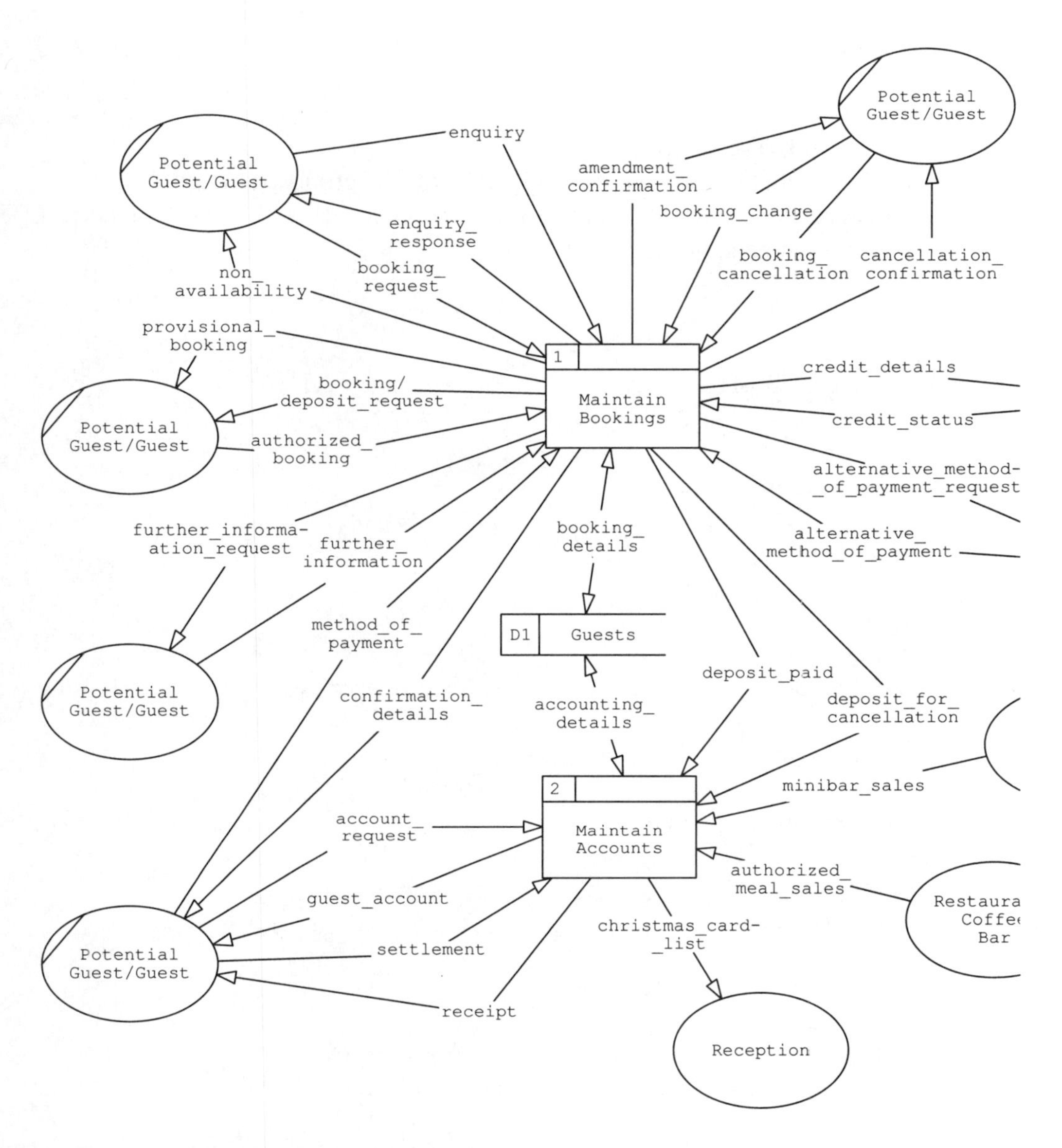

Figure E3.4c Required top level diagram.

the balanced top level diagram.

It is possible that you did not consider requirements 3 and 4 to be part of the maintain accounts subsystem. You may have created a new subsystem; however, your detailed processes should be the same, and you would need to ensure that you checked the balancing.

Chapter 4

Solution to Exercise 4.1

(a) Figure E3.4c shows the top level DFD for the proposed system.

The data flows from this diagram are identified below, including generic data flows where applicable.

Top level data flows

Data flows

```
account_request
accounting_details = [guest_purchase | guest_billing]
alternative_method_of_payment
alternative_method_of_payment_request
amendment_confirmation
authorized_booking
authorized_meal_sales
booking_cancellation
booking_change
booking_details = [guest_booking | provisional_details |
  booking_authorized |  booking_confirmed | amended_charge |
  guest_amendments | guest_cancellation]
booking_request
booking/deposit_request
cancellation_confirmation
christmas_card_list
confirmation_details
credit_details
credit_status
deposit_for_cancellation
deposit_paid
enquiry = [price_enquiry | room_enquiry | guest_enquiry]
enquiry_response = [price_response | room_response |
  guest_response]
further_information
further_information_request
guest_account
method_of_payment
minibar_sales
non_availability
provisional_booking
receipt
settlement
```

Terminators

Top level terminators.

```
Credit Card Company
Potential Guest/Guest
Reception
Restaurant/Coffee Bar
Room Bursar
```

Data stores

Top level data stores.

```
Guests
```

The following entries are taken from the lower level DFDs shown in Figures E3.4a and b, and Figures E3.3d–h.

DD entries for level 1 and 2 DFDs.

Data flows

```
amended_charge
amended_details
archived_booking
authorized_status
bill_details
booking_amendments
booking_authorized
booking_confirmed
booking_pending = * hidden on DFD *
  * Process 'Request Further Information' to data store
       'Bookings Pending' *
  * Process 'Record Further Information' to data store
       'Bookings Pending' *
cancellation_details
checked_booking
confirmation
credit_pending = * hidden on DFD *
  * Process 'Verify Payment Method' to data store 'Credits
       Pending' *
  * Data store 'Credits Pending' to process 'Verify Credit
       Status' *
  * Process 'Verify Credit Status' to data store 'Credits
       Pending' *
  * Process 'Record Alternative Payment Method'
              to data store 'Credits Pending' *
date_archived
deleted_archive
deleted_booking
details_complete
details_incomplete
guest_amendments
```

guest_billing = [bill_details | deleted_booking]
guest_booking
guest_cancellation
guest_enquiry
guest_payment
guest_purchase
guest_response
name_and_address
payment_details
price_enquiry
price_response
provisional_charge
provisional_details
provisional_room
provisional_status
rate
room_availability
room_booking = [provisional_room | authorized_status]
room_details = [room_availability | provisional_status]
room_enquiry
room_response
room_selected
sales_transaction
valid_booking

Data stores

Bookings Pending
Credits Pending
Guest Archives
Room Rates
Rooms
Sales Accounts

Process specifications

Process 1.1.1: Check Room Rate
Process 1.1.2: Check Room
Process 1.1.3: Check Booking Status
Process 1.2.1: Check Availability
Process 1.2.2: Calculate Room Charge
Process 1.2.3: Record Provisional Booking
Process 1.3.1: Verify Booking
Process 1.3.2: Request Further Information
Process 1.3.3: Record Booking/Payment
Process 1.3.4: Record Further Information
Process 1.4.1: Verify Payment Method

```
Process 1.4.2: Verify Credit Status
Process 1.4.3: Record Alternative Payment Method
Process 1.5: Confirm Booking
Process 1.6: Amend Booking
Process 1.7: Recalculate Charge
Process 1.8: Cancel Booking
Process 2.1: Record Guest Purchase
Process 2.2.1: Prepare Account
Process 2.2.2: Receive Payment
Process 2.3: Post Accounts
Process 2.4: Prepare Christmas Card List
Process 2.5: Delete Archive
```

(b) In this particular example, the DD entries have already been identified in part (a) of this exercise. On inspection of the data flows, it is obvious that there will be much repetition of data elements. The data structures chosen are shown first. Note that the data structures you have chosen could be very different but still be correct. You need to check this answer in its entirety against yours. The data elements are shown at the end of the exercise.

The following assumptions have been made:

1. Unique numbers are assigned to guests, rooms, bookings and minibar items (although an item of the same number may be found in more than one room).
2. A booking may involve more than one room, and each room could be required for a different length of time.
3. Different room rates may be charged for a room type at different times of the year.
4. Only the details of the person making the booking are stored.
5. All minibar and refreshment bills are added to the bill for the booking.
6. Accounts are always settled in full.

Data Structures

Data structures

```
book_struct = booking_no, {date_room_struct, booking_status,
                           room_charge}, total_charge
date_room_struct = date_required_from, date_required_to, room_no
date_type_struct = date_required_from, date_required_to, room_type
deposit_struct = deposit_required, deposit_received
guest_credit_struct = credit_card_no, expiry_date, credit_position
guest_namadd_struct = guest_name, guest_address
guest_pers_struct = guest_namadd_struct, guest_tel_no
minibar_struct = item_no, quantity_consumed, date_sold, item_price
refreshment_struct = refreshment_description, date_consumed, price
```

The DD entries below are taken from Figures E3.4a, d and e.

Data flows

Data flows

Note that the entries `guest_booking` and `provisional_details` are defined further, as they are within the scope of the DD required. The others are omitted as they fall outside the scope of this question, but of course would be in a complete DD.

```
booking_details = [guest_booking | provisional_details |
  booking_authorized | booking_confirmed | amended_charge |
  guest_amendments | guest_cancellation]
booking/deposit_request = guest_namadd_struct, book_struct,
                deposit_required
booking_request = guest_pers_struct, {date_type_struct}
enquiry = [price_enquiry | room_enquiry | guest_enquiry]
enquiry_response = [price_response | room_response | guest_response]
guest_booking = {booking_no, {date_room_struct},
              booking_status, booking_date}
guest_enquiry = guest_namadd_struct, {date_type_struct}
guest_response = {date_type_struct, booking_status}
non_availability = guest_namadd_struct, {date_type_struct,
                 booking_status}
price_enquiry = date_type_struct
price_response = date_type_struct, rate
provisional_booking = guest_namadd_struct, book_struct
provisional_charge = guest_pers_struct,
          {date_type_struct, room_no, room_charge}
provisional_details = guest_no, guest_pers_struct, booking_no,
              {date_room_struct, booking_status}
provisional_status = {room_no, {booking_date, booking_no,
                 booking_status}}
rate = * Data element 'rate' *
room_availability = {date_type_struct, room_no,
                  booking_status}
room_details = [room_availability | provisional_status]
room_enquiry = date_type_struct
room_response = date_type_struct, availability_text
room_selected = guest_pers_struct, {date_type_struct, room_no}
```

Terminator

Terminator

```
Potential Guest/Guest
```

Data stores. Complete entries for the data stores are included.

Data stores

```
Guests = guest_no, guest_pers_struct,
         {booking_no, {date_room_struct, {minibar_struct}},
         {refreshment_struct}, booking_status, booking_date,
            payment_method,
         [guest_credit_struct | deposit_struct]}
```

As stated in the assumptions, it is possible for a guest to have more than one booking at any one time; also, that a booking may be for a number of rooms. The minibar sales are associated with each room, though the refreshment bills are attributed to the guest who made the booking. Details relating to the status and date of the booking, and the payment method, are related to the booking. The booking status may be free, provisional, authorized or confirmed. It is accepted for example that different payment methods may be used for different bookings.

```
Room Rates = {room_type, {start_date, end_date, rate}}
Rooms = {room_no, room_type, {date, booking_no, booking_status}}
```

Data elements

```
availability_text
booking_date
booking_no
booking_status
credit_card_no
credit_position
date
date_consumed
date_required_from
date_required_to
date_sold
deposit_received
deposit_required
end_date
expiry_date
guest_address
guest_name
guest_no
guest_tel_no
item_no
item_price
payment_method
price
quantity_consumed
rate
refreshment_description
room_charge
room_no
room_type
start_date
total_charge
```

Process Specifications

Process specifications

```
Process 1.1.1: Check Room Rate
BEGIN
ACCEPT price_enquiry
RETRIEVE rate FROM Room Rates FOR
        date_required_from, date_required_to, room_type
CREATE price_response
END
```

```
Process 1.1.2: Check Room
BEGIN
ACCEPT room_enquiry
CHECK availability IN Rooms FOR
        date_required_from, date_required_to, room_type
CREATE room_response
END
```

Alternatively, `RETRIEVE availability FROM Rooms FOR date_required_from, date_required_to, room_type`

```
Process 1.1.3: Check Booking Status
BEGIN
ACCEPT guest_enquiry
FOR each separate enquiry
  RETRIEVE guest_booking FROM Guests
  CHECK same booking details IN Rooms
  CREATE an iteration of guest_response
ENDFOR
END
```

```
Process 1.2.1: Check Availability
BEGIN
ACCEPT booking_request
FOR each separate request
  RETRIEVE room_availability FROM Rooms
  IF OK
   CREATE an iteration of room_selected
  ELSE
   CREATE an iteration of non_availability
  ENDIF
ENDFOR
END
```

```
Process 1.2.2: Calculate Room Charge
BEGIN
ACCEPT room_selected
FOR each separate booking requirement
  RETRIEVE rate FROM Room Rates FOR
```

```
      date_required_from, date_required_to, room_type
  CALCULATE room_charge
  CREATE an iteration of provisional_charge
ENDFOR
END

Process 1.2.3: Record Provisional Booking
BEGIN
ACCEPT provisional_charge
ALLOCATE booking_no
total_charge = 0
SET booking_status TO 'provisional'
FOR each separate booking requirement
  WRITE provisional_status TO Rooms
  total_charge = total_charge + room_charge
ENDFOR
CALCULATE deposit_required
CREATE provisional_booking
IF new guest
  ALLOCATE guest_no
ENDIF
WRITE provisional_details TO GUESTS
CREATE booking/deposit_request
END
```

Chapter 5

Solution to Exercise 5.1

Likely entities	*Identifiers*
Applicant	applicant_no
Course	course_code
College	college_code
Qualification	qualification_code

Assumptions

1. Applicants' names might not be unique, so a unique applicant_no has been allocated.
2. Each qualification is allocated a unique code.

You might have identified the Local Education Authority as a likely entity. For the entities in the 'Likely entities' list above, if we ask the question 'How many occurrences will there be of this entity type in the system?', the answer will be, 'As many occurrences of `Applicant` as there are applicants who apply for courses; as many occurrences of `Course` as there are courses run by the colleges . . .', and so on. However, if we ask the same question of the entity type LEA, the answer is 'Just one'. Thus we do not need to include the LEA as an entity type in this system; the information that we want to store about the LEA is about its colleges, courses and applicants and their qualifications. Note that if the system were destined to hold information about a number of LEAs, then the answer to the question posed above would be different, and there would be a need to include it as an entity type within the model.

Solution to Exercise 5.2

Relationships

An applicant may apply for several courses.
An applicant may have several qualifications.
Each course is run at only one college.

See Figure E5.2

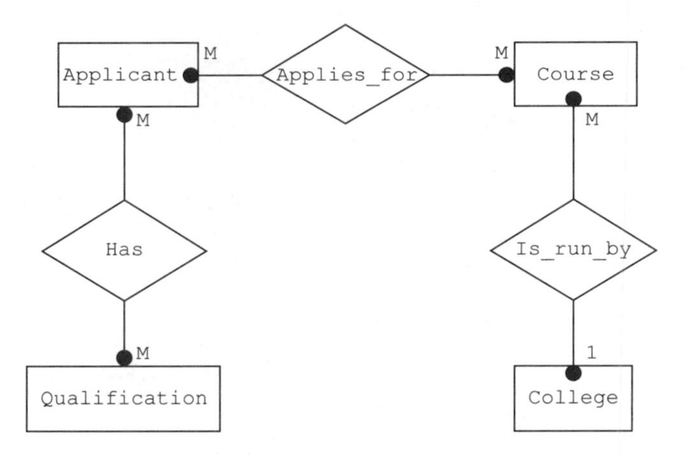

Figure E5.2 Solution to Exercise 5.2.

Assumptions (continued)

3. An applicant might not have any standard qualifications.
4. Details of standard qualifications are stored, regardless of whether any applicants have obtained them.
5. A course might not have received any applications.
6. Each college must run at least one course.

The degrees of the relationships in your solution should be the same as those shown in Figure E5.2. However, assumptions have to be made about the membership classes of some of the relationships, and you might have made different assumptions. For example, you might have assumed that every candidate must have at least one standard qualification, and so made

Applicant obligatory in its relationship with Qualification. As stated earlier, the vital issue is to ensure that your diagram is consistent with your assumptions.

Solution to Exercise 5.3
See Figure E5.3.

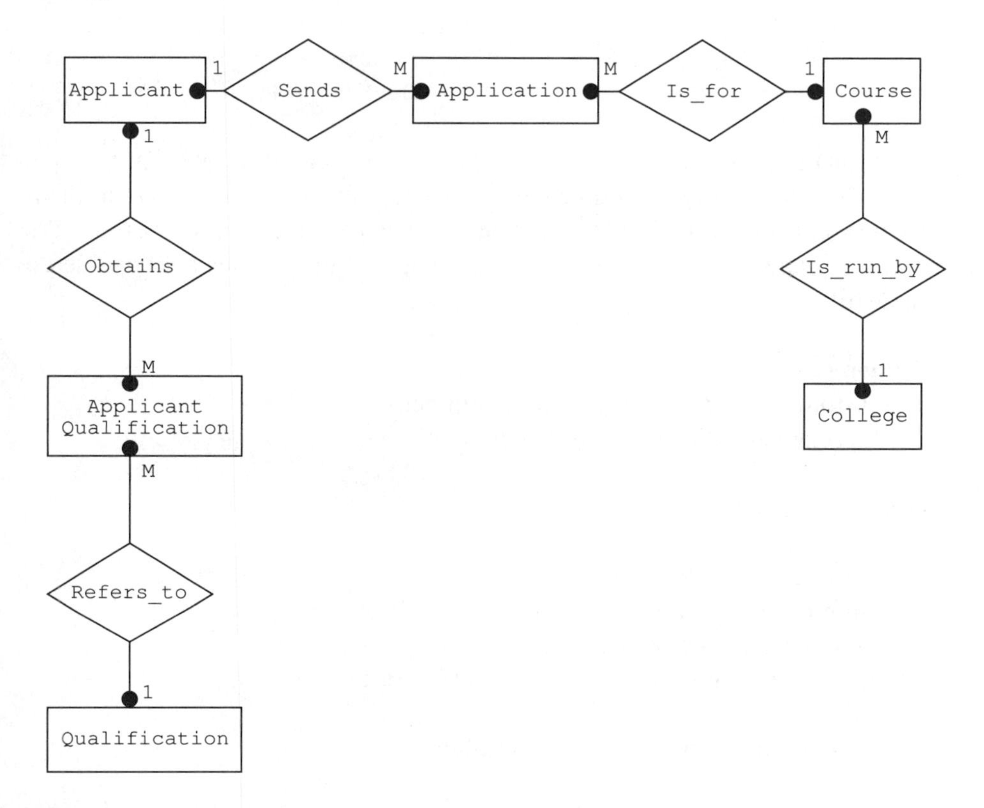

Figure E5.3 Solution to Exercise 5.3.

A similar argument could apply to Applicant Qualification, as an applicant could have taken the same qualification more than once; however, it is likely in this case that only the attempt resulting in the highest grade will be recorded.

The identifier of the entity Applicant Qualification will be applicant_no and qualification_code. However, will applicant_no and course_code be sufficient to uniquely identify an occurrence of the entity type Application? If an applicant may apply twice for the same course, then the date_of_application will be needed, too. Alternatively, a unique application_no can be assigned as the identifier of Application.

Solution to Exercise 5.4

Guideline steps 1 and 2:

Likely entities	*Identifiers*
Clothing Item	item_code
Customer	customer_no
Supplier	supplier_no
Customer Order	customer_order_no
Clothing Requisition	clothing_requisition_no

Manufacturers and textile importers have been treated as one entity type, Supplier; however, it would be equally acceptable at this point to treat them as two entity types. The decision as to whether one or two entities are required can be made only when the attributes of the entities are considered in detail.

See Chapter 6.

Assumptions

1. Customers are allocated a unique number.
2. Suppliers are allocated a unique number.

Guideline step 3:
Relationships

A customer sends in a customer order.
A customer order is for clothing item(s).
A supplier supplies clothing items.
Clothing items are ordered by the mail order company on a clothing requisition.
A clothing requisition is sent to a supplier.

Guideline step 4: See Figure E5.4a.

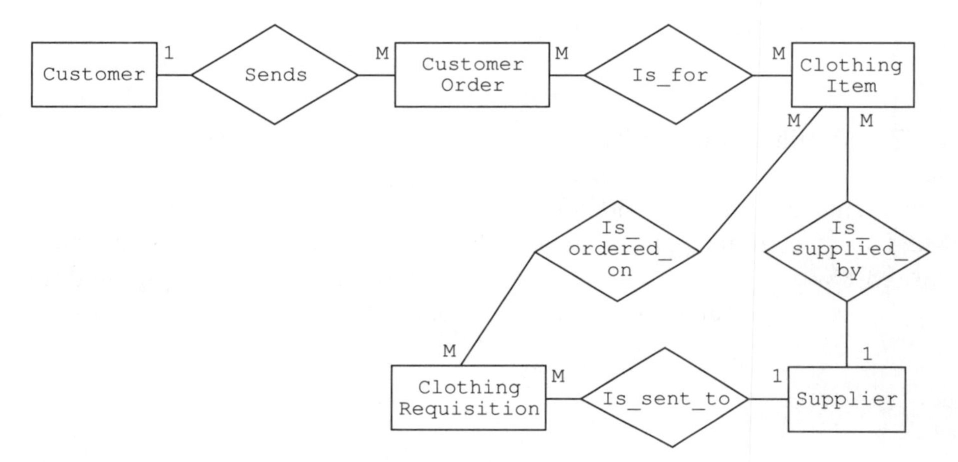

Figure E5.4a Guideline step 4.

Guideline step 5: See Figure E5.4b.

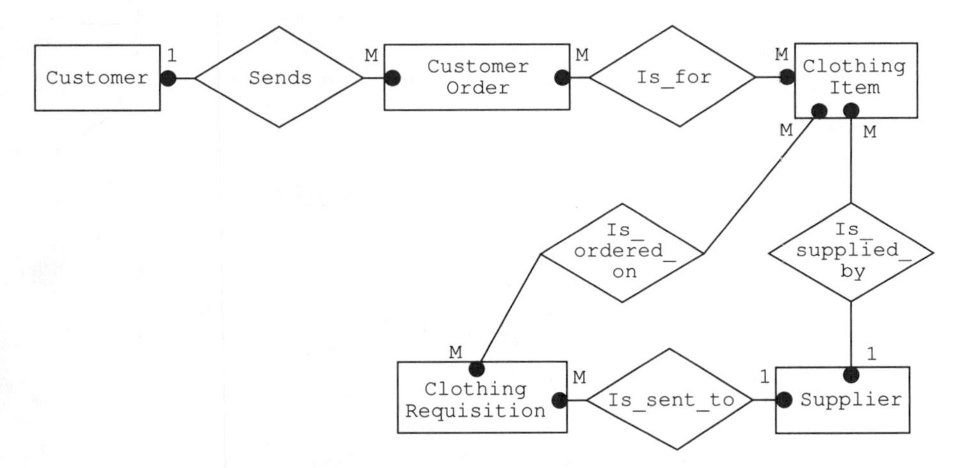

Exercise E5.4b Guideline step 5.

Assumptions (continued)

3. The mail order company holds on file both the customers who have sent in an order, and potential customers (who might have requested a mail order catalogue, for example, but not yet ordered any goods).
4. Not every clothing item may have been ordered by a customer (for example, new or unpopular items).
5. Details are kept only of suppliers who supply items of clothing (i.e. details of potential suppliers are not kept).
6. All items of clothing are obtained from a clothing requisition.
7. Details are only kept of clothing items that have been ordered.

You might have made some different assumptions, for example that details of potential suppliers *are* kept; make sure that your diagram accurately reflects your assumptions.

Guideline step 6: There are two M:M relationships in Figure E5.4b; see Figure E5.4c for the diagram after their decomposition.

Likely entities	*Identifiers (continued)*
`Customer Orderline`	`customer_order_no, item_code`
`Clothing Requisitionline`	`clothing_requisition_no, item_code`

Note that if during guideline step 5 you decided that details of potential suppliers are to be kept as well as actual suppliers, two relationships are needed between `Clothing Item` and `Supplier` (see Figure E5.4d). A supplier might be the actual supplier of some items, but the potential supplier of others.

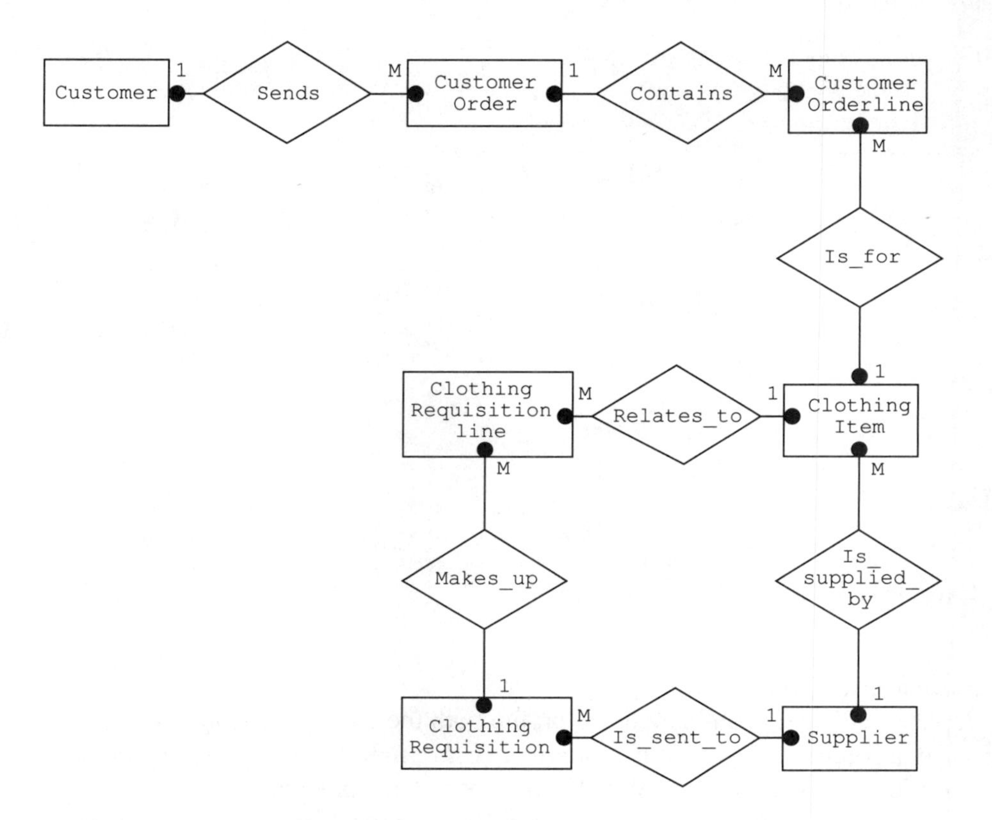

Figure E5.4c Guideline step 6.

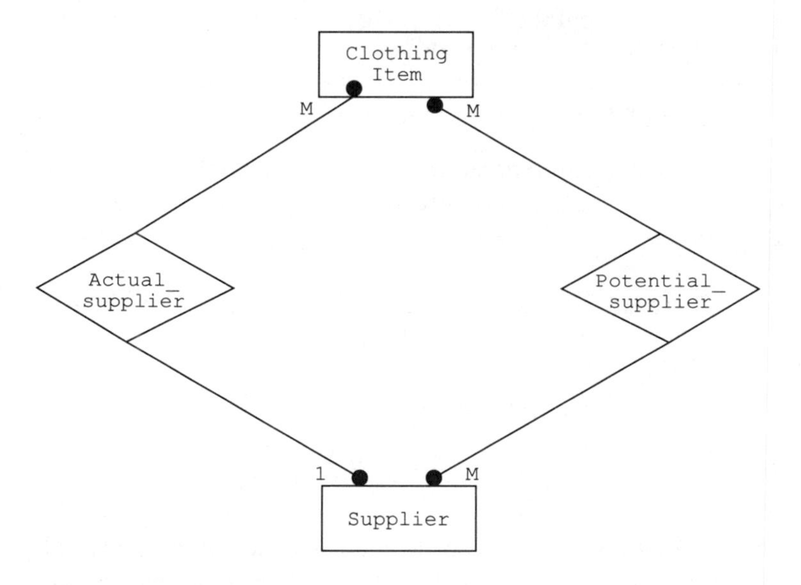

Figure E5.4d E-R diagram to support the storage of data about potential suppliers.

There is still just one actual supplier of each item of clothing; however, each item of clothing might have more than one potential supplier. Another M:M relationship has now been created. The relationship `Potential Supplier` will need an identifier; for example, `item_code` and `supplier-no`.

Solution to Exercise 5.5

There is a fan trap between `Invoice`, `Client` and `Property`. As Figure E5.5a illustrates, it is not possible to tell to which property an invoice relates. Figure E5.5b presents a solution.

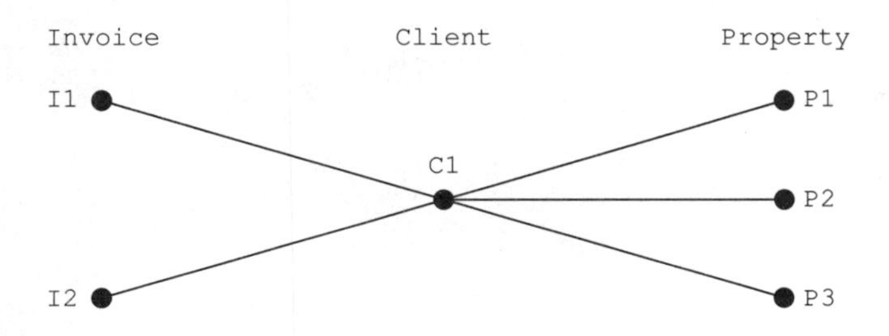

Figure E5.5a Occurrence diagram to illustrate the fan trap.

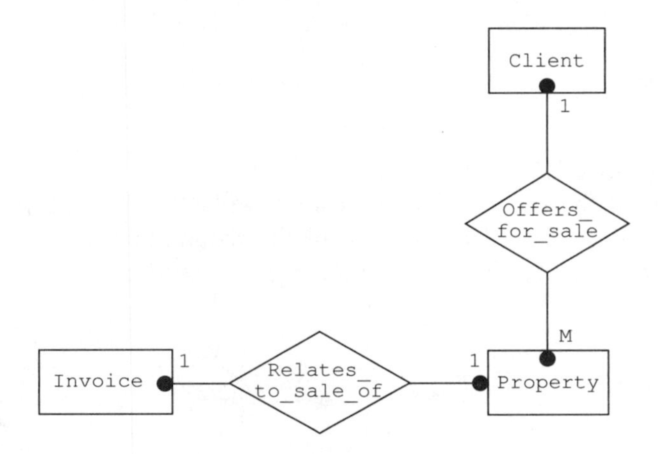

Figure E5.5b Solution to Exercise 5.5.

There is also a fan trap between `Viewing`, `Potential Buyer` and `Requirement`. However this one is not significant as there is no need to make a connection between the entities `Viewing` and `Requirement`.

Note that the decomposition of the M:M relationship between `Property` and `Requirement` would result in a third fan trap, between `Viewing`, `Property` and the new entity. Again this is not a significant trap, but this example does highlight how the decomposition of M:M relationships helps to identify potential fan traps.

Solution to Exercise 5.6

The entity `Module` is non-obligatory in its relationship with `Student`; it is possible for a module not to be chosen by any student. Thus it is not possible to determine for every module on which course it may be studied nor all of the modules offered by a particular course. In fact, it would only be possible to satisfy these two requirements if every year students on a course between them chose every module offered on that course.

Thus there is a chasm trap in Figure 5.19, which can be resolved by adding a relationship between `Module` and `Course`; see Figure E5.6.

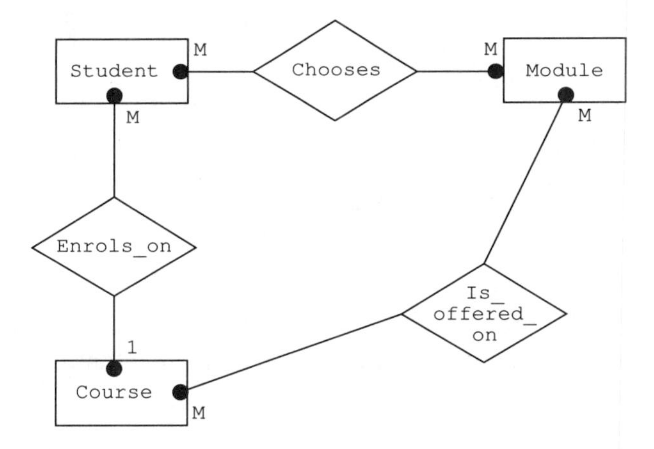

Figure E5.6 Resolution of the chasm trap in Figure 5.19.

Solution to Exercise 5.7

Decomposition of the two M:M relationships will result in two potential fan traps; see Figure E5.7. However, all of the requirements noted earlier can still be satisfied, so these potential traps are not significant.

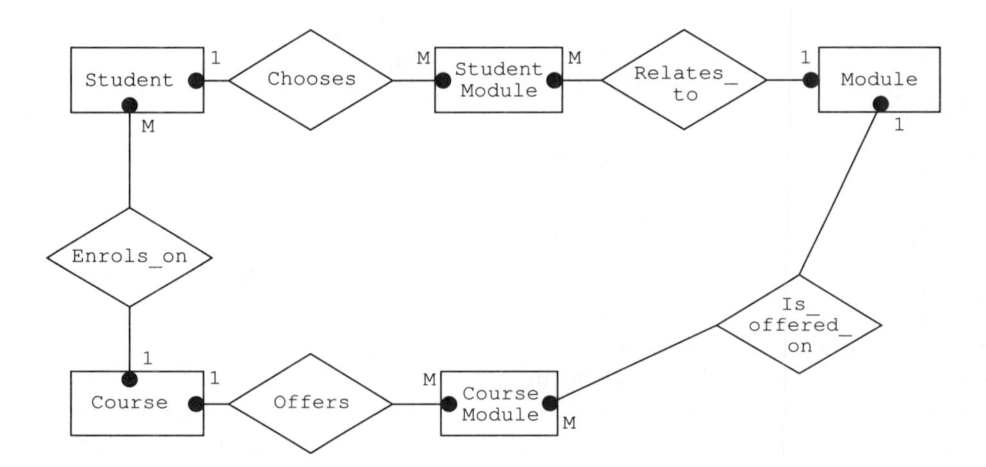

Figure E5.7 Decomposition of the M:M relationships in Figure E5.6.

Chapter 6

Solution to Exercise 6.1

The order of rows in the table in Figure 6.7 is significant, and there are two identical rows. Figure E6.1 illustrates the necessary amendments. As it is possible for more than one applicant to have the same name, a unique `applicant_no` has been introduced.

Applicant

applicant_no	applicant_name	course_no
A100	Wexford	C152
A101	Morse	C155
A101	Morse	C201
A102	Bond	C397
A103	Marple	C721
A104	Bond	C397

Figure E6.1 Solution to Exercise 6.1.

Solution to Exercise 6.2

The values `A101`, `Bond` and `C397` are duplicated but not redundant data. The values `Morse` and `Wine making` are redundant; one of the occurrences of each can be eliminated with no loss of information.

Solution to Exercise 6.3

Figure E6.3 shows the two table occurrences. The table types are:

```
Applicant   (applicant_no, applicant_name)
Application (applicant_no, course_no)
```

Solution to Exercise 6.4

An applicant may have more than one qualification and may apply for more than one course, thus the table supplied contains two repeating groups:

```
Applicant (applicant_no, applicant_name, applicant_address,
           date_of_birth, (qualification_code), (course_code))
```

Applicant

applicant_no	applicant_name
A100	Wexford
A101	Morse
A102	Bond
A103	Marple
A104	Bond

Application

applicant_no	course_no
A100	C152
A101	C155
A101	C201
A102	C397
A103	C721
A104	C397

Figure E6.3 Solution to Exercise 6.3.

It may be split as follows:

Applicant (applicant_no, applicant_name, applicant_address,
date_of_birth)

Applicant Qualification (applicant_no, qualification_code)

Application (applicant_no, course_code)

Solution to Exercise 6.5

Lecturer_no is a determinant of lecturer_name.
Lecturer_no and module_no together are a composite determinant of no_of_weeks.

Solution to Exercise 6.6

See Figure E6.6.

It has been assumed that delivery_address is determined by customer_order_no, and that a customer may request a whole order to be delivered to an alternative address. A customer would be able to use different delivery addresses for different orders (for example, if they regularly order goods for friends or relatives).

There are two other possibilities. Delivery_address could be determined by customer_no; in this case the enterprise rule would be that a customer may nominate just one delivery address that is different from their billing address. Alternatively, delivery_address could be determined by both customer_order_no and item_code so that different items could be delivered to different addresses (for example, if orders for the customer, for friends and for relatives were combined).

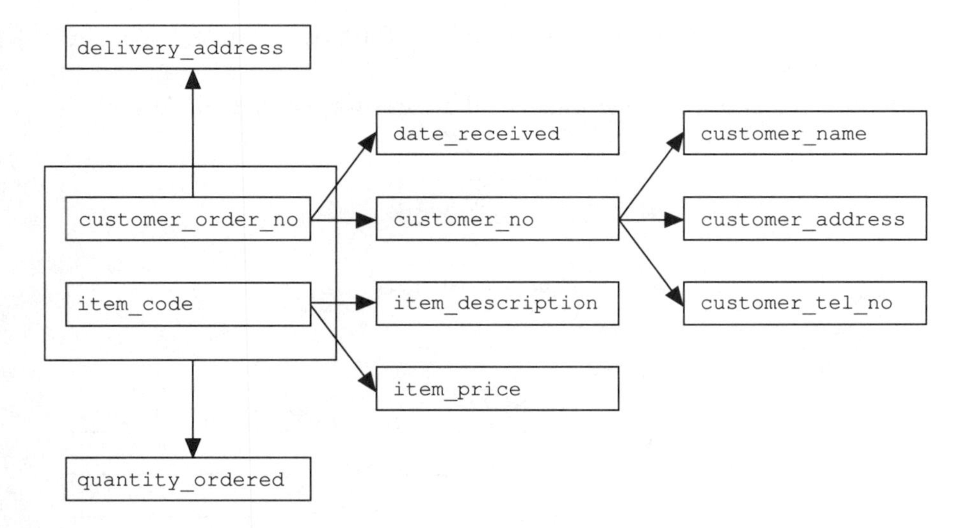

Figure E6.6 Solution to Exercise 6.6.

Solution to Exercise 6.7
See Figure E6.7.

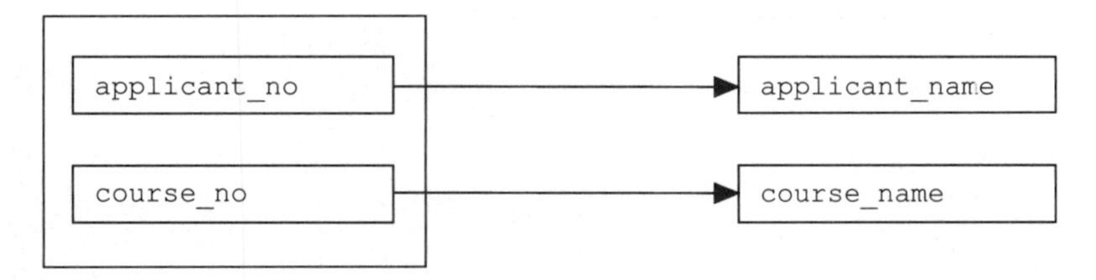

Figure E6.7 Solution to Exercise 6.7.

The determinancy diagram in Figure E6.7 illustrates that neither of the non-key attributes is determined by the whole of the composite identifier. Thus two further tables must be created to hold the details of applicants and courses. The `Application` table is still required; it represents the M:M relationship between `Applicant` and `Course`. The 2NF tables will therefore be as follows:

```
Application    (applicant_no, course_no)
Applicant      (applicant_no, applicant_name)
Course         (course_no, course_name)
```

Solution to Exercise 6.8
Assuming that an applicant is asked for just one referee, the tables are all in 1NF. If it is the case that each applicant has more than one referee, then the attributes `referee_name`, `referee_status` and `referee_address` form a

repeating group and must be removed to form a separate `Applicant Referee` table.

Figures E6.8a–c show determinancy diagrams for the three tables.

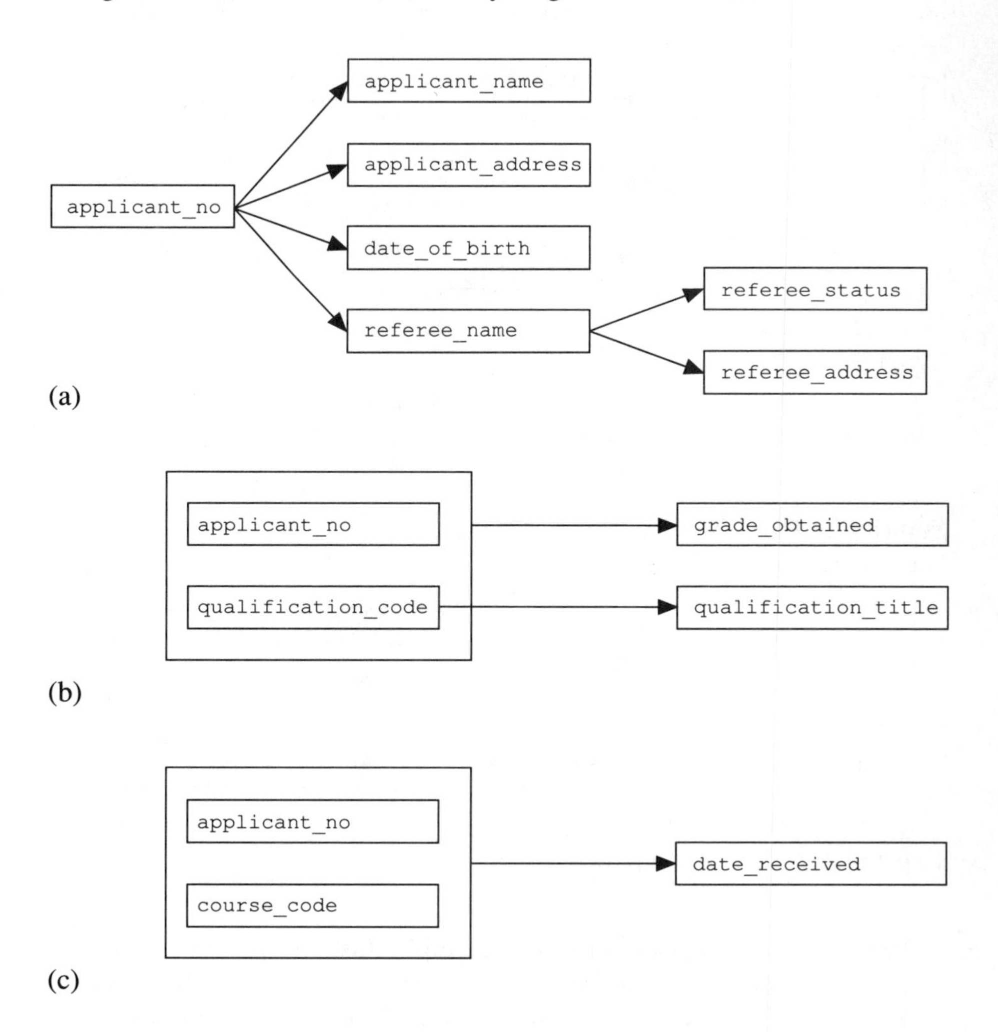

Figure E6.8 Determinancy diagrams for (a) `Applicant`, (b) `Applicant Qualification` and (c) `Application`.

It can be seen that `Applicant` is in 2NF but not in 3NF, `Applicant Qualification` is not in 2NF and `Application` is in 3NF. The complete set of 3NF tables is as follows:

```
Applicant (applicant_no, applicant_name, applicant_address,
                    date_of_birth, referee_name)
```

```
Referee  (referee_name, referee_status, referee_address)

Applicant Qualification (applicant_no, qualification_code,
                         grade_obtained)

Qualification  (qualification_code, qualification_title)

Application   (applicant_no, course_code, date_received)
```

If you assumed that an applicant has more than one referee, `referee_name` should be removed from `Applicant` and an additional table provided:

```
Applicant Referee (applicant_no, referee_name)
```

Solution to Exercise 6.9

Guideline steps 1 and 2:

Likely entities	*Identifiers*
Guest	guest_no
Room	room_no
Booking	booking_no
Guest Purchase	booking_no, purchase_no
Minibar Stock	room_no, item_no

Tables

```
Guest           (guest_no,
Room            (room_no,
Booking         (booking_no,
Guest Purchase  (booking_no, purchase_no,
Minibar Stock   (room_no, item_no,
```

Assumptions

1. Each guest is assigned a unique `guest_no`.
2. Each room is assigned a unique `room_no`.
3. Each booking is assigned a unique `booking_no`.
4. A guest may make a number of purchases; a purchase includes items from the minibar in the room, meals in the hotel restaurant or drinks at the hotel bar.
5. Purchases made by a guest are assigned a running number within that guest's booking number.
6. The minibar in each room contains a number of different items, each of which is assigned a unique `item_no`. Each item may be found in more than one room.

Guideline step 3:
Relationships
A guest makes a booking.
A booking can be made for one or more rooms.
A guest may make a number of guest purchases.
Minibar stock is held within a room.

Guideline step 4: See Figure E6.9a.

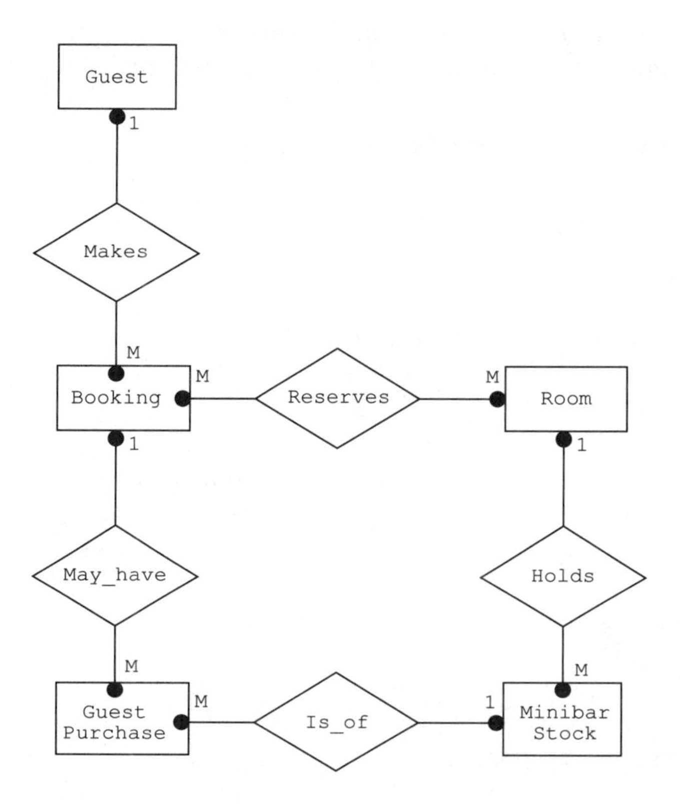

Figure E6.9a Guideline step 4.

Assumptions (continued)

7. The entity `Guest` is non-obligatory in its relationship with `Booking`, so details of potential guests can be stored in the same table.
8. Not all rooms have a minibar.
9. A booking does not have to involve any guest purchases.

Guideline step 5: See Figure E6.9b.
There is one M:M relationship in Figure E6.9a. Figure E6.9b shows the relationship after decomposition.

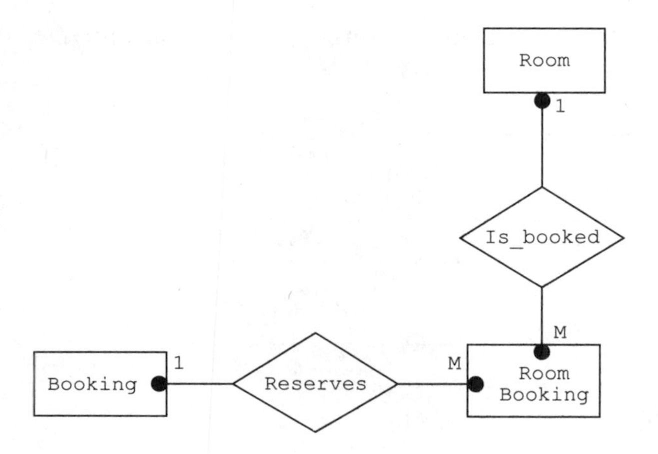

Figure E6.9b Guideline step 5.

Additional table

Room Booking (booking_no, room_no,

Guideline step 6:
Tables

Guest (guest_no,
Room (room_no,
Booking (booking_no, guest_no,
Guest Purchase (booking_no, purchase_no, room_no, item_no,
Minibar Stock (room_no, item_no,
Room Booking (booking_no, room_no,

Guideline step 7: Note that the attributes that have already been included as identifiers of the tables listed so far are omitted from the list below:

Attributes

guest_name	deposit_required
guest_address	deposit_received
guest_tel_no	room_type
booking_date	room_rate
credit_position	quantity_in_stock
credit_card_no	item_description
expiry_date	item_price
date_required_from	quantity_consumed
date_required_to	date_sold
booking_status	refreshment_description
payment_method	date_consumed
price	

Guideline step 8: Figure E6.9c contains a determinancy diagram for the attributes that are determined by booking_no and guest_no.

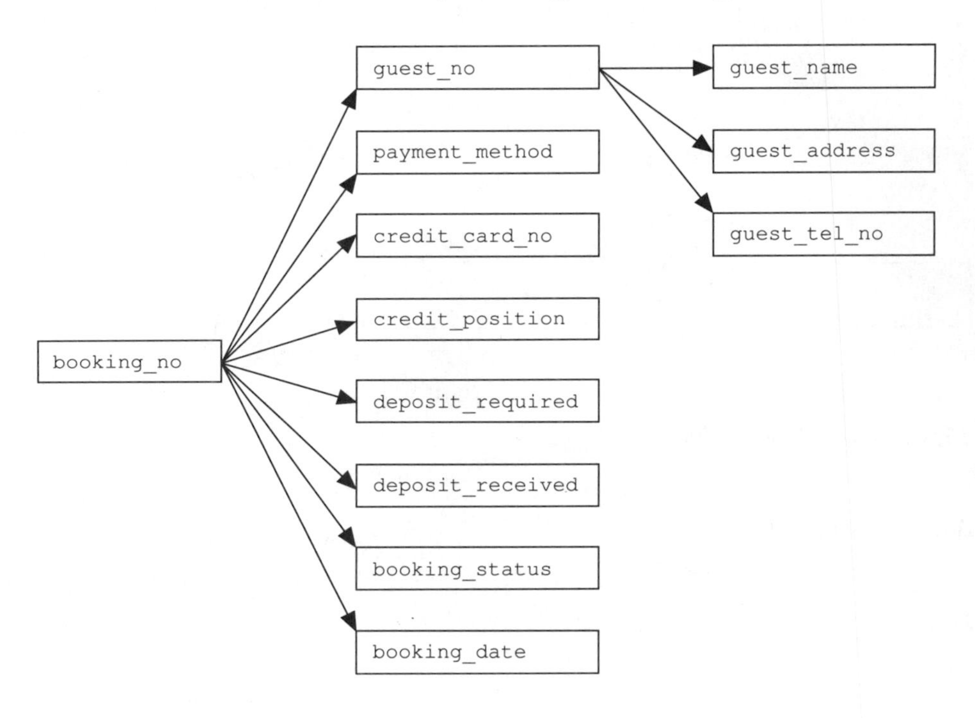

Figure E6.9c Guideline step 8.

It is assumed that the payment_method is determined by the booking_no rather than by the guest_no; a regular guest could use a different payment method on different occasions. A guest might also use a different credit card for each booking. Hence the attributes relating to credit details are determined by booking_no. Since a deposit is only required if a credit card is not used, an occurrence of Booking will *either* store credit details *or* deposit details. To avoid having null values in Booking, the entities Guest Credit and Guest Deposit have been created to store this data.

A second determinancy diagram illustrates the need for some more tables (see Figure E6.9d).

Refreshment_description refers to any meals taken by the guest in the restaurant and any drinks at the hotel bar. The Guest Purchase table has been split into two tables, Minibar Purchase and Refreshment, as different attributes need to be stored about each type of purchase.

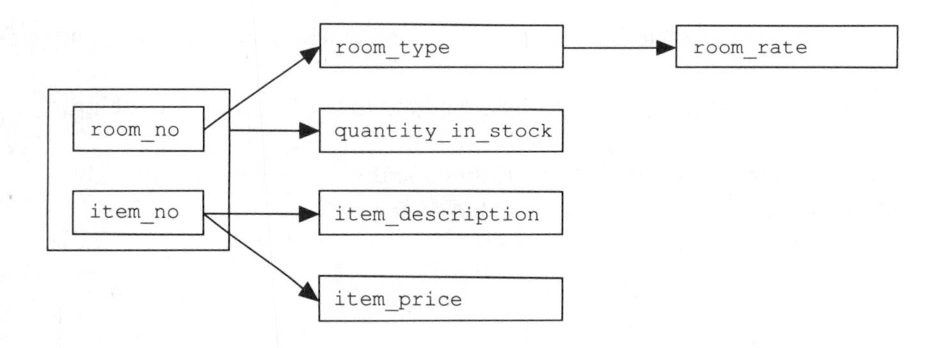

Figure E6.9d Guideline step 8 (continued).

Tables

Guest (guest_no, guest_name, guest_address, guest_tel_no)
Room (room_no, room_type)
Booking (booking_no, guest_no, booking_date, payment_method, booking_status)
Guest Credit (booking_no, credit_card_no, expiry_date, credit_position)
Guest Deposit (booking_no, deposit_required, deposit_received)
Minibar Purchase (booking_no, purchase_no, room_no, item_no, quantity_consumed, date_sold)
Minibar Stock (room_no, item_no, quantity_in_stock)
Room Booking (booking_no, room_no, date_required_from, date_required_to)
Item (item_no, item_description, item_price)
Rate (room_type, room_rate)
Refreshment (booking_no, refreshment_sequence_no, refreshment_description, date_consumed, price)

Assumptions (continued)

10. Payment_method is determined by the booking_no.
11. A guest could use a different credit card for each booking.
12. Room_type indicates whether the room is a double with *en suite*, a double with hot and cold, a single with *en suite*, a single with hot and cold or a family room.
13. A booking may involve more than one room, and each room could be required for a different length of time.
14. Refreshment_description indicates the type of refreshment taken, such as lunch, dinner, a bar bill, and so on.
15. Meals taken in the hotel restaurant and drinks taken at the hotel bar may be added to the guest's bill.
16. Only the details of one person making the booking are stored.

17. One bill is made out for each booking, even when several rooms are involved.
18. All minibar and refreshment bills are added to the bill for the booking.

Guideline step 9: See Figure E6.9e. Entities and relationships for the tables added in guideline step 8 have been added.

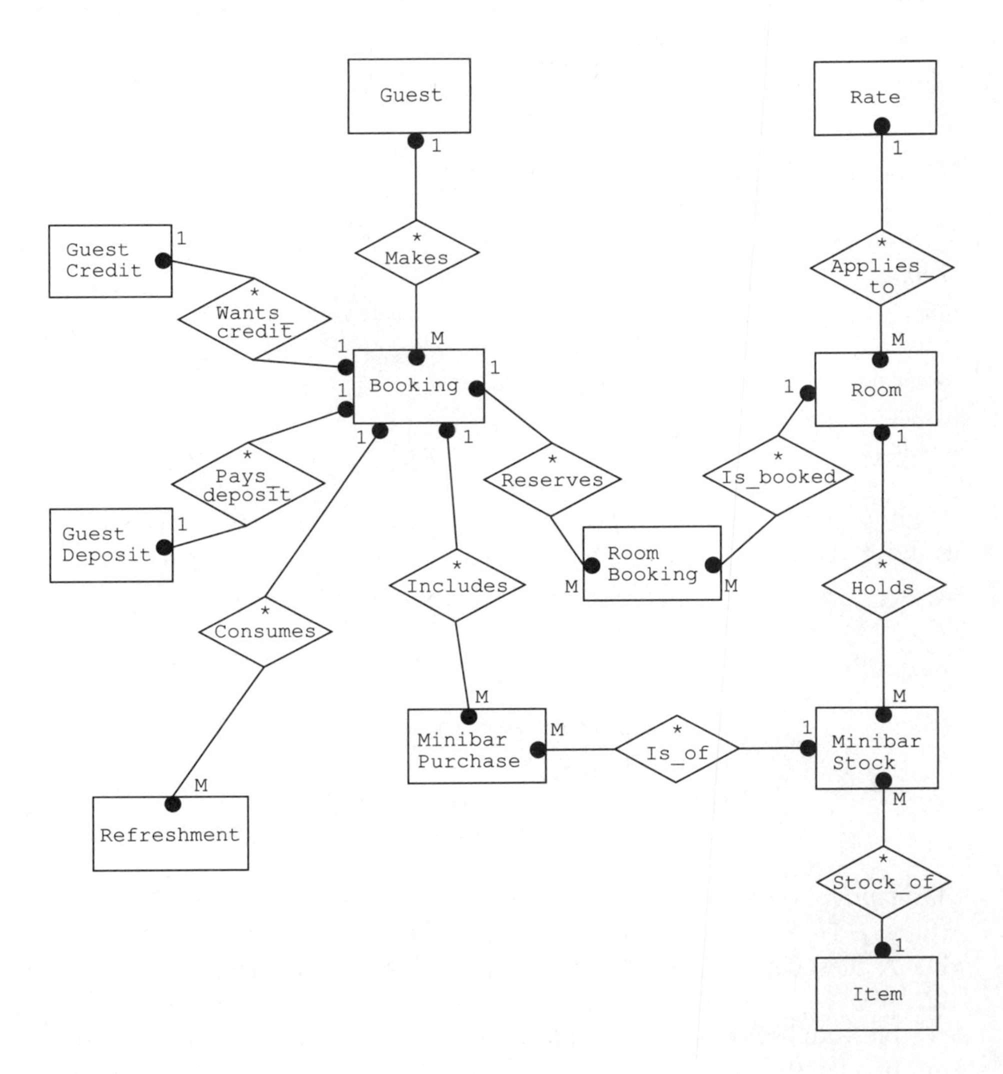

Figure E6.9e Guideline step 9.

Guideline step 10: There is a fan trap between `Refreshment`, `Booking` and `Room Booking`; it has been assumed that the bill for a refreshment is assigned to the booking.

See assumption 18.

The significance of the fan trap is that if more than one room is involved in a booking, it cannot be determined which room was responsible for a refreshment bill. If this kind of detail is required, then the entity `Refreshment` should have a M:1 relationship with `Room Booking` instead of with `Booking`. The `Refreshment` table should be modified as follows:

```
Refreshment      (booking_no, room_no, refreshment_sequence_no,
                  refreshment_description, date_consumed, price)
```

Note that as the entity `Minibar Purchase` refers to the purchase of items from the minibar in a particular room, it is possible to determine which `Room Booking` was responsible for making the purchase.

Guideline step 11: If the fan trap described in guideline step 10 is assessed as significant, the amendments to the diagram and to the `Refreshment` table described in that step will need to be made. The diagram should then be re-examined to ensure that no further connection traps have been created by the modifications.

The suggested amendment will not create any further connection traps.

Chapter 7

Solution to Exercise 7.1

Figure 7.4 contains two errors. Firstly, the selection box R cannot appear on its own. If the intention is to denote that N consists of either R or nothing, then a null box should be introduced, as in Figure E7.1a. Alternatively, if it was intended that N should consist of R and only R, then the selection notation is a mistake, and box R can be omitted altogether, as in Figure E7.1b.

Secondly, selection and iteration have been combined on the same level of a branch of the tree. If it is intended that P consists of either T or U, followed by an iteration of V, then two structure boxes need to be introduced to represent this sequence, as in Figure E7.1a. An alternative interpretation is that P consists of either T or U or an iteration of V, in which case the diagram needs to amended as in Figure E7.1b.

Note that incorrect use of the notation leads to ambiguities; the introduction of structure boxes to maintain the integrity of the notation eliminates any doubt about how the diagram is to be interpreted.

Solution to Exercise 7.2

(i) This is not permissible, as a student must be assigned to a personal tutor before the deletion event.
(ii) This is permissible.
(iii) This is permissible.
(iv) This is not permissible; there must be a deletion event, either graduation, failure or withdrawal.

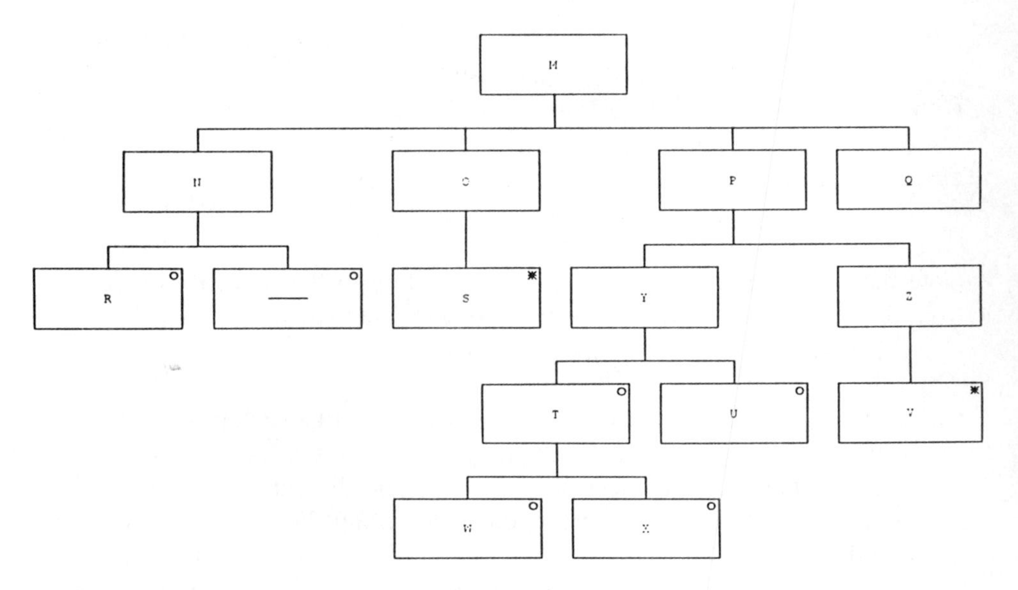

Figure E7.1a Solution to Exercise 7.1.

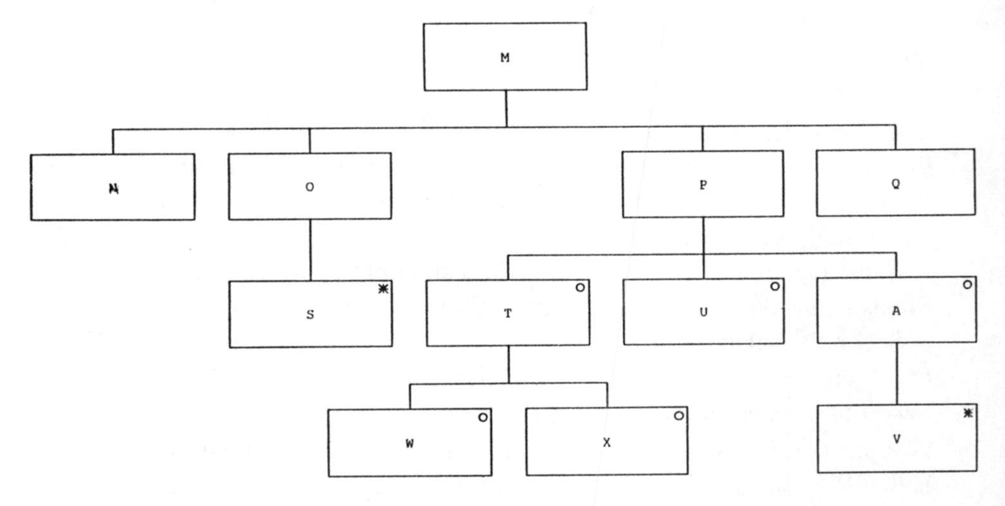

Figure E7.1b Alternative solution to Exercise 7.1.

Solution to Exercise 7.3
See Figure E7.3.

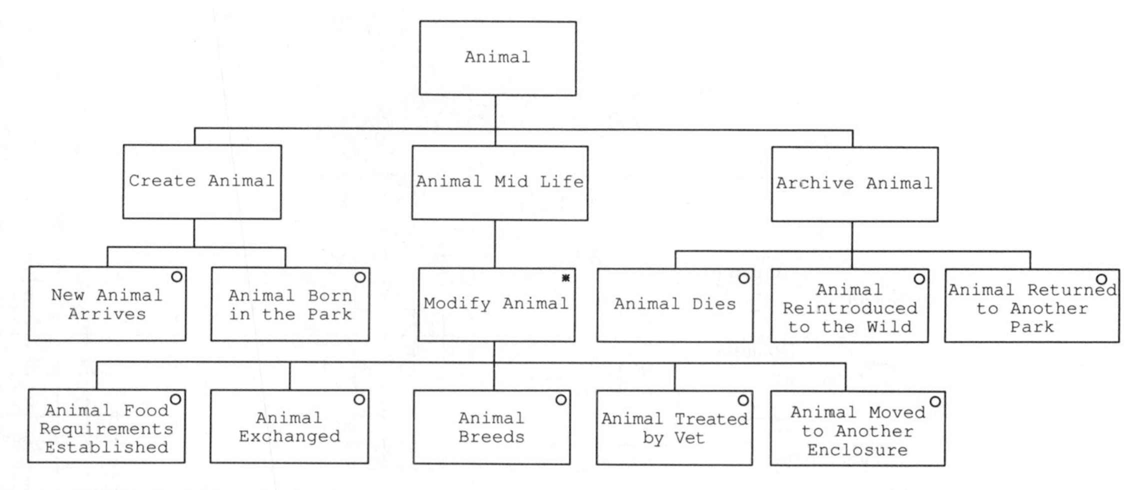

Figure E7.3 Solution to Exercise 7.3.

Solution to Exercise 7.4

See the EEM in Figure E7.4.

In developing this EEM some assumptions have been made:

Assumptions

1. When a provisional booking is made an occurrence of `Guest Deposit` is created, as this contains the attribute `deposit_required` which the guest will need to know to confirm the booking.
2. When a booking is cancelled, all details of the booking are deleted.
3. When a guest settles their bill the appropriate booking and guest purchase details are deleted. However the guest details are retained for 18 months (see Chapter 3, Exercise 3.4).

Some of the entities have no events that create, modify or delete them. Room details will be input when the system is set up and are unlikely to change very often. The rate for the room will change from time to time, and items offered for sale in the minibars are also likely to change. Once again, omissions from earlier models have been highlighted. As these omissions do not affect `Booking`, the ELH for which will be developed in the next exercise, they will not be addressed here.

See Chapter 8.

Solution to Exercise 7.5

Guideline step 2: The following events form a sequence:

- Provisional booking made
- Booking confirmed
- Guest purchases refreshments
- Guest settles bill.

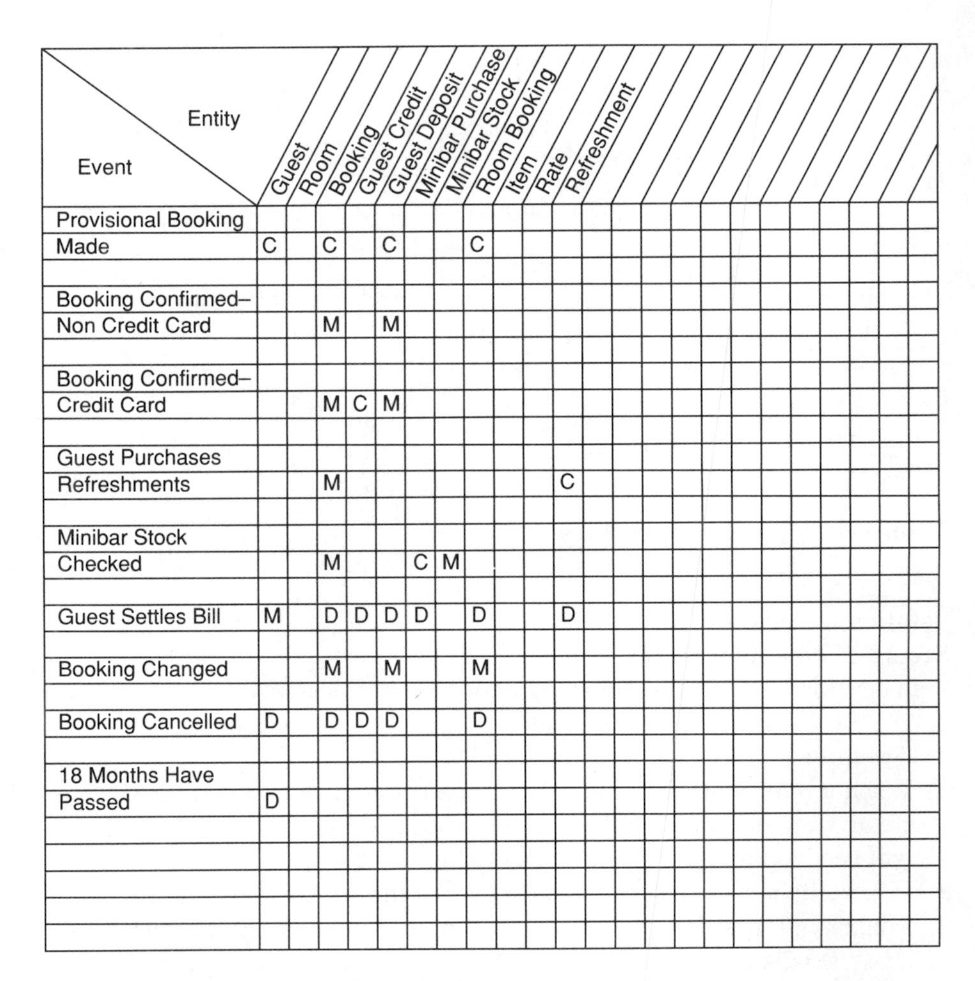

Event \ Entity	Guest	Room	Booking	Guest Credit	Guest Deposit	Minibar Purchase	Minibar Stock	Room Booking	Item	Rate	Refreshment
Provisional Booking Made	C		C		C			C			
Booking Confirmed– Non Credit Card			M		M						
Booking Confirmed– Credit Card			M	C	M						
Guest Purchases Refreshments			M								C
Minibar Stock Checked			M			C	M				
Guest Settles Bill	M		D	D	D	D		D			D
Booking Changed			M		M			M			
Booking Cancelled	D		D	D	D			D			
18 Months Have Passed	D										

Figure E7.4 Solution to Exercise 7.4.

The following events are selections:

- Booking confirmed – non credit card; or
- Booking confirmed – credit card.

The following events are iterations:

- Guest purchases refreshments;
- Minibar stock checked.

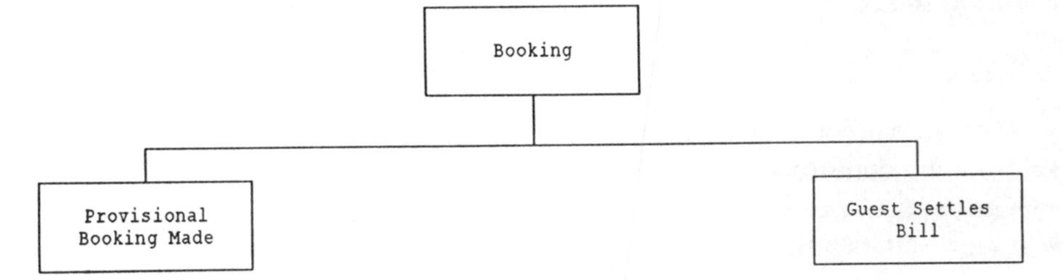

Figure E7.5a Guideline step 3.

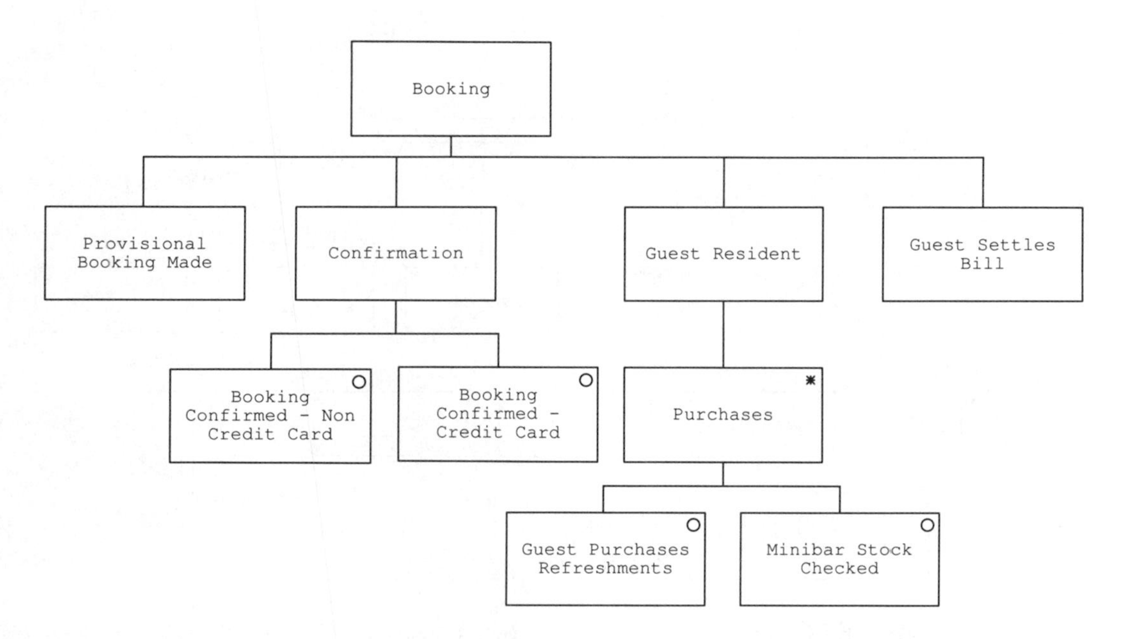

Figure E7.5b Guideline step 4.

The event `Booking Cancelled` is being omitted at this point.

Guideline step 3: There is one creation and one deletion event (see Figure E7.5a).

Guideline step 4: See Figure E7.5b.

Guideline step 5: If your solution is different to Figure E7.5b, check that you have used the notation correctly. Does your ELH allow a permissible combination of events? If not, identify any errors and repeat guideline steps 2–5.

The ELH in Figure E7.5b does not support the case where a guest is requested to offer an alternative method of payment, the credit check on the credit card offered having failed. See Chapter 8 for a discussion of the reconciliation of inconsistencies between the models.

Solution to Exercise 7.6

See Figure E7.6.

A booking could be changed or cancelled either before or after confirmation, so a parallel structure has been used. The construction of the `Alteration` branch allows a booking to be changed zero, one or many times. It allows a booking to be cancelled not at all or just once; a quit and resume structure has been used to ensure that, on cancellation, the life of an occurrence of `Booking` comes to an end. It is also permissible for a booking to be changed before it is cancelled.

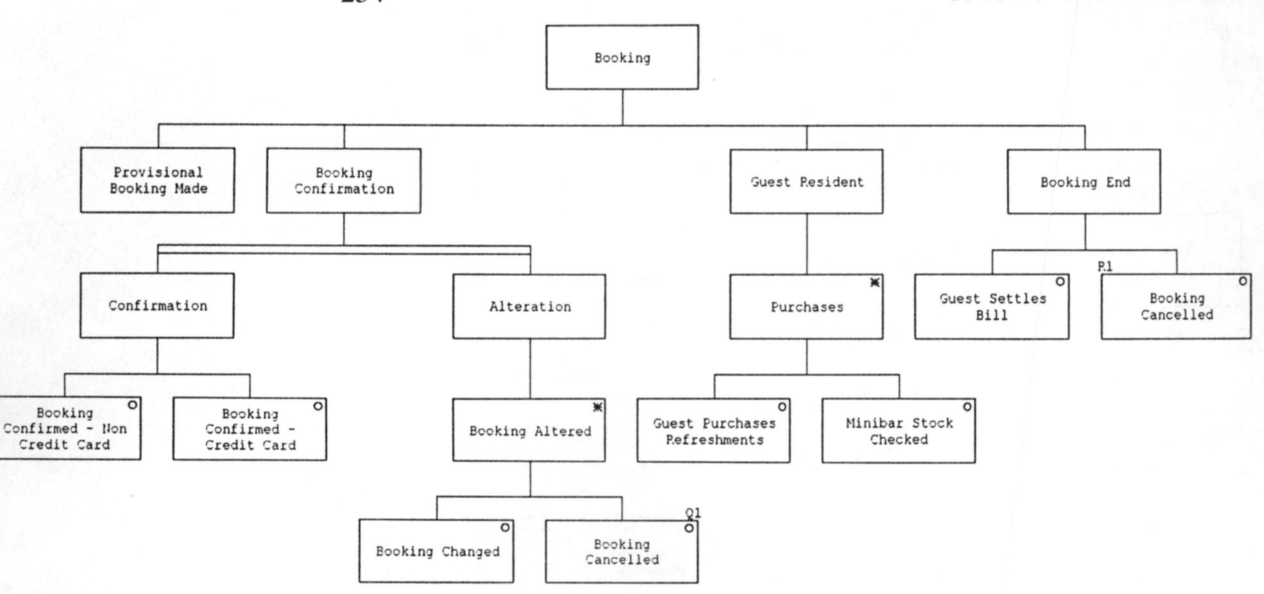

Figure E7.6 Solution to Exercise 7.6.

Solution to Exercise 7.7

The iteration asterisk denotes 'zero or more times', so an occurrence of `Animal` might never be modified. Thus, it is valid for it to be in any of the states 2–5 prior to a deletion event.

Solution to Exercise 7.8

See Figure E7.8.

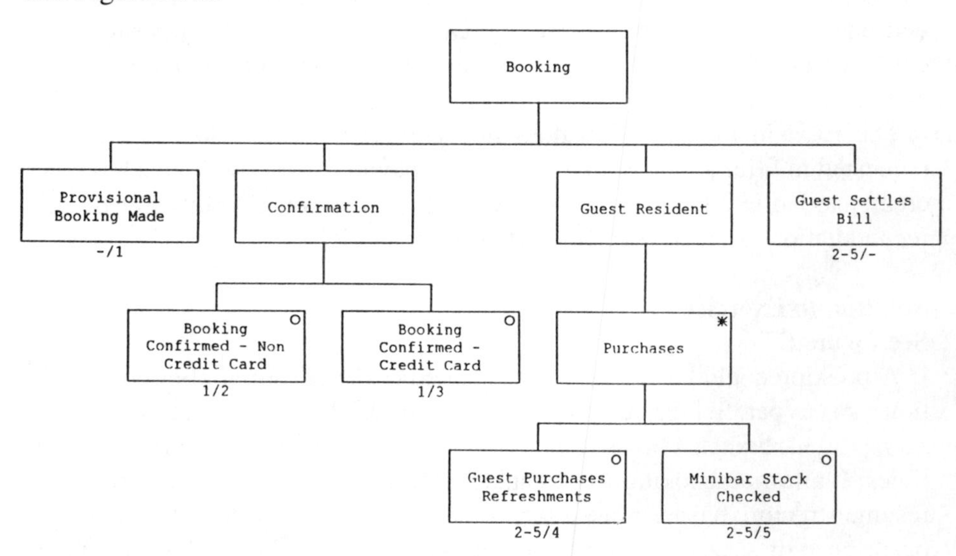

Figure E7.8 Solution to Exercise 7.8.

Chapter 8

Solution to Exercise 8.1

(a) and (b)

Data stores	*Entities*
`Bookings Pending`	
`Credits Pending`	
`Guests`	`Guest`
	`Booking`
	`Guest Credit`
	`Guest Deposit`
	`Minibar Purchase`
	`Refreshment`

The data store shows the dates and rooms for a particular booking. Within the entities, reference may be made to this data held in the entity `Room Booking`.

`Guest Archives`	
`Room Rates`	`Room Rate`

The entity shows one rate for a room whereas the data store shows that different rates apply for different time periods.

`Rooms`	`Room`
	`Room Booking`

The entity `Room Booking` incorporates the dates for each booking in a from/to format. The data store holds the details for a room separately for each day specified within a particular booking

`Sales Accounts`	
	`Item`
	`Minibar Stock`

(c)

The data stores `Bookings Pending` and `Credits Pending` are transient data stores where data is held temporarily, therefore no entities are required.

SSADM identifies transient data stores with a different identifier 'T' within a DFD. This alternative has not been used in this text.

The details about the bookings are handled differently. The organization of the data store facilitates the enquiries for free dates. Both models support the processing required. This would probably be resolved at a later time in the design stage, when processing efficiencies are considered.

The E-R model supports the maintenance of the minibar stock, whereas the DFD records details of items used; which is required? How does the

E-R model support the processing of the sales accounts; exactly what is the requirement? Both these issues would need to be checked with the user.

The requirements for the system state that the guest details should be archived and kept for 18 months; also that Christmas cards are to be sent to guests who have stayed in the hotel for a week or longer in the current calendar year. The DFD supports this by archiving all of the guest details as soon as their account is settled, and uses the archived file to print the Christmas card list. The E-R model does not distinguish between current and past guests. This issue would need to be revisited during design.

Solution to Exercise 8.2

(a)

Events on EEM/ELHs	*Data flows on DFDs*
`Provisional Booking Made`	`booking_request`
`Booking Confirmed - Non Credit Card)`	`authorized_booking`
`Booking Confirmed - Credit Card )`	
	`further_information?`
	`alternative_method_of_payment?`
`Guest Settles Bill`	`settlement`
`Minibar Stock Checked`	`minibar_sales`
`Guest Purchases Refreshments`	`authorized_meal_sales`
`Booking Changed`	`booking_change`
`Booking Cancelled`	`booking_cancellation`
`18 months have passed`	`date_archived`

The data flow `further_information` is not supported on the ELH. If clarification only was required there may be no changes to any entities. If the original information supplied with the booking was incomplete then changes may be required. The action required should be resolved with the user and the model(s) changed accordingly.

The data flow `alternative_method_of_payment` is not supported on the ELH. The action required should be resolved with the user and the model(s) changed accordingly.

See Solution to Exercise 8.1c.

(b)

Room Rates – read-only.

Sales accounts – write-only.

Chapter 9

Solution to Exercise 9.1

(a) Top level DFD for the required system.

There is rarely time in an examination to develop a full set of DFDs, so it is common to be asked to provide a top level DFD and perhaps to explode one of the processes. The next question provides a hint about one of the processes that must appear in the solution to this question.

Note the importance of reading the whole set of tasks before beginning work.

The development of a solution to these questions requires reference to the guidelines for developing a top level DFD, and to the points to keep in mind when drawing a levelled set of DFDs.

See Chapters 2 and 3. Note the comments made in Chapter 9.

Firstly, the inputs and outputs of the system and the source and recipient of each one are identified:

Inputs and outputs	*Source*	*Recipient*
new_plant_details	Owner	
seeds_sown	Gardener	
plants_removed_for_sale	Gardener	
due_ready_date_changed	Gardener	
new_greenhouse_details	Owner	
amended_greenhouse_details	Owner	
daily_temp_and_humid_values	Gardener	
watering_completed	Gardener	
special_notes	Gardener	
invalid_temp_or_humid_value		Supervising Gardener
reports		Owner
order		Supplier
delivery_note	Supplier	
invoice	Supplier	
invoice_query		Supplier
query_reply	Supplier	
checked_invoice		Owner
payment_authorization	Owner	
payment_details		Supplier

Assumptions

1. Hilary and Alec have been called the owners as a logical view is required; the model developed will apply no matter who may run GreenField in the future.
2. The owners decide what new plants should be offered.
3. The owners provide the details of new greenhouses, and decide upon the changes to the temperature and humidity values of existing greenhouses.

The terminators and their associated data flows may now be drawn around the outside of the page.

The next guideline step in Chapter 2 ('Guidelines for drawing a current physical top level DFD') is to identify and draw for each data flow a process which will receive/generate the data. As we are developing a top level DFD of a levelled set, albeit that we are only going to show the lower level of just one of the processes, it is helpful to bear in mind guideline step 4 from Chapter 3 ('Guidelines for drawing a levelled set of DFDs') which suggests the combination of processes that belong to the same logical function from the user's point of view or that access the same data. We already know, from question (b), that one process will be `Maintain Greenhouses`. In searching for other logical functions, `Maintain Plants` and `Maintain Supplies` are suggested by the input and output flows. For the required system, the ability to generate reports is required and this suggests another process. With the addition of the necessary data stores the complete top level DFD for the required system is illustrated in Figure E9.1a.

In accordance with guideline step 9 from Chapter 2, the diagram should be checked for consistency and completeness.

(b) Explode the process dealing with the maintenance of greenhouses

The DFD for process 2, `Maintain Greenhouse Record`, is shown in Figure E9.1b. A process has been drawn to receive or to generate each of the process's input and output data flows.

(c) DD entries

The DD entries have been completed for process 2.5, `Check Temperature and Humidity Values`.

(i) Data Flows

```
temp_or_humid_details = {greenhouse_no, max_temp, min_temp,
    max_humid, min_humid, date, temp, humid}
invalid_temp_or_humid_value = greenhouse_no, date, [ max_temp,
    min_temp, temp | max_humid, min_humid, humid ], text
```

(ii) Process Specification

```
BEGIN
* Time trigger, each day *
FOR all greenhouses
RETRIEVE temp_or_humid_details FROM Greenhouse
COMPARE temp with max_temp and min_temp
IF temp GREATER THAN max_temp OR temp LESS THAN min_temp
  text = 'invalid temperature'
  CREATE invalid_temp_or_humid_value
ENDIF
COMPARE humid with max_humid and min_humid
IF humid GREATER THAN max_humid OR humid LESS THAN min_humid
```

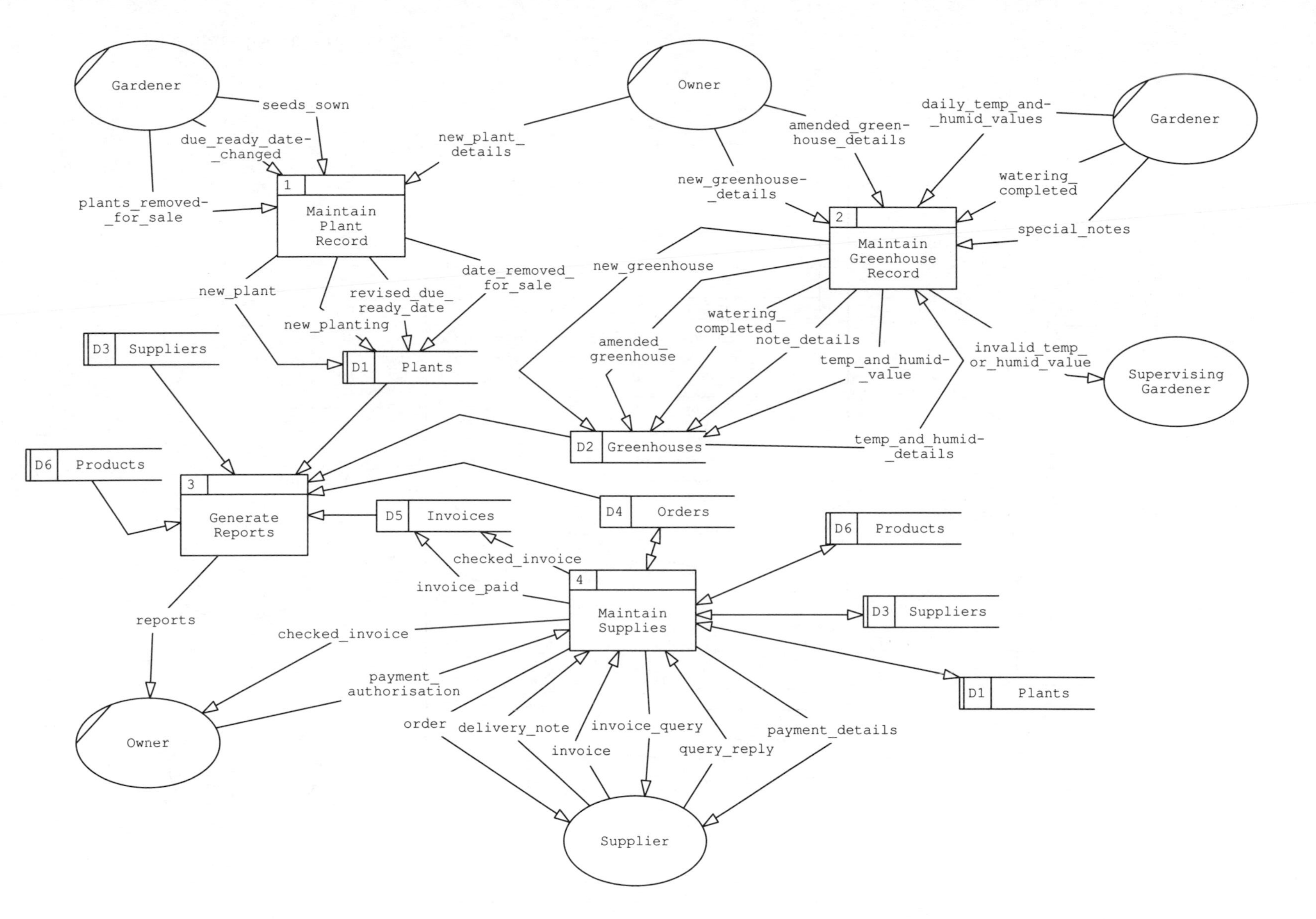

Figure E9.1a Top level DFD.

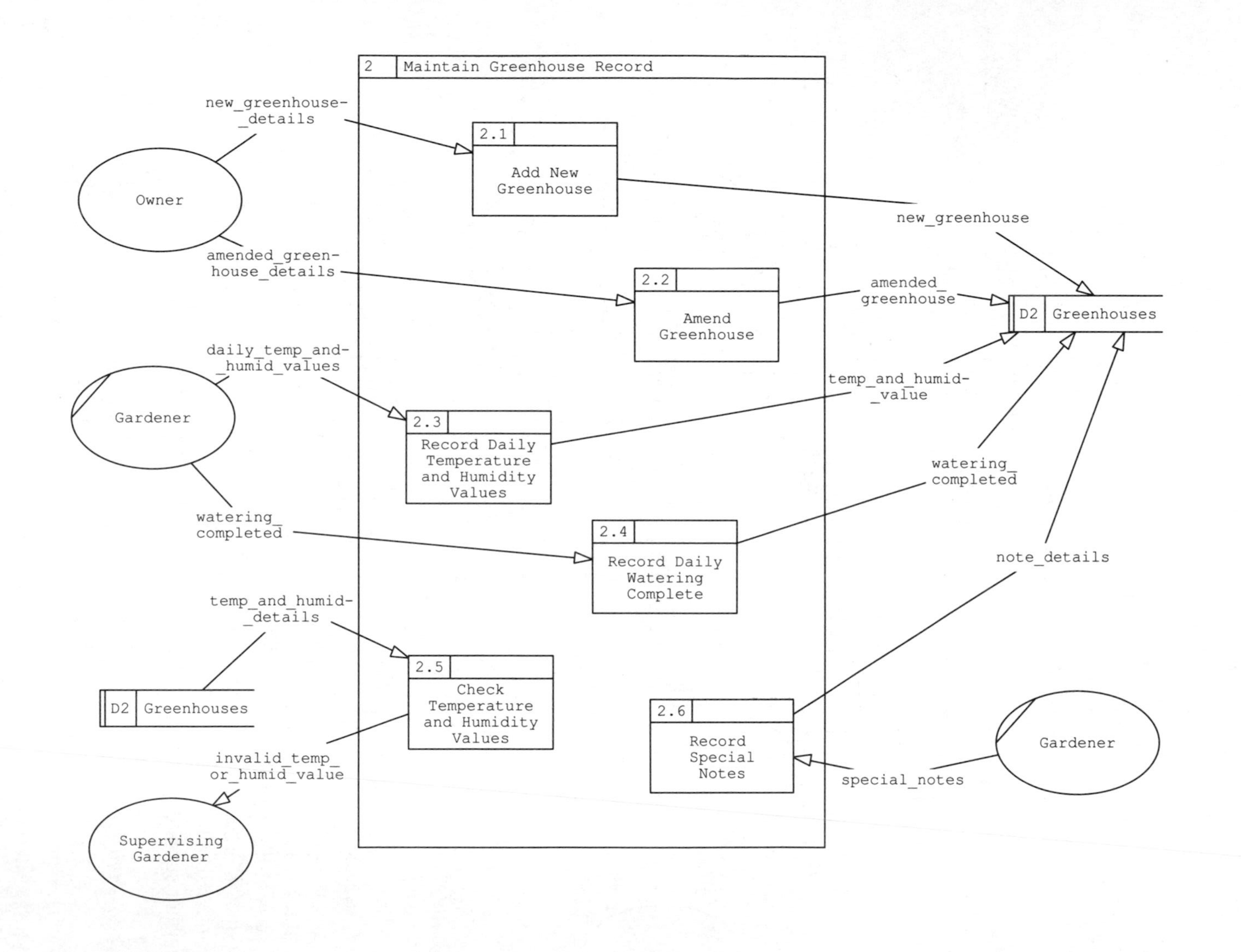

Figure E9.1b Process 2.

```
  text = 'invalid humidity'
  CREATE invalid_temp_or_humid_value
ENDIF
ENDFOR
END
```

(d) E-R model

Guidelines for developing an E-R model are contained in Chapter 6.

Guideline steps 1 and 2:

Likely entities	*Identifiers*
Greenhouse	greenhouse_no
Plant	plant_no
Gardener	gardener_no
Supplier	supplier_no
Order	order_no
Invoice	invoice_no
Product	product_no

Tables

```
Greenhouse     (greenhouse_no,
Plant          (plant_no,
Gardener       (gardener_no,
Supplier       (supplier_no,
Order          (order_no,
Invoice        (invoice_no,
Product        (product_no,
```

Guideline step 3:

Relationships

A plant is grown in a greenhouse.
A gardener supervises a greenhouse.
An order is sent to a supplier.
A supplier sends an invoice.
A plant is ordered on an order.
A product is ordered on an order.
A supplier supplies products.
A supplier supplies plants.

Guideline step 4: See Figure E9.1c.

Assumptions (continued)

4. A plant may be grown in a greenhouse or it may be purchased from a supplier.

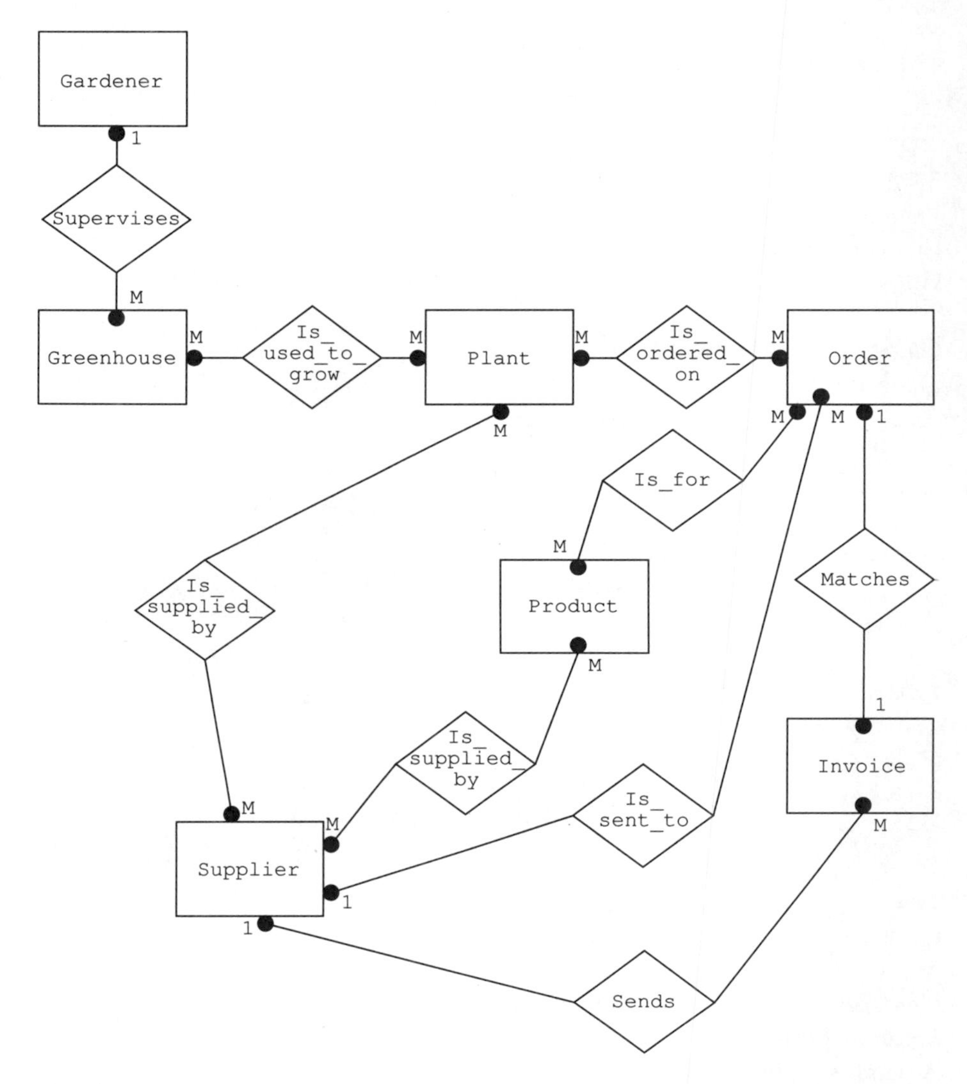

Figure E9.1c Guideline step 4.

5. A greenhouse does not have to be used to grow plants; it can be used only to display plants to the public.
6. Each greenhouse must be supervised by a gardener.
7. Each gardener may supervise more than one greenhouse.
8. The same data is kept about gardeners and supervising gardeners, so they are treated as one entity.
9. Details are kept of potential suppliers as well as of suppliers that have been sent an order.
10. A supplier sends an invoice for each order received.

Guideline step 5: See Figure E9.1d. The figure shows that an order must

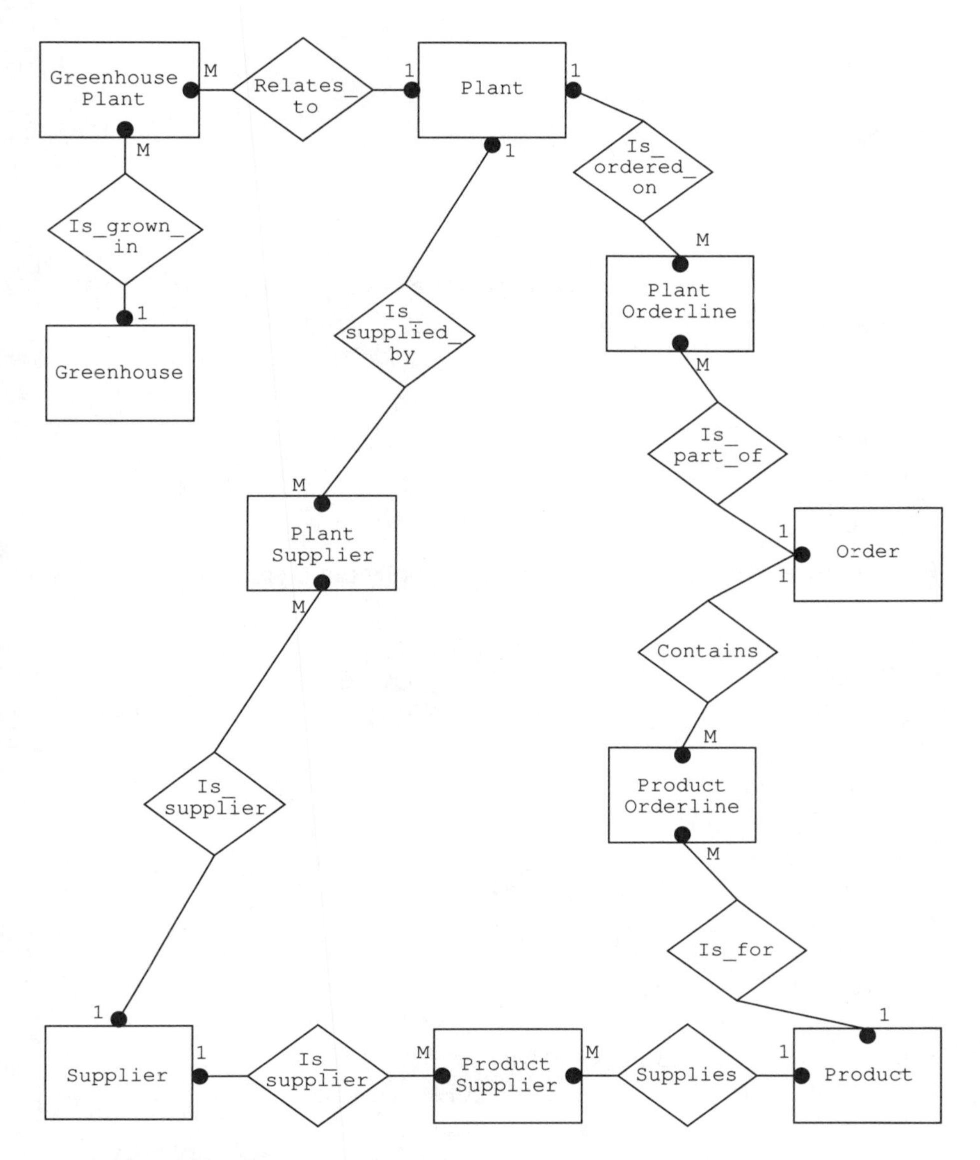

Figure E9.1d Guideline step 5.

contain an orderline, and that an orderline can be either a plant orderline or a product orderline. If a plant orderline and a product orderline are found to contain the same attribute types, this part of the diagram may be redrawn, as shown in Figure E9.1e. For the present, they will be treated as two separate entities.

Question (e) provides the name for one of the new entity types, `Greenhouse Plant`.

Additional Tables

```
Greenhouse Plant        (plant_no, greenhouse_no,
Plant Orderline         (order_no, plant_no,
```

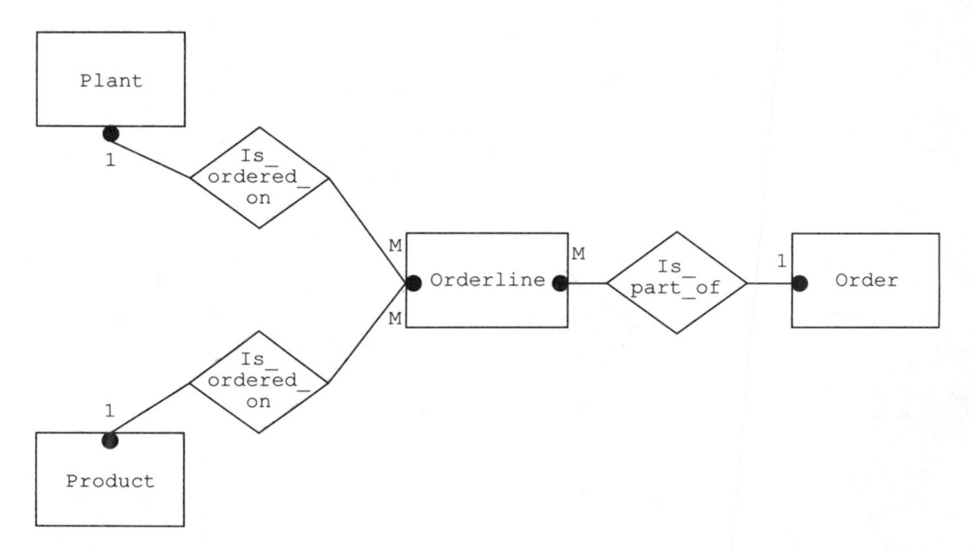

Figure E9.1e An orderline contains either a plant or a product.

```
Product Orderline      (order_no, product_no,
Plant Supplier         (supplier_no, plant_no,
Product Supplier       (supplier_no, product_no,
```

Guideline step 6:

Tables

```
Greenhouse        (greenhouse_no, supervising_gardener_no,
Plant             (plant_no,
Gardener          (gardener_no,
Supplier          (supplier_no,
Order             (order_no, supplier_no,
Invoice           (invoice_no, supplier_no, order_no,
Greenhouse Plant       (plant_no, greenhouse_no,
Plant Orderline        (order_no, plant_no,
Product Orderline      (order_no, product_no,
Plant Supplier         (supplier_no, plant_no,
Product Supplier       (supplier_no, product_no,
```

Guideline step 7: Note that attributes that have already been included as identifiers of the tables listed so far are omitted from the list below.

Attributes

```
max_temp                 watering_instructions
min_temp                 date_sown
max_humid                no_of_trays
min_humid                date_due_ready
temp                     date_removed_for_sale
```

humid
reading_date
gardener_name
gardener_address
gardener_tel_no
special_note
latin_name
order_date
quantity_ordered
supplier_name
supplier_address
supplier_tel_no
common_name

Guideline step 8: Each greenhouse has its temperature and humidity readings taken each day; as these values would form a repeating group within Greenhouse, a new entity, Greenhouse Daily Record, is created.

Assumptions (continued)
11. Only one set of readings is recorded for each greenhouse each day.

Tables

Greenhouse (greenhouse_no, supervising_gardener_no, max_temp, min_temp, max_humid, min_humid)
Plant (plant_no, latin_name, common_name, watering_instructions)
Gardener (gardener_no, gardener_name, gardener_address, gardener_tel_no)
Supplier (supplier_no, supplier_name, supplier_address, supplier_tel_no)
Order (order_no, supplier_no, order_date)
Invoice (invoice_no, supplier_no, order_no)
Greenhouse Plant (plant_no, greenhouse_no, date_sown, no_of_trays, date_due_ready, date_removed_for_sale)
Plant Orderline (order_no, plant_no, quantity_ordered)
Product Orderline (order_no, product_no, quantity_ordered)
Plant Supplier (supplier_no, plant_no)
Product Supplier (supplier_no, product_no)
Greenhouse Daily Record (greenhouse_no, reading_date, temp, humid, special_note)

Plant Orderline and Product Orderline contain similar attributes. If further analysis were to confirm that there are no additional attributes to be stored for either entity or that any additional attributes needing to be stored apply to both of them, they could be combined to form one entity, Orderline. Plant Supplier and Product Supplier could be combined with the same proviso.

Guideline step 9: See Figure E9.1f.

As there are no connection traps the E-R model is complete.

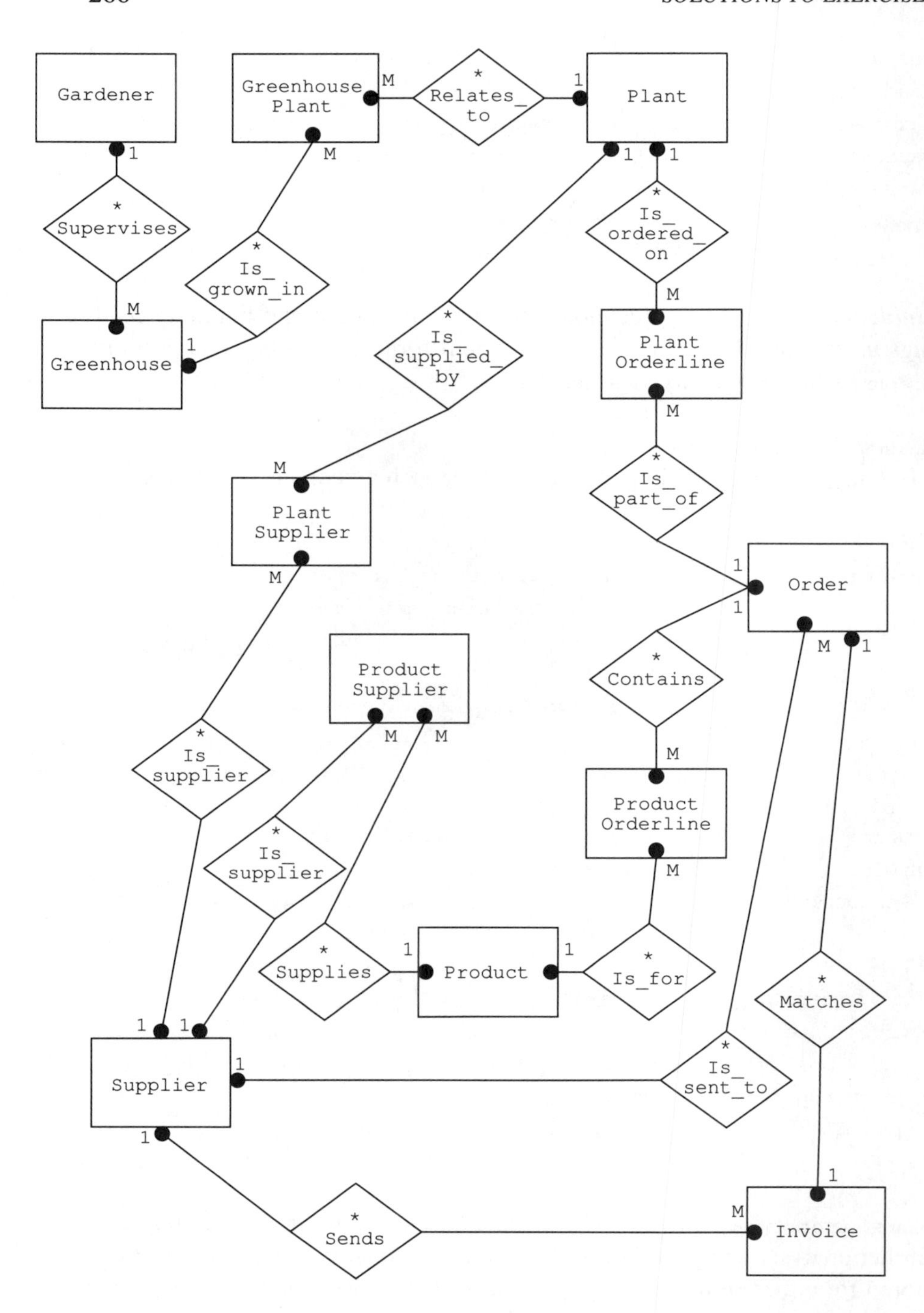

Figure E9.1f Guideline step 9.

(e) ELH for `Greenhouse Plant`

Guidelines for developing an ELH are contained in Chapter 7. Since only one ELH is required, it is not necessary to construct an event/entity matrix, but the events that affect the entity `Greenhouse Plant` need to be identified. The events are:

- Seeds Sown
- Due Ready Date Changed
- Plants Removed For Sale
- Plants Die
- One Year Has Passed (the records are required to be kept for one year).

Guideline step 2: The following events form a sequence:

- Seeds Sown
- Due Ready Date Changed
- Plants Removed For Sale
- One Year Has Passed.

The following events are a selection:

- Plants Removed For Sale, or
- Plants Die.

The following event is an iteration:

- Due Ready Date Changed.

Guideline steps 3 and 4: Figure E9.1g contains the ELH for `Greenhouse Plant`, including state indicators.

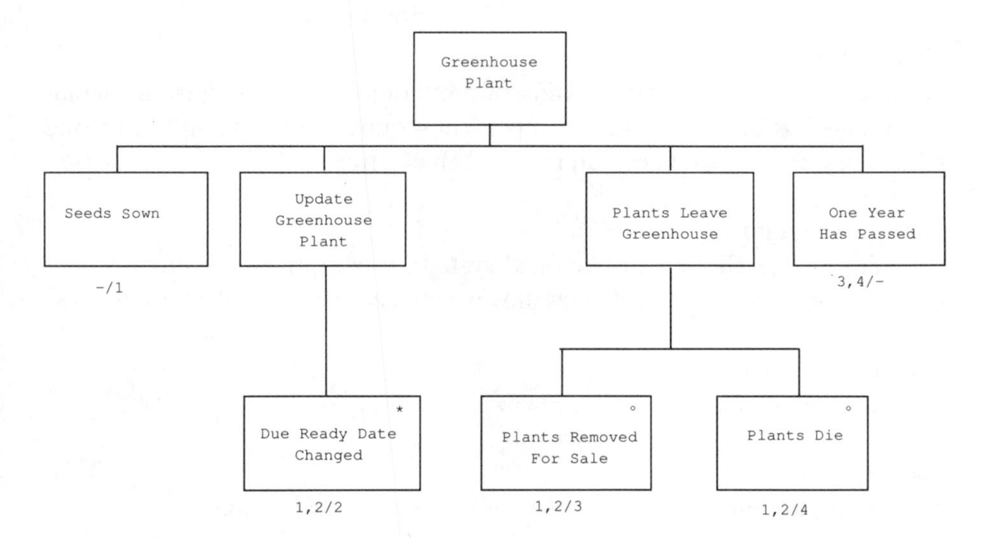

Figure E9.1g ELH for `Greenhouse Plant`.

Glossary

Attribute
A data element that is associated with an entity or a relationship.

Balancing
Method of checking that a set of levelled DFDs are complete and consistent.

Boyce/Codd Normal Form (BCNF)
A table is in BCNF if it conforms to the rule 'Every determinant must be a candidate identifier'.

CASE Construct
May be used to represent selection in Structured English.

CASE Tools
Computer-Aided System or Software Engineering Tools comprise software that supports the development of computer systems. Front-end CASE tools support the activities within the analysis and design phases of systems development.

Connection Trap
A misunderstanding of the meaning of a relationship and its misrepresentation on an E-R diagram, which can prevent required information from being held. They are of two types; fan trap and chasm trap.

Context Diagram
Provides an overall view of a defined system. It comprises one process box, the terminator(s) and the data that moves between them and the system.

Data Dictionary (DD)
A repository of data about data.

Data Element
A meaningful item of data that will not be decomposed further.

Data Flow
Shows where data moves between different components on a DFD. Known as 'data in motion'. Comprises one or more pieces or elements of data.

Data Flow Diagram (DFD)
A diagrammatic way of showing the passage of data through a system, and depicts any changes made to that data and what data is stored. Used to model the process view of a system.

Data Store
On a DFD shows where data is held either temporarily or more permanently. Usually comprises a collection of packets of data each of which may consist of a number of data elements. Known as 'data at rest'.

Data Structure
A collection of data elements that are regularly used together.

Decision Tables
Used to represent in tabular form the conditional logic of processes and the action(s) taken depending on those conditions. Alternative to decision trees and Structured English.

Decision Trees
Used to represent alternative actions that may result from different combinations of circumstances. Alternative to decision tables and Structured English.

Degree
A property of a relationship; it may be 1:1, 1:M or M:M.

Determinant
An attribute A is said to be the determinant of attribute B if a given value of A is associated with just one value of B at any given time.

Dictionary Balancing
Used to check balancing of DFDs when generic data flows are used.

Effect
The change caused by an event, such as the creation, deletion or modification of an entity occurrence.

Enterprise Rule
A rule that relates to an organization's data model, defining among other issues the degree and the membership class of a relationship.

Entity
Something about which an organization wishes to collect and store data. It is capable of being uniquely identified.

Entity Life History (ELH)
Used to model the effect of events on an entity over time.

Entity-Relationship (E-R) Diagram
Illustrates the entities within a system and the relationships between them. Used to model the data view of a system.

Entity-Relationship (E-R) Model
Comprises an E-R diagram and a set of well-normalized tables.

Event
Triggers a process to update data. It should be noted that the event is not the process itself, but whatever initiates or activates the process.

Foreign Key
An attribute or a collection of attributes within a table that exists elsewhere in the information system as a primary key.

Function
A business activity recognized by a user.

Functional Primitive
Processes on a DFD that are not broken down any further. Also known as 'bottom level' or 'elementary' processes. There should be a process specification in the DD for each functional primitive.

Identifier (Primary Key)
An attribute or a collection of attributes that distinguishes a single entity occurrence from every other entity occurrence of that type.

Iteration
Repetition of logical steps until a predefined condition is obtained.

Levelling
Process whereby a set of DFDs is produced allowing us to view the system at different levels of detail. Also known as 'partitioning'.

Logical
Independent of implementation – shows what is done.

Membership Class
Indicates whether every occurrence of an entity will take part in a specified relationship, when it is said to be 'obligatory', or whether only some occurrences will take part, when it is said to be 'non-obligatory'.

Normalization
A technique for deriving a set of tables that is free from redundancy.

Partitioning
Process whereby a set of DFDs is produced allowing us to view the system at different levels of detail. Also known as levelling.

Physical
Implementation dependent – shows what is done and how it is done.

Primary Key
See Identifier.

Process
An action or series of actions which produces a change or development. Used to show the transformation of data from one form or state to another on a DFD.

Process Specification
In DD shows detailed logic or logical steps within a process. Often represented in Structured English. Known as 'elementary process description' in SSADM.

Programming Constructs
A program component which may only be entered at the beginning of the component and exited at the end. There are three allowable types of construct: sequence, selection and iteration.

Prototype
A working model of part(s) of an information system which emphasizes specific aspects of that system; for example, human-computer interface and functionality.

Prototyping
An iterative approach with user participation used to determine systems requirements.

Redundant Data
Data that has been unnecessarily duplicated.

Relation
A mathematical concept derived from set theory; a two-dimensional structure for describing an entity or relationship and its attributes.

Relationship
An association between two or more entities.

Resource Flow
Shows the flow of resources on a current physical DFD or resource flow diagram (in SSADM).

Resource Store
Shows the storage of resources on a current physical DFD or resource flow diagram (in SSADM).

Selection
A choice of one of two or more alternatives.

Sequence
Consists of one or more logical steps applied one after another with no interruptions.

SSADM
Structured Systems Analysis and Design Method.

State Indicators
Added to the leaf boxes on an ELH to show the valid state or states that an entity occurrence must be in for the event to take place (the 'valid prior' value(s)) and the state that an entity occurrence will be in after the event has taken place (the 'set to' value).

Structured English
Used to show logical steps within a process specification. Alternative to decision tables and decision trees.

Table
Often used as a synonym for 'relation'.

Terminator
On a DFD, a data source or destination which is outside the boundary of a defined system. Usually represents a person or department, another system or another organization. Other names – external entity, source (originator) or sink (recipient).

Top level DFD
DFD showing all the major processes being modelled. Level below the context diagram.

Well-normalized
A table is well-normalized if it is in BCNF.

References

Ashworth, C. and Slater, L. (1993) *An Introduction to SSADM version 4*. McGraw-Hill, New York.

Atzeni, P. and De Antonellis, V. (1993) *Relational Database Theory.* Benjamin/Cummings, New York.

Avison, D. and Fitzgerald, G. (1988) *Information System Development: Methodologies, Techniques and Tools*. Blackwell Scientific, Oxford.

Avison, D. and Wood-Harper, A.T. (1990) *Multiview*. Blackwell Scientific, Oxford.

Blethyn, S.G. and Parker, C.Y. (1990) *Designing Information Systems*. Butterworth-Heinemann, Oxford.

Cameron, J.R. (1989) *JSP and JSD,* 2nd edn. IEEE Press, Washington, DC.

Cardenas, A.F. (1985) *Database Management Systems* 2nd edn. Allyn & Bacon.

Checkland, P. (1981) *Systems Thinking, Sstems Practice*. Wiley, Chichester.

Checkland, P. and Scholes, J. (1990) *Soft Systems Methodology in Action*. Wiley, Chichester.

Chen, P. P.-S. (1976) The Entity-Relationship model – toward a unified view of data. *ACM Transactions on Database Systems*, **1** (1), 9–36.

Daniels, A. and Yeates, D. (1988) *Basic Systems Analysis,* 3rd edn. Pitman, London.

Date, C.J. (1990) *An Introduction to Database Systems*. 5th. edn. Addison-Wesley, Reading,MA.

DeMarco, T. (1979) *Structured Analysis and System Specification*. Yourdon Press, New York.

Downs, E., Clare, P. and Coe, I. (1992) *Structured Systems Analysis and Design Method: Application and Context*. 2nd. edn. Prentice-Hall, Englewood Cliffs, NJ.

Flynn, D.J. (1992) *Information Systems Requirements: Determination and Analysis*. McGraw-Hill, New York.

Gane, C. and Sarson, T. (1977) *Structured Systems Analysis*. Prentice-Hall, Englewood Cliffs, NJ.

Goldsmith, S. (1993) *A Practical Guide to Real-time Systems Development*. Prentice-Hall, Englewood Cliffs, NJ.

Graham, I. (1991) *Object Oriented Methods*. 2nd edn. Addison-Wesley, Wokingham.

Howe, D.R. (1989) *Data Analysis for Database Design.* 2nd, edn. Edward Arnold, London.

Ingevaldsson, L. (1979) *JSP: A Practical Method of Program Design.* Chartwell-Bratt Ltd, London.

Jackson, M.A. (1975) *Principles of Program Design.* Academic Press, New York.

Kendall, K.E. and Kendall, J.E. (1992) *Systems Analysis and Design.* 2nd. edn. Prentice-Hall, Englewood Cliffs, NJ.

Kent, W. (1983) A simple guide to five normal forms in relational database theory. *Communications of the ACM*, **26** (2), 120–5.

Korth, H.F. and Silbershatz, A.(1991) *Database System Concepts,* 2nd edn. McGraw-Hill, New York.

Maier, D. (1983). *The Theory of Relational Databases*. Pitman, London.

Mannila, H. and Raiha, K.-J. (1992) The *Design of Relational Databases.* Addison-Wesley, Reading, MA.

Martin, J. (1976) *Principles of Data-base Management.* Prentice-Hall, Englewod Cliffs, NJ.

Martin, J. and McClure, C. (1988) *Structured Techniques. The Basis for CASE*. Prentice-Hall, Englewood Cliffs, NJ.

Maude, T.I. and Willis, G. (1991) *Rapid Prototyping*. Pitman, London.

McClure, C. (1989) *CASE is Software Automation.* Prentice-Hall, Englewood Cliffs, NJ.

Miller, G. (1956) The magical number seven plus or minus two: some limits on our capacity for processing information. *Psychological Review*, **63**, 81–97.

Mumford, E. (1983) *Designing Human Systems.* Manchester Business School, UK.

Patel, D., Hayes, A. and Sun, Y. (1994) *Object-oriented Analysis and Design*. Chapman & Hall, London.

Senn, J.A. (1989) *Analysis and Design of Information Systems,* 2nd edn. McGraw-Hill, New York.

Skidmore, S. (1994) *Introducing Systems Analysis,* 2nd edn. NCC Blackwell, Oxford.

Skidmore, S. and Wroe, B. (1990) *Introducing Systems Design.* NCC Blackwell, Oxford.

Skidmore, S., Farmer, R. and Mills, G. (1992) *SSADM Version 4: Models and Methods*. NCC Blackwell, Oxford.

SSADM Reference Manual. (1990) Volumes 1 to 4. CCTA. NCC Blackwell, Oxford.

Spurr, K. and Layzell, P. (eds) (1992) *CASE: Current Practice Future Prospects*. Wiley, Chichester..

Sully, P. (1993) *Modelling the World with Objects.* Prentice-Hall, Englewood Cliffs, NJ.

Vonk, R. (1990) *Prototyping the Effective Use of CASE Technology.* Prentice-Hall, Englewood Cliffs, NJ.

Ward, P.T. and Mellor, S.J. (1985) *Structured Development for Real-time Systems*. 3 volumes. Yourdon Press, New York.

Yourdon, E. (1989) *Modern Structured Analysis*. Prentice-Hall, Englewood Cliffs, NJ.

Index

Page numbers in **bold** refer to figures